SATELLITE COMMUNICATION

By

Sapna Katiyar

Assistant Professor

Deptt. of Electronics & Comm. Engg.

Greater Noida Institute of Technology,

Greater Noida (UP)

S.K. KATARIA & SONS®

Publisher of Engineering Books

4885/109, Prakash Mahal, Dr. Subhash Bhargav Lane,

Opposite Delhi Medical Association, Daryaganj, New Delhi-110002

Phone: +91-11-43551243, Mobile: +91-9871775858

e-mail: katariabook@gmail.com, katariabooks@yahoo.com

Website : www.skkatariaandsons.com

(e-books, subscriptions and bulk buying also available on our website)

Published by :

S.K. Kataria & Sons®

4885/109, Prakash Mahal, Dr. Subhash Bhargav Lane,
Opposite Delhi Medical Association, Daryaganj, New Delhi-110002
Phone: +91-11-43551243, Mobile: +91-9871775858
e-mail: katariabook@gmail.com, katariabooks@yahoo.com

First Edition: 2007-2008; Fourth Edition: 2012; Reprint: 2024

Printed at : AISA Enterprises, Delhi

PREFACE

Satellite Communication is one of the most impressive spin-offs from space program, and made a major contribution to the international communications. Satellite communication systems are now an integral part of major wide area telecommunication networks throughout the world. The purpose of this book is to introduce the various elements of satellite Communication system and to develop the principles of system design. The understanding of a complete satellite system requires coverage of broad range of topics.

The engineering aspect of Satellite Communication combines diverse topics like antennas, radio wave propagation, data communication, modulation, detection, coding, filtering, orbital mechanics and satellite subsystems. Satellite Communication engineering emphasizes the relevant material from these areas that is important to the book's subject matter and derives equations that the reader can follow and understand.

The aim of this book is to present fundamental principles common to the majority of communication systems in a simple and concise manner.

Satellites play an important role in telephone communication, TV and radio programs distribution and certain other communications. This is a major field of study and has its own intensive literature.

This book has been written as a text book to meet the requirements of Satellite Communication suitable for engineering students. It deals with the fundamental aspects of satellite communication. The subject matter is divided into several chapters covering duly recognized area of theory and applications. Problems and examples have been included to illustrate the applicability of the concepts described in this book. It also contains some recent results of propagation modeling and system design studies. At the end of each chapter summary is given, it will help the reader to review the important points. Though the subject of Satellite Communication is too vast, author has made all the efforts to keep the text simple. Major topics covered under each chapter are as follows :

Chapter 1 gives history of satellite with frequency allocations and current status of satellite communication.

Chapter 2 describes the fundamentals of satellite orbits, orbital parameters, and principle features of geostationary orbit and basic ideas of orbital mechanics.

Chapter 3 explains the design consideration of satellite subsystems. The outline issues related to the design of communication satellites and earth stations are discussed. It also explains the mechanism of satellite station keeping and satellite attitude.

Chapter 4 describes the link budget method essential to the design of satellite communication links It focuses on the configuration of the satellite communication link, method for calculating the receive power, noise generated by the communication link and method for calculating the ratio of carrier signal power to noise power density.

Chapter 5 discusses propagation effects and their impact on system designs. These effects are severe at frequencies above 10 GHz. It also describes the propagation characteristics like rainfall attenuation and fading.

Chapter 6 discusses the earth station design requirements with their subsystems.

Chapter 7 describes various satellite systems like Intelsat (The International Telecommunication Satellite Organization) and Inmarsat (The International Maritime Satellite Organization)

Chapter 8 explains the different application areas of the satellite.

Chapter 9 describes the development of Radio and MSC, overview, concepts and applications of GMSC for maritime, land and aeronautical applications. It also explains the electromechanical characteristics with basic relations and classification of antenna.

The author welcomes reader's feedback and suggestions which would be helpful to improve the book in the coming editions.

SAPNA KATIYAR

AUTHOR

ACKNOWLEDGEMENT

I take this opportunity to express my gratitude to Mr. Ashu Goyal (Chairman, ABESIT Group of Institutions), Prof. Nitij Gupta (Executive Director, ABESIT Group of Institutions), Dr. Monika Gupta (Director, ABESIM), who have given me an opportunity to serve and contribute to the development of Electrical, Electronics and Communication Deptt.

I am very grateful to Prof. A.Q. Ansari and Prof. Ibraheem, Deptt. of Electrical Engineering, Jamia Millia Islamiya, New Delhi for their time to time much needed valuable guidance.

I would like to express my sincere thanks to Mr. R.K. Yadav (HOD, E&C), Ms. Usha Sharma, Ms. Kalpana Hazarika (ITS Engg. College, Gr. Noida), Mr. Arvind Kumar, Ms. Dipti Priyadarshni, Ms. Praina Pandita (CET-IILM Gr. Noida), Mr. Sunil Kumar, Mr. Safdar Tanveer, Mr. Yogesh Kumar, Mr. Priyesh Tiwari, Mr. Manish, Mr. Abdul Rahman, Ms. Arpana, Ms Rachana Singh, Ms. Usha Chauhan HOD, E&C, (PCCS, Gr. Noida) and my all collegues of ABESIT group of Institutions, Ghaziabad for their continuous help and valuable suggestions.

I am grateful to my family members for their constant motivation and guidance that has helped me to complete this project. I would also express my gratefulness for their patience and understanding during the time I devoted to writing the book.

I take this opportunity to express my thanks to Mr. Sanjeev Kumar Kataria of M/s S.K. Kataria and Sons for publishing this book in such a short time. I am also thankful to Mr. Sharda Singh and Mr. Sachin for their assistance.

Last but not the least my thanks are due to all my students who have been a real source of academic persuit and encouragement to me.

SAPNA KATIYAR
AUTHOR

Dedicated to My Husband

"Abhai"

CONTENTS

1

CHAPTER

INTRODUCTION

INSIDE THIS CHAPTER

1.1. THE ORIGIN OF SATELLITE COMMUNICATIONS

Scientists from different countries conceives various ideas for communications through space along with the technological breakthroughs in different fields of science. In 1945, Clarke, in his article **'Extra Terrestrial Relays: Can Rocket Stations give Worldwide Radio Converge?'** published in Wireless World outlined the basic technical consideration involved in the concept of satellite communications.

The idea of satellite communication was conceived by **Arthur C. Clarke** a famous British Science fiction writer in 1945. He had already pointed out that a satellite in a circular equatorial orbit with a radius of about 42,242 km wound have an angular velocity that matched the earth's. So it would always remain above the same spot on the ground and it could receive the relay signals from most of the hemisphere. Three satellites spaced 120 degree apart could cover the whole world with some overlap provided, the message could be relayed between the satellites and thus reliable communications between any two points in the world was possible.

He had also stated that the electrical power for the satellite would be obtained by the conversion of the son's radiation by means of solar cells. Even the rocket technology was available to put a satellite into a low orbit; synchronous orbit was not achieved until 1963.

1.2. BRIEF HISTORY OF SATELLITE COMMUNICATION

- In 1945 an RAF electronics officer and member of the British Interplanetary society, Arthur C. Clarke, wrote a short article on wireless world that described the use of satellite in 24 hours orbits high above the world's land masses to distribute television programmes. His article apparently had little lasting effects in spite of Clarke's repeating the story in his 1951 THE EXPLORATION OF

SPACE. Perhaps the first person to carefully evaluate the various technical options in satellite communications and evaluate the financial prospects was John R. Pierce of AT&T's Bell Telephone Laboratories Who, in a 1954 speech and 1955 article, elaborated the utility of a communications "mirror" in space, a medium orbit "repeater". In comparing the communications capacity of a satellite, which he estimated at 1,000 simultaneous telephone calls and the communications capacity of the first trans-atlantic telephone cable (TAT-1), which could carry, 36 simultaneously telephone calls at a cost of 30-50 million dollars. Price woundered if a satellite would be worth a billion dollars.

- After the 1957 launch of Sputnik I, many considered the benefits, profits, and prestige associated with satellite communications. Because of congressional fear of "duplication". NASA confined itself to experiments with "mirrors" or "passive" communications satellites (ECHO), while the Department of Defence was responsible for "repeater" or "active" satellite which amplify the received signals at the satellite - providing much higher quality communications.
- In 1960 AT&T filled with the Federal Communications Commission (FCC) for permission to launch an experimental communications satellite with an view to rapidly implementing an operational system. The U.S. government reacted with surprise—there was no policy in place to help execute the many decisions related to the AT&T proposal. By the middle of 1961, NASA had awarded a competitive contract to RCA to build a medium orbit (4000 miles high) active communication satellite (RELAY); AT&T was building its own medium orbit satellite (TELESTAR) which NASA would launch on a cost-reimbursable basis; and NASA had awarded a sole-source contract to Hughes Aircraft Company to build a 24-hours (20,000 mile high) satellite (SYNCOM). The military programme, ADVENT, was cancelled a year later due to complexity of spacecraft and cost over runs.
- By 1964, two TELESTARS, two RELAYS and two syncoms had operated successfully in space. This timing was fortunate because the communications satellite corporation (COMSAT), formed as a result of the communications satellite Act of 1962, was in process of contracting their first satellite. For a variety of reasons, including costs, COMSAT ultimately chose to reject the joints AT&T/RCA offer of a medium-orbit satellite incorporating the best of TELESTAR and RELAY. They chose the 24-hour orbit (geosynchronous) satellite offered by Hughes Aircraft Company for their first two systems and a TRW geosynchronous satellite for their third system.
- On April 6, 1965 COMSATs first satellite, EARLY BIRD, was launched from cape canaveral, Global satellite communication had begun.

1.3. ELEMENTS OF A SATELLITE COMMUNICATION

Two major elements of satellite communication systems are:

1. **Space segment:**
   - Communication satellite
   - Satellite bus subsystems
   - Satellite payloads
   - Launch vehicle

2. **Ground segment:**
   - Transmit equipment
   - Receive equipment
   - Antenna system

The basic elements of satellite communication system are earth stations, terrestrial system and users. The basic structure of a satellite communication system is shown in Fig. 1.1.

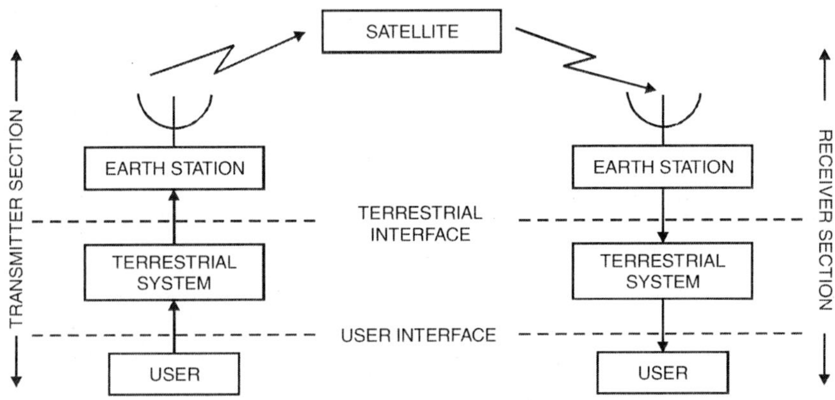

Fig. 1.1. Basic structure of a satellite communication system.

- This consist of many earth stations on the ground and these are linked with a satellite in space. The user is connected to the earth station through a terrestrial network and this terrestrial network may be a telephone switch or dedicated link to a earth station.
- The user generates a baseband signal that is processed through a terrestrial network and transmitted to a satellite at the earth station.
- The satellite consists of a large number of repeaters in space, that receives the modulated RF carrier in its uplink frequency spectrum from all the earth stations in the network, amplifier these carriers and retransmits them back to the earth stations in the down link frequency spectrum.
- To avoid the interference downlink frequency spectrum should be different from uplink frequency spectrum.
- The signal at the receiving earth stations in processed to get back the base band signal, it is sent to the user through a terrestrial network.

Commercial communication satellites use a frequency band of 500 MHz bandwidth near 6 GHz for uplink transmission and another 500 MHz bandwidth near 4 GHz for downlink transmission. While an uplink of 5.725 to 7.075 GHz and a downlink of 3.4 to 4.8 GHz is used. Since **6/4 GHz band is also used in many countries** for microwave links (terrestrial communications), so the problem of mutual interference may occurs. The 500 MHz allocation of frequency band is usually divided into 12 channels of approximately 40 MHz each. This allows each of 12 transponders to carry one TV channel or 1500 analog FM voice circuits.

- Modern communication satellite also employ frequency reuse to increase the number of transponders in the alloted bandwidth.
- 14/12 GHz band is also used in commercial communication satellites because of over crowding of geostationary satellites at 6/4 GHz band. Here uplink is of 12.75 to 14.8 GHz and downlink of 10.7 to 12.3 GHz. This band may be extensively used in future.
- A third band is the 30/20 GHz band where 2.5 GHz bandwidth has been allocated with a uplink of 27.5 to 31 GHz and downlink of 18.1 to 21.2 GHz.

1.3.1. Earth Station Transmitter

The basic block diagram of an **earth station transmitter** is shown in Fig. 1.2.

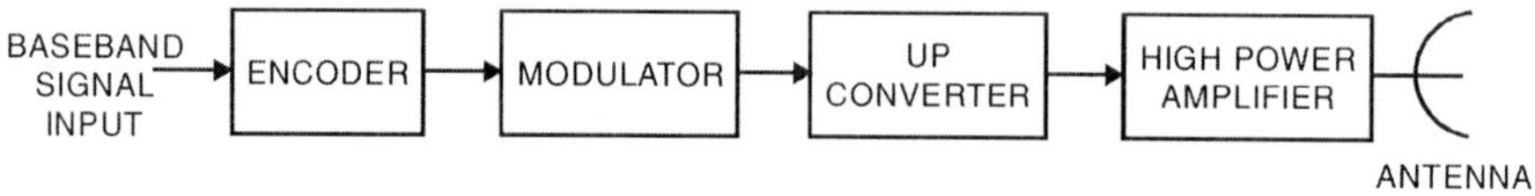

Fig. 1.2. Block diagram of an earth station transmitter.

The baseband signal from the terrestrial network is processed through the encoder and modulator, then it is converted to uplink frequency. Finally it is amplified by high power amplifier and directed towards the appropriate part of antenna.

1.3.2. Earth Station Transmitter

The block diagram of an **earth station receiver** is shown in Fig. 1.3.

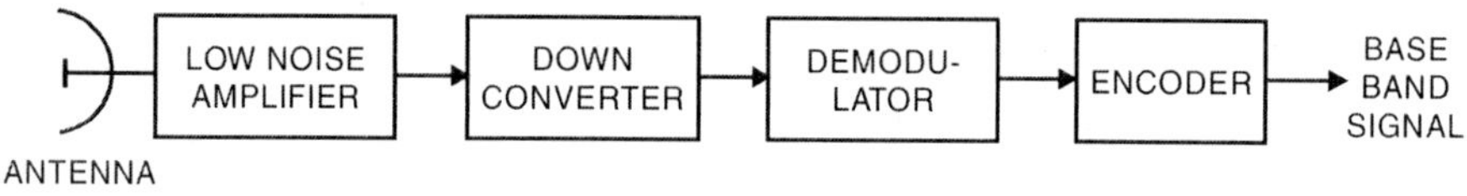

Fig. 1.3. Block diagram of an earth station receiver.

The signal received from the satellite is processed through low noise amplifier and then it is down converted, demodulated by demodulator and decoded by decoder, thus the original baseband signal is obtained.

In case of failure to maintain uninterruptable power some critical components will be installed with automatic switch over. The isolation of high power amplifier and low noise amplifier is of much concern in the design considerations of an earth station. Some **mutual interference effects** may be there:

- These includes interference caused by side lobes, if antenna interferes with an adjacent satellite or an earth station equipment.
- Other sources of interference are microwave relay links, sun transit and intermodulation produces generated in the transponder.

Before 1983 the spacing between the two geostationary satellites was 4° of an equatorial arc and the smallest earth station antenna diameter was 5 m. Now-a-days the spacing allowed between two adjacent satellite in space is 2° along the equitorial arc. The closer spacing has allowed twice as many satellite to occupy the same orbital arc, therefore now a days every earth station antenna is designed accordingly.

1.3.3. Advantages and Disadvantages of 6/4 GHz

Some of the advantages and disadvantages of 6/4 GHz are given below :

Advantages of 6/4 GHz Band :

(1) In this band there is no absorption by the rain.

(2) It has the fewest propagation problems. Attenuation is low below 10 GHz at an elevation angle of 5° or more.

(3) There is no change of polarization when the waves pass through ionosphere.

(4) RF components for these band were being used for terrestrial relay links also and hence easily available.

(5) Sky noise is low at 4 GHz and therefore signal to noise ratio at the receiving antenna is not much deteriorated.

(6) Broad beamwidth allowed large coverage area for receiving.

(7) It is easier to generate high power at 4 GHz at the satellite than at 11 GHz.

Disadvantages of 6/4 GHz Band :

(1) Bandwidth is limited to 500 MHz which can be extended to 1000 Mhz at the most by using orthogonal polarization scheme.

(2) Interference from other users is more. Rather the band is congested.

(3) Power cannot be concentrated in very small area on the earth. Higher frequencies can produce narrower beams.

(4) Direct reception in home TVs is not easily possible because of the need of big sized parabolic dishes. Higher bands are required for the same.

1.3.4. Reason for the Downlink Frequency to be Lower than the Uplink Frequency

Frequency of downlink is smaller than that of the uplink, several factors are there :

(1) **Output power amplifier in transponder :** It is the most important factor because the final power amplifier in the transponder generates more power at lower frequencies than at higher frequencies.

(2) **Effective area of the receiving antenna :** The effective capture area of the downlink antenna should be more so as to receive more energy at the earth's receiving antenna. Effective area of an antenna is directly proportional to the square of the frequency. Thus, the frequency should be kept lower.

(3) **Path loss :** It is less at lower frequencies than at higher frequencies.

(4) **Beam width :** It should be wider as a satellite would send energy to a large number of earth stations. Lower the frequency wider is the beam width.

1.4. SATELLITE FREQUENCY ALLOCATION AND BAND SPECTRUM

Allocating frequencies to satellite services is a complicated process which requires international coordination and planning. This is carried out under the auspices of the international Telecommunication union. To facilitate frequency planning, the world is divided into three regions :

Region 1 : Europe, Africa, what was formerly the Soviet Union and Mongolia.

Region 2 : North and South America and Greenland

Region 3 : Asia, Australia and the South West pacific.

Within these regions, frequency bands are allocated to various satellite services, although a given service may be allocated different frequency band in different regions. Some of the services provided by satellites.

(*a*) Fixed Satellite Services (FSS)

(*b*) Broadcasting Satellite Service (BSS)

(*c*) Mobile Satellite Service

(*d*) Navigation Satellite Service

(*e*) Metrological Satellite Service

There are six frequency bands that have been allocated for the use with satellite communication and these are given as

Table 1.1. Frequency Bands for Satellite Communication.

S. No.	Band	Downlink Band (MHz)	Uplink Band (MHz)
1.	UHF—Military	250–270 (Appx.)	292–312 (Appx.)
2.	C Band—Commercial	3700–4200	5925–6425
3.	X Band—Military	7250–7750	7900–8400
4.	Ku Band—Commercial	11700–12200	14000–14500
5.	Ka Band—Commercial	17700–21200	27500–30000
6.	Ka Band—Military	20200–21200	43500–45500

Table 1.2. Frequency Band Designation.

S. No.	Frequency range, GHz	Band designation
1.	0.1–0.3	VHF
2.	0.3–1.0	UHF
3.	1.0–2.0	L
4.	2.0–4.0	S
5.	4.0–8.0	C
6.	8.0–12.0	X
7.	12.0–18.0	Ku
8.	18.0–27.0	K
9.	27.0–40.0	Ka
10.	40.00–75	V
11.	75–110	W
12.	110–300	mm
13.	300–3000	μm

Table 1.3. Frequency Allocations for Fixed Satellite Service and Broadcasting Satellites.

S. No.	Frequency	Fixed satellite service	Broadcasting satellite
1.	2500–2535 MHz	Down Regions II & III	Down
2.	2535–2655	Down Region II	Down
3.	2655–2690	Up Regions II and III	Down
4.	3400–3700	Down	
5.	3700–4200	Down	
6.	4500–4800	Down	
7.	5725–5850	Up Region I	
8.	5850–5925	Up	
9.	5925–7075	Up	
10.	7250–7450	Down	
11.	7450–7550	Down	
12.	7900–8025	Up	
13.	8025–8400	Up	
14.	10.7–11.7 GHz	Down	Up Region I
15.	11.7–12.1	Down Region II	Down Regions I and II
16.	12.1–12.2	Down Region II	Down
17.	12.2–12.3	Down Region II	Down Regions I and II
18.	12.3–12.5		Down Regions I and II
19.	12.5–12.75	Up/Down	Down Region II
20.	12.75–13.25	Up	
21.	14–14.5	Up	
22.	14.5–14.8		Up
23.	17.3–17.7		Up
24.	17.7–18.1		Up
25.	18.1–18.6	Down	
26.	18.6–18.8	Down	
27.	18.8–20.2	Down	
28.	20.2–21.2	Down	
29.	22.5–23.0		Down Regions II and III
30.	27–27.5	Up Regions II and III	
31.	27.5–29.5		
32.	29.5–31	Up	
33.	37.5–39.5	Down	
34.	39.5–40.5	Down	
35.	40.5–42.5		Down
36.	42.5–43.5	Up	Down
37.	47.2–49.2		Up
38.	49.2–50.2	Up	
39.	50.4–51.4	Up	
40.	71.0–75.5	Up	
41.	81–84	Down	
42.	84–86		Down
43.	92–95	Up	
44.	102–105	Down	
45.	149–164	Down	
46.	202–217	Up	
47.	231–241	Down	
48.	265–275	Up	

Table 1.4. Frequency Allocation for General Mobile Satellite Service.

S. No.	Frequency	General Mobile
1.	806–890 MHz	Regions II and III
2.	1530–1535	Limited use
3.	1544–1545	Down
4.	1545–1559	Down
5.	1645.5–1646.5	Up
6.	19.7–21.2 GHz	Down
7.	29.5–31.0	Up
8.	39.5–40.5	Down
9.	43.5–47	Up
10.	66–71	Up/Down
11.	71–74	Up/Down
12.	81–84	Up/Down
13.	95–100	Up/Down
14.	134–142	Up/Down
15.	190–200	Up/Down
16.	252–265	Up/Down

1.5. SATELLITE ORBITS

Orbital path is defined as the path in which satellite goes round the earth. The distance from the center of earth to the satellite is known as **orbital radius.** An orbit is characterized by attributes :

- Altitude
- Inclination.

The high of the satellite from the surface of the earth is known as **altitude** and any point in the orbit the angle of rotation of the satellite with horizon is known as **inclination** At altitude of 2000-5000 km and 15,000-30,000 km, there are belts of ionized particles known as **'Inner Van Allen Belt'** and **'Outer Van Allen Belt'** respectively. Satellite orbits are important in these range as communication would be impossible. Radius varies in case of eccentric orbit while it is constant in circular orbit.

Satellite Orbits are classified in two broad categories *i.e.*,

- Non-Geostationary Orbit (NGSO)
- Geo Stationary Orbit (GSO)

Satellite orbits are also classified in terms of the orbital height. These are :

- Low Earth Orbit (LEO)
- Medium Earth Orbit (MEO)
- Geosynchronous Earth Orbit (GEO)

1.5.1. Non-Geostationary Orbit (NGSO)

Early ventures with satellite communications used satellites in Non-geostationary low earth orbits due to the technical limitations of the launch vehicles in placing satellites in higher orbits. With the advancement of launch vehicles and satellite technologies, once the Geo Stationary Orbit (GSO) was achieved, majority of the satellite for telecommunications started using GSO due to its many advantages.

Satellite orbits with orbital height of approximately 1000 km or less are known as **Low Earth Orbit** (LEO). LEOs tend to be in general circular in shape. Satellite orbits with orbital heights of typically in the range of 5000 km to about 25,000 km are known as **Medium Earth Orbit (MEO)/Intermediate Circular orbit** (ICO). MEO and ICO are often used synonymously, but MEO classification is not restricted to circular orbits. Satellites in Highly Elliptical Orbit (HEO) are suitable for communications in the higher latitudes. Russian Molnya satellites have highly inclined elliptical orbits with a perigee of about 1000 km, apogee of 40,000 km, inclination of 63.435 deg and orbital period of 12 hours. In Geosynchronous Earth Orbit (GEO) the satellite is in equatorial circular orbit with an altitude of 35,786 km and orbital period of 24 hours. Three satellites in GEO placed 120° apart over equator cover most of the world for communications purposes.

1.5.2. Geostationary Orbit (GSO)

There is only one geostationary orbit possible around the earth, lying on the earth's equatorial plane and the satellite orbiting at the same speed as the rotational speed of the earth on its axis. For a Satellite to have an orbital period equal to that of earth's rotation *i.e.,* a sidereal day (23 Hrs 56 min. 4.09 sec.) an altitude of 35,786 km is required. Such a satellite orbiting at a velocity of 3.075 km/sec remains fixed relative to any point on earth or geostationary.

1.5.3. Geostationary Earth Orbit (GEO)

- These satellites are in orbit 35,863 km above the earth's surface along the equator.
- Objects in Geostationary orbit revolve around the earth at the same speed as the earth rotates. This means GEO satellites remain in the same position relative to the surface of earth.

Characteristics

1. A satellite in a geostationary orbit appears to be stationary with respect to the earth.
2. Three conditions are required for an orbit to be stationary.
   - The satellite must travel eastward at the same rotational speed as the earth.
   - The orbit must be circular.
   - The inclination of the orbit must be 0.
3. The radius of the geostationary orbit = 42164 km.

Advantages

- A GEO satellite's distance from earth gives it a large coverage area, almost a fourth of the earth's surface.
- GEO satellites have a 24 hour view of a particular area.
- These factors make it ideal for satellite broadcast and other multipoint applications.
- User terminals do not have to track the satellite.
- Only a few satellites can provide global coverage.
- Maximum life-time (15 years or more).
- Above Van Alien Belt Radiation.
- Often the lowest cost system and simplest in terms of tracking and high speed switching.

Disadvantages

- A GEO satellite's distance also cause it to have both a comparatively weak signal and a time delay in the signal, which is bad for point to point communication.
- GEO satellites, centered above the equator, have difficulty broadcasting signals to near polar regions.
- Transmission latency or delay of 250 millisecond to complete up/down link.
- Satellite antennas must be of larger aperture size to concentrate power and to create narrower beams for frequency reuse.
- Poor look angle elevations at higher latitudes.

1.5.4. Low Earth Orbit (LEO)

- LEO satellites are much closer to the earth than GEO satellites, ranging from 500 to 1,500 km above the surface.
- LEO satellites don't stay in fixed position relative to the surface, and are only visible for 15 to 20 minutes each pass.
- A network of LEO satellites is necessary for LEO satellites to be useful.
- At LEO, a satellite can circle the Earth in approximately one and a half hours.

Advantages

- A LEO satellite's proximity to earth compared to a GEO satellite gives it a better signal strength and less of a time delay, which makes it better for point to point communication.
- A LEO satellite's smaller area of coverage is less of a waste of bandwidth.
- Low latency or transmission delay.
- Higher look angle (especially in high-latitude regions).
- Less path loss or beam spreading.
- Easier to achieve high levels of frequency re-use.
- Easier to operate to low-power/low-gain ground antennas.

Disadvantages

- A network of LEO satellites is needed, which can be costly.
- LEO satellites have to compensate for Doppler shifts cause by their relative movement.
- Atmospheric drag effects LEO satellites, causing gradual orbital deterioration.
- Larger number of satellites (50 to 70 satellites). Thus higher launch costs to deploy, build, and operate.
- Harder to deploy, track and operate. There is higher TTC&M costs even with cross links.
- Shorter in-orbit lifetime due to orbital degradation.

1.5.5. Medium Earth Orbit (MEO)

- A MEO satellite is in orbit somewhere between 8,000 km and 18,000 km above the earth's surface.
- MEO satellites are similar to LEO satellites in functionality.
- MEO satellites are visible for much longer periods of time than LEO satellites, usually between 2 to 8 hours.
- MEO satellites have a larger coverage area than LEO satellites.
- These orbits are primarily reserved for communications satellites that cover the North and South Pole.
- Unlike the circular orbit of the geostationary satellites, MEO's are placed in an elliptical (oval-shaped) orbit.

Advantages

- A MEO satellite's longer duration of visibility and wider footprint means fewer satellites are needed in a MEO network than a LEO network.
- Less latency and delay than GEO (but greater than LEO).
- Improved look angle to ground receivers in higher latitudes.
- Fewer satellites to deploy and operate and cheaper TTC&M systems than LEO (but more expensive than with GEO)
- Longer in-orbit lifetime than LEO systems (but less than GEO)

Disadvantages

- A MEO satellite's distance gives it a longer time delay and weaker signal than a LEO satellite, though not as bad as a GEO satellite.
- More satellites to deplay than GEO (10 to 18 vs. 3 to 4).
- Ground antennas are generally more expsnsive and complex because of the need to track satellites. Or, one must use lower-gain, quasi-omni antennas.

  Increased exposure to Van Allen Belt radiation.

Table 1.5 shows the various characteristics of each of these four types of satellite.

Table 1.5. Performance Characteristics of Different Altitude Schemes.

Types of Satellite	LEO	MEO	GEO	HEO
Altitude	500–1500 km	5000–10,000 km	36,000 km	15,000–30,000 km
Rotation period	90 mins	5–12 hour	24 hours	Less than 24 hours
Time in sight	15 mins	2–4 hours	Always	8 hours
Example	Iridium	GPS	VSAT	Molnya
Uses	Mobile communication and for surveying	Global communication such as e-mail, FAX, telephony	Global communication on such as TV and radio transmission, data transmission	Communication amount the polar countries

1.6. ADVANTAGES AND DISADVANTAGES OF SATELLITE COMMUNICATION

Communication through satellite has several advantages and disadvantages. Until the evolution of satellite communication, long distance communication through space could be done by using cascaded radio relays, very low frequency (below 30 kHz) and high frequency or short waves radio (3–30 MHz). But the cascaded radio relays were limited to overland spans. So the satellite has given a major contribution in the field of communication, it is capable of transmitting high capacity over long distances either overland or water. Also because of its unique geometry, it is inherently a broadcast medium with a ability to transmits simultaneously from one point to an arbitrary number of other points within its converge area.

1.6.1. Advantages

1. Point to multipoint communication is possible whereas terrestrial relay are point to point. This is why satellite relays are wide area broadcast.
2. Circuits for the satellite can be installed rapidly. Once the satellite is in position, earth stations can be installed and communication may be established within some days or even hours.
3. During critical conditions each earth stations may be removed relatively quickly from a location and reinstalled somewhere else.
4. Mobile communications can be easily achieved by the satellite communication because of its flexibility in interconnecting mobile vehicles.
5. Satellite communication has the economical advantage *i.e.,* the satellite cost is independent of the distance.
6. As compared to fibre optic cable, the satellite communication has the advantage of the quality of transmitted signals and the location of earth stations. The sending and receiving information is independent of distance.

1.6.2. Disadvantages

1. With the satellite in position the communication path between the terrestrial transmitter and the receiver is approximately 75000 km long.

2. There is a delay of 1/4 sec between the transmission and reception of a signal because the velocity of electromagnetic wave is 3×10^5 km/sec. This delay produces ECHO which is actually caused by an imperfect impedance matching.
3. The time delay reduces the efficiency of satellite in data transmission and long file transfer, which carried out over the satellites.
4. Over-crowding of available bandwidths due to low antenna gains.
5. High atmospheric losses above 30 GHz limits the carrier frequencies.

1.7. SATELLITE APPLICATIONS

Applications are innumerous but can be broadly classified as :

- In communication such as T.V., telephony, data transfer such as e-mail and internet etc. are mostly done through different communication satellite these days.
- Remote sensing and earth observation can be done with the help of lower earth orbit (LEO) satellite.
- Metrological applications such as wheather survey, to study different layers and amount of ozone content in the atmosphere.
- Military applications like short distance local communication from any camp to another, to study the location of the enemy, etc.

The number of operational and planned satellite communications system in growing very rapidly.

The military have developed the global positioning system (GPS), but now people are using these satellite services to determine their exact latitude, longitude and attitude wherever they are in the world GPS satellites can be used for navigation.

1.8. TYPES OF SATELLITES

The satellite may be classified into two categories :

(1) Passive satellite.

(2) Active satellite.

The major difference between these two is that wheather the communication relay involves passive reflection or an active electronic system. An **active satellite** is one which has transmitting equipment abroad, such as a transponder. It is a device which receives a signal from earth, amplifies it, and retransmits the same signal back to the earth (either immediately or after a delay).

A **passive satellite** merely reflects or scatters incident radiations from the earth, a portion of the radiation being reflected or scattered back in the direction of the earth.

One may provide all active satellite with omnidirectional transmitting antennas (radiating equally in all directions) or directive antennas (radiating most of the energy toward the earths). Directive antennas would require very less transmitted power but would require antenna stabilization so as to directed the radiated energy towards the earth.

1.8.1. Passive Satellites

The principle of communication by passive satellite is based on the properties of scattering of electromagnetic waves from different surface areas. Thus an electromagnetic wave incident on a passive satellite is scattered back towards the earth and a receiving station can receive the scattered wave. The passive satellites used in the early years of satellite communications were both artificial as well as natural.

Although passive satellites were simple, the communications between two distant places were successfully demonstrated only after overcoming many technical problems. The large attenuation of the signal while traveling the large distance between the transmitter and the receiver via the satellite was one of the most serious problems. The disadvantages of passive satellites for communications are:

- Earth Stations required high power (10 kW) to transmit signals strong enough to produce an adequate return echo.
- Large Earth Stations with tracking facilities were expensive.
- Communications via the Moon is limited by simultaneous visibility of the Moon by both the transmit and the receive stations along with the larger distance required to be covered compared to that of closer to earth satellite.
- A global system would have required a large number of passive satellites accessed randomly by different users.
- Control of satellites not possible from ground.

1.8.2. Active Satellites

In active satellites, which amplify and retransmit the signal from the earth have several advantages over the passive satellites.

The advantages of active satellites are:

- Require lower power earth station.
- Less costly.
- Not open to random use.
- Directly controlled by operators from ground.

Disadvantages of active satellites are:

- Disruption of service due to failure of electronics components on-board the satellites.
- Requirement of on-board power supply.
- Requirement of larger and powerful rockets to launch heavier satellites in orbit.

World's first active satellite **SCORE** (Satellite Communication by Orbiting Relay Equipment) was launched by US Airforce in 1958 at orbital height of 110 to 900 miles.

1.8.3. Digital Satellite Communication

With the developments in digital electronics technology, access techniques and possibility of expanded communication network, digital signalling techniques have been introduced in communication by the satellite. Some of the advantages are :

1. Increased efficiency of multiplexing multiple signals or handling digital messages in packets for convenient switching.

2. Relative insensitivity of digital circuits for retransmission.
3. Potential for extremely low error rate and high fidelity through error detection and correction.
4. Communications privacy.
5. Flexibility of digital hardware implementation and it permits the use of microprocessors, miniprocssors, digital swiching and the use of large scale integrated circuits (LSI).

So digital transmission techniques have gained increased usage for satellite communication, microwave relay or waveguide transmission.

In analog communication system frequency division multiplexing—frequency modulation–frequency division multiple access (FDM–FM–FDMA) has been widely used and provides good quality satellite links but the number of earth stations are limited. A digital satellite systems such as that using quaternary phase shift keying–time division multiple access (QPSK–TDMA) can accommodate a large number of earth stations with a small loss in transponder capacity and thus the system can quickly respond to traffic variation. Efficiency can be further increased by demand assignment and digital speech interpolation. Digital communication techniques use onboard switching and processing, multiple spot beams and beam hopping, therefore such digital systems are capable of serving a mixture of large, medium and small earth stations with high efficiency. The recent technique of code division multiple access has allowed to employ microearth stations (0.5 m antenna) at an extremely low cost with a good service quality. ISDN system will further enhance the importance of digital communication networking with satellites.

Summary

1. Satellites in geostationary orbit remain permanently at an attitude of about 37,870 km.
2. The first communication satellite was launched in the early 1970s.
3. Satellite can receive, process, amplify and retransmits signals for most of a hemisphere.
4. Communication satellite uses a frequency band of 6/4 GHz and this band is also used in many countries for microwave links.
5. The spacing between the two geostationary satellite is 2° along the equatorial arc.
6. For frequency allocation the world is divided into three regions and the services provided by the satellites are fixed satellite services, broadcasting satellite services, mobile services, navigation services and metrological satellite services.
7. Point to multipoint communication is possible by satellite.
8. Active satellite is having a transmitting equipment which passive satellite reflects or scatters incident radiations from the earth.
9. Efficiency of a satellite communication can be increased by multiple access techniques.
10. Some of the satellite applications are satellite communication satellite, remote sensing, military, wheather and positioning satellites.
11. A system of three geostationary satellites spaced 120° apart along the equator would be able to provide global coverage.
12. Passive satellites reflect signals they receive to earth or other satellites. They have no transmitter or energy signals.

13. The device used to convert the digital data into a usable form by the communications channel and back to digital data is known as modem. It is an acronym for modulator demodulator.
14. Commercial communication satellites use a frequency band of 500 MHz bandwidth near 6 GHz for uplink transmission and another 500 MHz bandwidth near 4 GHz for downlink transmission.
15. Modern communication satellite also emply frequency reuse to increase the number of transponders in the allotted bandwidth.
16. 6/4 GHz band is most popular because of less propagation problem.
17. The major disadvantage of 6/4 GHz band is its more interference from other users.
18. The major advantages of satellite communication is that the satellite costs are independent of the distance.
19. Digital transmission techniques have gained increased usage for satellite communication, microwave relay or waveguide trnasmission.
20. Advantages over terrestrial systems are :
    - They can cover large areas.
    - Inherent broadcast capability.
    - Easy deployment and configuration.
    - High speed network access.
21. Mobile communications can be easily achieved by the satellite communication because of its flexibility in interconnecting mobile vehicles.

Review Questions

1. How did the idea of geostationary satellite originates and grow ?
2. Explain the various frequency bands used in satellite communication. Compare the merits and demerits of different bands.
3. To cover the entire globe, three communication satellites are sufficient. Why ?
4. How does a communication satellite differ from a communication relay ?
5. What are the basic elements of a satellite communication system ? Explain with a suitable block diagram. **[UPTU 2008-09]**
6. Draw the block diagram of an earth station transmitter and explain the function of each block in brief.
7. Why the uplink frequency is different from the downlink frequency ? Also give the reason to keep higher uplink frequency.
8. Discuss the advantages and disadvantages of the 6/4 GHz band used in satellite communication.
9. Describe briefly the advantages and disadvantages offered by satellite communications.
10. What do you meant by active and passive satellites ?
11. List the advantages of a digital satellite communication.
12. Discuss briefly the current status of a satellite communication.
13. Why is uplink frequency kept different from downlink frequency ? What are the most frequently used values for them ?
14. Differentiate between active and passive satellite.
15. Briefly discuss the satellite frequency spectrum.

16. Discuss the various applications of satellite communication.

17. What do you understand by digital satellite communication ?

18. List the various classes of services provided by satellites.

19. Prove that for covering the globe three communication satellites are sufficient.

20. Why does microwave play an important role in satellite communication ?

21. Give a brief idea about mobile satellite to communication.

22. What is the need of satellite communication modern world ?

23. In what way satellite has an edge in communication compared to other methods ?

24. What are the future trends in satellite communication ? **[UPTU 2007-08]**

25. Make a comperative study between LEO, MEO, GEO and HEO types of satellites.

Objective Type Questions

1. Originally broadband long distance communications were made possible by the advent of

(*a*) Geostationary satellites

(*b*) HF radio

(*c*) Telegraph cables

(*d*) Repeater amplifiers

2. Microwave links are preferred to coaxial cables for transmission because

(*a*) They have less overall distortion

(*b*) They have greater bandwidths

(*c*) Of their relative immunity to impulse noise

(*d*) They are cheaper.

3. Satellite used for intercontinental communications are known as

(*a*) COMSAT

(*b*) MARISAT

(*c*) DOMSAT

(*d*) INTELSAT

4. Satellite used for wheather forecast applications are known as

(*a*) GORIZONT (*b*) TIROS–N

(*c*) COMSAT (*d*) SPOT

5. Satellite communication links are preferred over submarines cables because

(*a*) Of their multiple access ability

(*b*) They are feester

(*c*) They involve lesser costs

(*d*) None of these

6. Total band of frequencies allocated for satellite communications is spread over

(*a*) 100 MHz (*b*) 4 GHz

(*c*) 6 GHz (*d*) None of these

7. The downlink frequency in the C-band transponder is about

(*a*) 6 GHz (*b*) 4 GHz

(*c*) 14 GHz (*d*) 11 GHz

8. "In a satellite system, the uplink frequency is greater than that of downward link." Is it true ?

(*a*) Yes (*b*) Not always

(*c*) No (*d*) None of these

9. Rain attenuation and sky noise is low at

(*a*) 4 GHz (*b*) 6 GHz

(*c*) 12 GHz (*d*) 14 GHz

10. In satellite communication 6/4 GHz band has a disadvantage

(*a*) There is no absorption by the rain

(*b*) Sky noise is low

(*c*) More interference from other users

(*d*) Fewest propagation problems.

ANSWERS

1. (*a*)	**2.** (*a*)	**3.** (*d*)	**4.** (*b*)	**5.** (*a*)
6. (*d*)	**7.** (*b*)	**8.** (*a*)	**9.** (*a*)	**10.** (*c*)

❑❑❑

2

CHAPTER

Orbital Mechanism and Launching of Satellite

INSIDE THIS CHAPTER

2.1. INTRODUCTION

The orbit of a satellite is geostationary and it has to be maintained geostationary at any cost. **Mainly there are two problems regarding the satellite :**

1. Launching and putting the satellite into geostationary orbit.
2. Maintaining the satellite into its orbit *i.e.* station keeping.

There are some standards recommended by **International Telecommunication Union (ITU)** and **International Consultive Committee (CCIR).** They prescribe the parking slots for individual countries and this helps in avoiding the unncessary losses or traffic capacity. The communication satellite is a very large investment and its life is limited. Therefore its full utilisation is very essential.

The communication satellites move around the earth as planets do around the sun and therefore the Kepler's laws are applicable to them also.

2.1.1. Kepler's First Law

It states that the path followed by a satellite around the primary will be an ellipse. An ellipse has two focal points shown as F_1 and F_2 in Fig. 2.1.

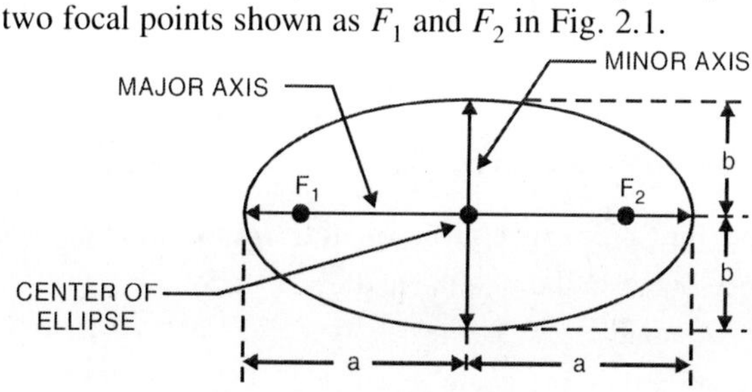

Fig. 2.1.

The center of the mass of the two body system, termed the **bary center**, is always centered on one of the foci. The semimajor axis is denoted by 'a' and semiminor axis is by 'b'. The eccentricity 'e' is given by

$$e = \frac{\sqrt{a^2 - b^2}}{a} \quad ...(2.1)$$

The eccentricity and semimajor axis are two of the orbital parameters specified for satellite (spacecraft) orbiting the earth.

For $0 < e < 1 \rightarrow$ Elliptical orbit

$e = 0 \rightarrow$ Circular orbit

2.1.2. Kepler's Second Law

It states that for equal time intervals, a satellite will sweep out equal areas in its orbital plane, focussed at the body center as shown in Fig. 2.2.

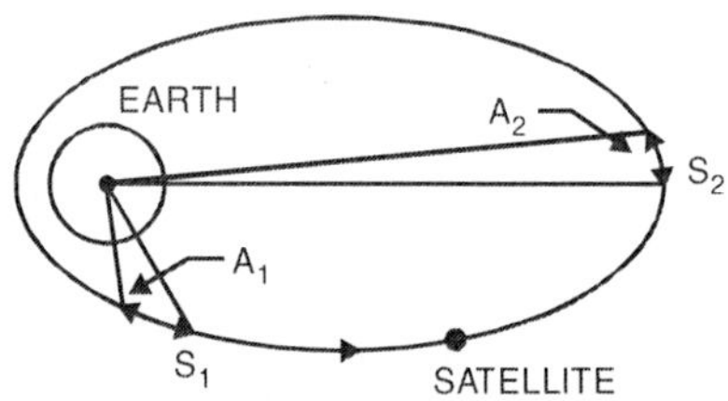

Fig. 2.2.

Assuming that the satellite travels distances S_1 and S_2 meters in 1 sec, then the areas A_1 and A_2 will be equal. The average velocity in each case is S_1 and S_2 meters per sec. and because of the equal area law, it follows that the velocity at S_2 is less than at S_1.

2.1.3. Kepler's Third Law

It states that the square of the periodic time of orbit is proportional to the cube of the mean distance between the two bodies. The mean distance is equal to semi-major axis, a

So,
$$a^3 = \frac{\mu}{n^2} \quad ...(2.2)$$

where n = Mean motion of the satellite in radians/sec.

μ = Earth's geocentric gravitational constant

$= 3.98 \times 10^{14}$ m^3/sec^2

with n in radians per second. The orbital period in sec is given by

$$T = \frac{2\pi}{n}$$

So
$$T = 2\pi\sqrt{\frac{a^3}{\mu}} \quad ...(2.3)$$

The Newton's gravitational law of forces also control the satellite orbits.

Some important points useful in the determination of orbit of satellites are :

- Distance between the satellite and earth station.
- Converge angle.
- Earth station pointing angle.

- Eclipse and solar interferences.
- Polarisation angle.
- Converge angle.
- Slant range.

Because of a continuous drift in the satellite orbit both in longitude and lattitude due to various disturbances, perfect stationary satellite orbit is not possible. So a satellite is constrained to remain within a window. Whose limits are defined by angular shift as seen from the center of the earth around the required nominal position, usually this window is 75 km on the sphere containing the geostationary satellite orbit.

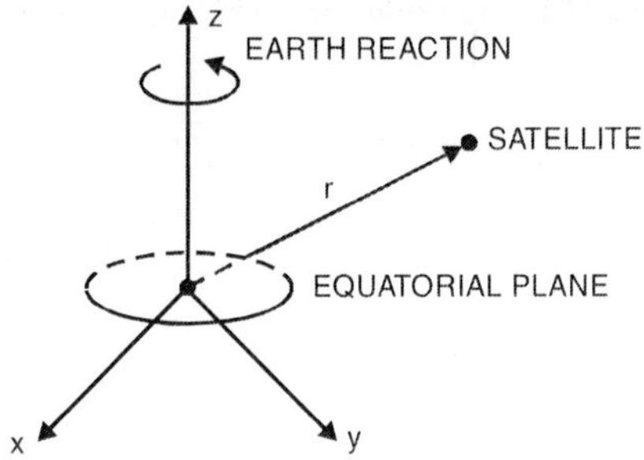

Fig. 2.3. Co-ordinate system used to describe the earth and a satellite.

2.2. ORBITAL MECHANICS

2.2.1. Equation of the Orbit

Let us consider the rectangular co-ordinate system. The origin is at the centre of earth, and the z-axis extends through the north geographic pole. The co-ordinate system is right handed, it is fixed in space and the earth rotates on the z-axis. It is assume that the center of mass of the earth satellite system coincides with the center of mass of the earth at the origin.

The vector r locates the moving satellite with respect to the centre of the earth. If the mass of a satellite is m. The gravitational force F on the satellite is given by

$$F = -\frac{GMm\hat{r}}{r^2} \quad ...(2.4)$$

where, M = Mass of earth

G = 6.672×10^{-11} Nm/kg^2 (Universal gravitational constant)

$\hat{r}$ = Unit vector in the r direction

Product term = $GM = 3.98 \times 10^5$ km^3/s^2 (Kepler's constant) 'μ'

According to the Newton's second law

$$F = m\frac{d^2r}{dt^2}\hat{r} \quad ...(2.5)$$

Equating the inertial force on satellite to the gravitational force, we have

$$-\frac{\mu m\hat{r}}{r^2} = m\frac{d^2r}{dt^2}\hat{r}$$

$$-\frac{\mu\hat{r}}{r^2} = \frac{d^2r}{dt^2}\hat{r} \quad ...(2.6)$$

This is a **second order vector linear differential equation** and its solution will involve six undetermined constants called the **orbital elements.** Writing it as

$$r = r\hat{r} \quad ...(2.7)$$

and rearranging Eq. (2.6), we have

$$\frac{1}{r}\cdot\frac{d^2r}{dt^2}+\frac{\mu r}{r^3} = 0 \qquad ...(2.8)$$

Motion defined by above equation is confined to a plane. Taking $r \times$ each term yields

$$r\times\frac{d^2r}{dt^2} = 0 \qquad ...(2.9)$$

Invoking the rule for finding the derivative of a product, it follows that

$$\frac{d}{dt}\left(r\times\frac{dr}{dt}\right) = \frac{dr}{dt}\times\frac{dr}{dt}+r\times\frac{d^2r}{dt^2} \qquad ...(2.10)$$

The **cross product** of any vector with itself is zero; hence Eq. (2.10) may be rewritten as

$$\frac{d}{dt}\left[r\times\frac{dr}{dt}\right] = 0 \qquad ...(2.11)$$

This is equivalent to

$$\boxed{r\times\frac{dr}{dt} = h} \qquad ...(2.12)$$

where, h = Constant

It is the **orbital angular momentum** of the satellite.

h can be a constant only if the orbit lies in a plane. So the problem of motion of satellite in three dimensions reduces to the problem of motion in a plane.

X_0 and Y_0 axis lie in the orbital plane and the z-axis is perpendicular to it. It is difficult to solve the Eq. (2.8) because of second derivative of r but it can be solved by expressing the equation in a rectangular co-ordinate system where the unit vectors are constant (Fig. 2.4). The first two co-ordinates lie in the plane and z_0 is normal to it. Expressing the Eq. (2.8) in terms of these co-ordinates yields

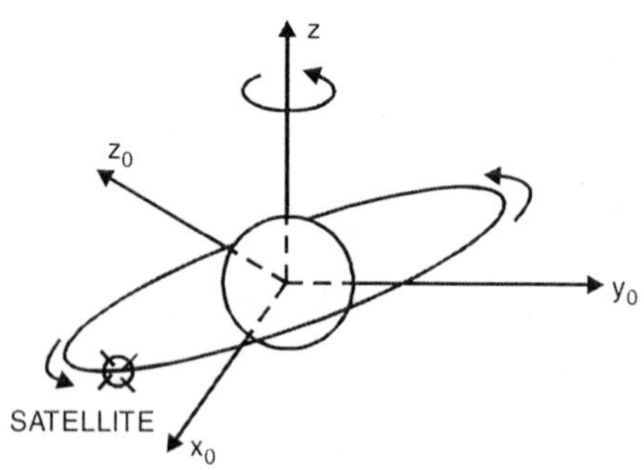

Fig. 2.4. Orbital plane co-ordinate system.

$$\hat{X}_0\left(\frac{d^2x_0}{dt^2}\right)+\hat{Y}_0\left(\frac{d^2y_0}{dt^2}\right)+\frac{\mu(x_0\hat{X}_0+y_0\hat{Y}_0)}{(x_0^2+y_0^2)} = 0 \qquad ...(2.13)$$

Above equation can be solved easily if it is expressed in polar co-ordinate system (r_0, ϕ_0).

Using the transformations, we have

$$x_0 = r_0\cos\phi_0 \qquad ...(2.14)$$

$$y_0 = r_0 \sin \phi_0 \qquad \text{...(2.15)}$$

$$\hat{X}_0 = \hat{r}_0 \cos \phi_0 - \hat{\phi}_0 \sin \phi_0 \qquad \text{...(2.16)}$$

$$\hat{Y}_0 = \hat{\phi}_0 \cos \phi_0 + \hat{r}_0 \sin \phi_0 \qquad \text{...(2.17)}$$

and equating the $\hat{r}_0$ components of Eq. (2.13), we have

$$\frac{d^2 r_0}{dt^2} - r_0 \left(\frac{d\phi_0}{dt}\right)^2 = -\frac{\mu}{r_0} \qquad \text{...(2.18)}$$

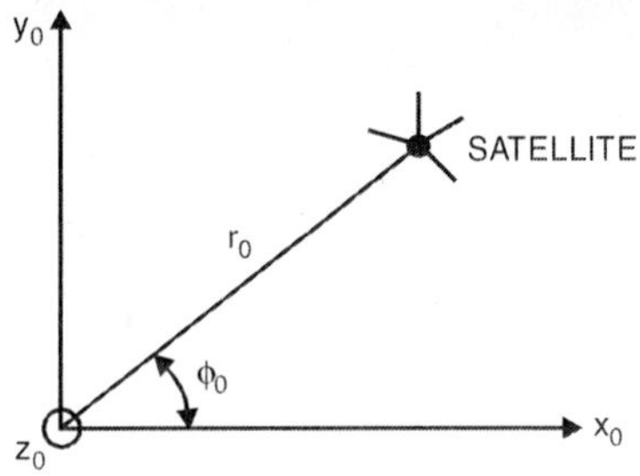

Fig. 2.5. Polar co-ordinates in the orbital plane.

Similarly equating $\hat{\phi}_0$ components, we have

$$r_0 \left(\frac{d^2\phi_0}{dt^2}\right) + 2\left(\frac{dr_0}{dt}\right)\left(\frac{d\phi_0}{dt}\right) = 0 \qquad \text{...(2.19)}$$

We may rewrite the Eq. (2.19) as

$$\frac{1}{r_0}\frac{d}{dt}\left(r_0^2 \frac{d\phi_0}{dt}\right) = 0 \qquad \text{...(2.20)}$$

and this is equivalent to

$$r_0^2 \frac{d\phi_0}{dt} = \text{a constant} = |\,h\,| = h \qquad \text{...(2.21)}$$

where h is the magnitude of the angular momentum vector in Eq. (2.12), writing it as

$$r_0 \left(\frac{d\phi_0}{dt}\right)^2 = \frac{h^3}{r_0^3} \qquad \text{...(2.22)}$$

Now substituting this expression into Eq. (2.18), we have

$$\frac{d^2 r_0}{dt^2} - \frac{h^2}{r_0^3} = -\frac{\mu}{r_0} \qquad \text{...(2.23)}$$

To find out the equation relating r_0 and ϕ_0, we have to eliminate t from Eq. (2.23). So we define a new variable 'u' by

$$u = \frac{1}{r_0} \qquad \text{...(2.24)}$$

So that,
$$\frac{dr_0}{d\phi_0} = -\frac{1}{u^2}\cdot\frac{du}{d\phi_0} \qquad \text{...(2.25)}$$

and using the relationship, we have

$$\frac{dr_0}{dt} = \left(\frac{dr_0}{d\phi_0}\right)\left(\frac{d\phi_0}{dt}\right) = \left(\frac{dr_0}{d\phi_0}\right)\left(\frac{h}{r_0^2}\right)$$

$$= -h.\left(\frac{du}{d\phi_0}\right) \qquad \text{...(2.26)}$$

Now transforming $\left(\frac{d^2 r_0}{dt^2}\right)$ in Eq. (2.23) to

$$\frac{d^2 r_0}{dt^2} = -h^2 u^2 \left(\frac{d^2 u}{d\phi_0^2}\right) \quad \text{...(2.27)}$$

We may rewrite Eq. (2.23) as

$$\frac{d^2 u}{d\phi_0} + u = \frac{\mu}{h^2} \quad \text{...(2.28)}$$

According to the standards, the solution of this differential equation is

$$\boxed{u = \frac{u}{h^2} + c\cos(\phi_0 - \theta_0)} \quad \text{...(2.29)}$$

where c and θ_0 are constants and it can be determined by the boundary conditions. Representing the above equation in terms of r_0, we have

$$r_0 = \frac{1}{\frac{\mu}{h^2} + c\cos(\phi_0 - \theta_0)}$$

$$= \frac{\left(\frac{h^2}{\mu}\right)}{1 + \left(\frac{h^2}{\mu}\right) c\cos(\phi_0 - \theta_0)}$$

$$\boxed{r_0 = \frac{P}{1 + e\cos(\phi_0 - \theta_0)}} \quad \text{...(2.30)}$$

For $e < 1$, this is the equation of an ellipse. Whose semilatus spectrum P is given by

$$\boxed{P = \frac{h^2}{\mu}} \quad \text{...(2.31)}$$

where, Eccentricity $(e) = \frac{h^2 c}{\mu}$

for $e = 0$, orbit is a circle with the earth at its center.

2.2.2. Describing the Orbit

The quantity θ_0 serves to orient the ellipse with respect to orbital plane axes x_0 and y_0 if θ_0 = zero, accordingly we can choose x_0 and y_0.

Then the **equation of orbit** will be

$$\boxed{r_0 = \frac{P}{1 + e\cos\phi_0}} \quad \text{...(2.32)}$$

Some of the **orbital parameters** are shown in Fig. 2.6.

The value of eccentricity e determines the type of orbit.

$$e = \begin{cases} 0 & \rightarrow \text{Circle} \\ <1 & \rightarrow \text{Ellipse} \\ 1 & \rightarrow \text{Parabola} \\ >1 & \rightarrow \text{Hyperbola} \end{cases}$$

The point in the orbit where the satellite is closest to the earth is known as **perigee** and the point where the satellite is farthest from the earth is known as **apogee**. To make θ_0 equal to zero, we have chosen x_0 axis so that the apogee and perigee lie along it.

The differential area swept by vector r_0 from the origin to the satellite in time dt is given by :

$$\boxed{dA = 0.5 r_0^2 \left(\frac{d\phi_0}{dt}\right) dt = 0.5h\, dt}$$

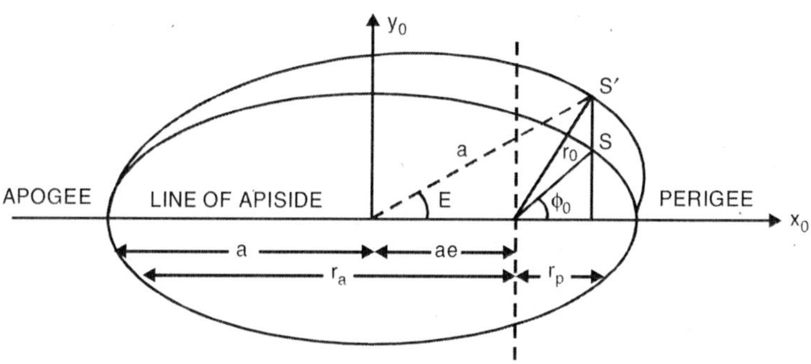

Fig. 2.6. Geometry of an elliptical orbit of communication satellite.

In the Fig. 2.6, *a* is the semi-major axis and *b* is semi-minor axis.

- Orbital period, $T = 2\pi\sqrt{\frac{a^3}{\mu}}$ ...(2.33)

$$\mu = GM = 3.98 \times 10^{14}\ \text{m}^3/\text{s}^2$$

- Satellite velocity for a circular orbit $= \sqrt{\frac{\mu}{a}}$ ...(2.34)
- Satellite velocity at an orbital point *S* (distance to the centre of earth *r*) is

$$v_s^2 = \frac{2\mu}{r} - \frac{\mu}{a} \quad \text{...(2.35)}$$

$$x = r_0 \cos\phi_0 = a(\cos E - e) \quad \text{...(2.36)}$$

$$y = r_0 \sin\phi_0 = a \sin E\,(1 - e^2)^{1/2} \quad \text{...(2.37)}$$

- Eccentricity, $e = \frac{c}{a} = \frac{r_a - r_p}{r_a + r_p}$ ...(2.38)
- Semi-major axis, $a = \frac{(r_a + r_p)}{2} = \frac{p}{(1 - e^2)}$ ...(2.39)
- Apogee distance, $r_a = a + c = a(1 + e)$ ...(2.40)

- Perigee distance, $r_p = a - c = a(1-e)$ ...(2.41)
- Locus parameter, $P = a(1-e^2) = \dfrac{2r_a r_p}{(r_a + r_p)}$ ...(2.42)
- Semi-minor axis, $b = a(1-e^2)^{1/2} = (r_a r_p)^{1/2}$ ...(2.43)

2.2.3. Locating the Satellite in the Orbit

The equation of orbit is,

$$r_0 = \frac{P}{1 + e\cos\phi_0}$$

and the semi-major axis is given as

$$a = \frac{P}{(1-e^2)}$$

So the equation of orbit may be rewritten as

$$r_0 = \frac{a(1-e^2)}{1 + e\cos\phi_0} \quad ...(2.44)$$

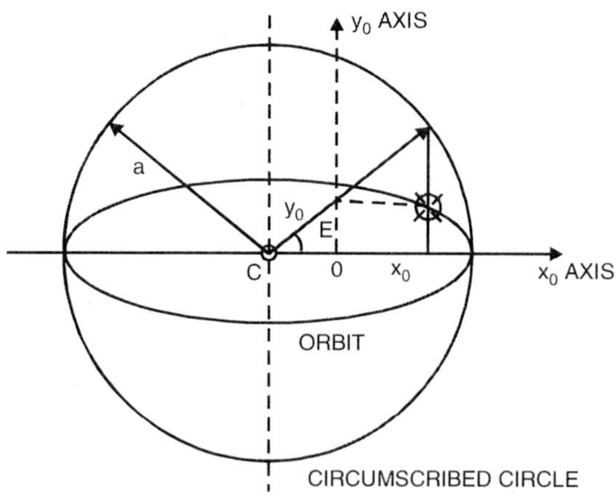

Fig. 2.7. Satellite position is specified in orbital plane co-ordinate system $b > (x_0, y_0)$.

The angle ϕ_0 is measured from x_0 axis and is called the **true enomaly.** Positive x_0 axis passes through its perigee, ϕ_0 measures the angle from the perigee to the instantaneous position of the satellite. The **rectangular co-ordinates of the satellite** are given by

$$x_0 = r_0 \cos\phi_0 \quad ...(2.45)$$

$$y_0 = r_0 \sin\phi_0 \quad ...(2.46)$$

The orbital period T is the time required for the satellite to complete one revolution and travel 2π radian. The **average angular velocity** is

$$\boxed{\eta = \frac{2\pi}{T} = \frac{\mu^{1/2}}{a^{3/2}} = \frac{1}{a}\left(\frac{\mu}{a}\right)^{1/2}} \quad ...(2.47)$$

If the time required for a spacecraft moving at this angular velocity to go around any circle is T sec., then the object will go around the circumscribed circle with a constant angular velocity η and it will complete one revolution as the satellite requires to complete one orbital revolution.

To find the equation of the orbit, we have to eliminate 't' from the equation of motion. But for the spacecraft location 't' is required.

We can relate the linear velocity v to r_0 and ϕ_0 by

$$v^2 = \left(\frac{dx_0}{dt}\right)^2 + \left(\frac{dy_0}{dt}\right)^2 = \left(\frac{dr_0}{dt}\right)^2 + r_0^2\left(\frac{d\phi_0}{dt}\right)^2 \quad \text{...(2.48)}$$

It can be shown that

$$v^2 = \frac{\mu}{a}\left(\frac{2a}{r_0} - 1\right) \quad \text{...(2.49)}$$

From Eqs. (2.21), (2.22) and (2.39)

$$r_0^2\left(\frac{d\phi_0}{dt}\right)^2 = \frac{h^2}{r_0^2} = \frac{\mu p}{r_0^2} = \frac{\mu a(1-e^2)}{r_0^2} \quad \text{...(2.50)}$$

then Eq. (2.48) will be

$$\frac{\mu}{a}\left(\frac{2a}{r_0} - 1\right) = \left(\frac{dr_0}{dt}\right)^2 + \left(\frac{\mu a}{r_0^2}\right)(1-e^2) \quad \text{...(2.51)}$$

and

$$\frac{dr_0}{dt} = \left\{\left(\frac{\mu}{ar_0^2}\right)\left[a^2e^2 - (a-r_0)^2\right]\right\}^{1/2} \quad \text{...(2.52)}$$

Solving the above equation for dt and multiplying by the mean angular velocity 'η', we have

$$\boxed{\eta . dt = \left(\frac{r_0}{a}\right)\frac{dr_0}{[a^2e^2 - (a-r_0)^2]^{1/2}}} \quad \text{...(2.53)}$$

Let us consider the geometry of the circumscribed circle as shown in Fig. 2.7. A line from the center of ellipse to the point A makes an angle E with the x_0 axis; E is called **eccentric anomaly** of the satellite and it is related to radius r_0 by

$$r_0 = a(1 - e\cos E) \quad \text{...(2.54)}$$

$$a - r_0 = ae\cos E \quad \text{...(2.55)}$$

Expressing the Eq. (2.53) in terms of E

$$\eta . dt = (1 - e\cos E)\, dE \quad \text{...(2.56)}$$

if $\quad t_p$ = Time of perigee

It is the time of closest approach to the earth when the satellite is crossing x_0 axis and time when E is zero. Integrating both sides of Eq. (2.56), we have

$$\boxed{\eta(t - t_p) = E - e\sin E} \quad \text{...(2.57)}$$

Left side of Eq. (2.57) is called **mean anomaly,** M

$$\boxed{M = \mu(t - t_p) = E - e \sin E} \qquad ...(2.58)$$

where, M = Arc length in radian

η = Mean angular velocity

2.2.4. Locating the Satellite with Respect to Earth

To locate the satellite at a point (x_0, y_0, z_0) in the recangular co-ordinate system of the orbital plane, we have to develop transformations that permits the satellite to be located from a point to the rotating surface of the earth.

We start with a fixed rectangular co-ordinate system (x_i, y_i, z_i) called the **geocentric co-ordinate system** whose origin is at the center of earth. It is shown in Fig. 2.8. z_i axis coincides with the earth's axis of rotation. X_i points towards a fixed location in space and is called the **first point of aries.** This is the direction of a line from the center of earth through the center of sun at the vertical equinox, at the instant when the subsolar point crosses the equator south to north. This co-ordinate system translates, as the earth revolves around the sun and it moves through space. X_i direction is the same as the earth's position. The plane (x_i, y_i) contains the earth's equator and is called **equitorial plane.**

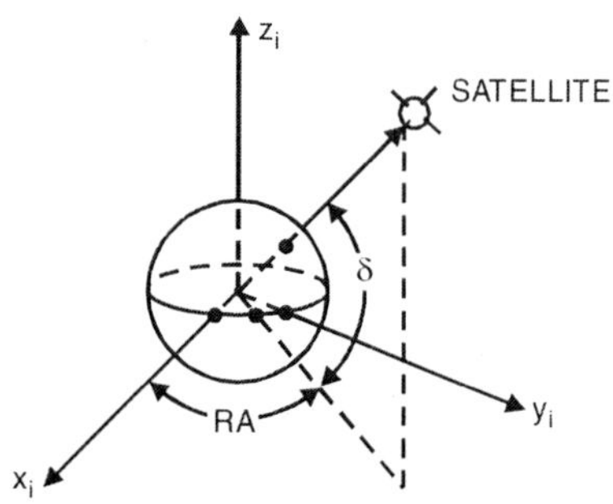

Fig. 2.8. The geocentric equitorial system.

The measurement of angular distance eastward in equitorial plane from the x_i axis is called **right ascension** and is denoted by symbol R_A. The points at which the orbit penetrates the equitorial plane are called **nodes.** Satellite moves upward through the equitorial plane at the **ascending node** and downward at any **descending node**. The right ascersion of ascending node is denoted by 'r'**.** The angle that the orbital plane makes with the equitorial plane is called **inclination** 'i' and all these quantities are shown in Fig. 2.9.

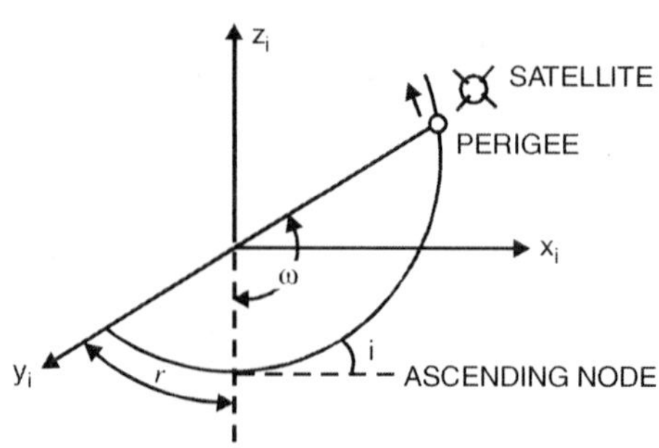

Fig. 2.9. Locating the orbit in geocentric equitorial system (Satellite penetrates the equitorial plane).

To locate the orbital co-ordinate system with respect to equitorial co-ordinate system, there is a need of ***w*** **argument of perigee.** This is the angle measured along the orbit from the ascending node to the perigee.

The orbital plane co-ordinate (x_0, y_0, z_0) of a satellite is related to the satellite co-ordinates (x_i, y_i, z_i) by a linear transformation and is given by

$$\begin{bmatrix} x_i \\ y_i \\ z_i \end{bmatrix} = \begin{bmatrix} (\cos r)(\cos w) - (\sin r)(\sin w)(\cos i) & -(\cos r)(\sin w) - (\sin r)(\cos i)(\cos w) & (\sin r)(\sin i) \\ (\sin w).(\cos w) + (\cos r)(\cos i).(\sin w) & -(\sin w)(\sin r) + (\sin r)(\cos i)(\cos w) & -(\cos w)(\sin i) \\ (\sin i)(\sin w) & (\sin i)(\cos w) & (\cos i) \end{bmatrix} \begin{bmatrix} x_0 \\ y_0 \\ z_0 \end{bmatrix} \quad ...(2.59)$$

One more co-ordinate transformation is required to locate the satellite w.r.t. a point on the rotating earth. So we need a rotating rectangular system (x_r, y_r, z_r) attached to the earth whose $x - y$ plane and z-axis corresponds to the geocentric equatorial system, as shown in Fig. 2.10. The rotating system turns at angular velocity Ω_e and T_e measured the elapsed time since x_r coincides with x_i axis.

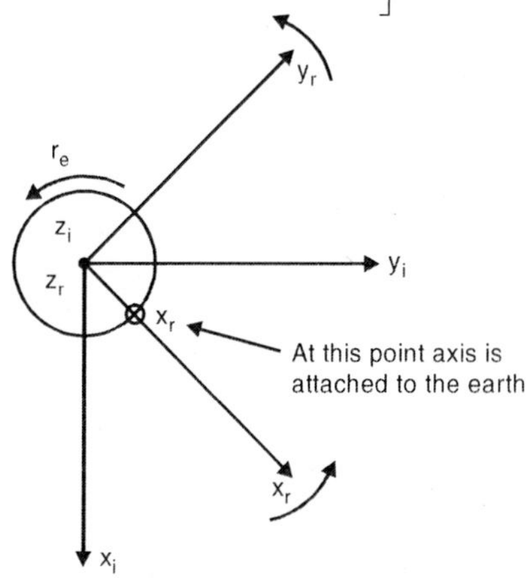

Fig. 2.10. The relationship between the rotating co-ordinate system (x_r, y_r, z_r) and the geocentric equatorial system (x_i, y_i, z_i).

The co-ordinates of satellite in the rotating system are related to the co-ordinates in the geocentric equatorial system and given by

$$\begin{bmatrix} x_r \\ y_r \\ z_r \end{bmatrix} = \begin{bmatrix} \cos(r_e T_e) & \sin(r_e T_e) & 0 \\ -\sin(r_e T_e) & \cos(r_e T_e) & 0 \\ 0 & 0 & 1 \end{bmatrix} \begin{bmatrix} x_i \\ y_i \\ z_i \end{bmatrix} \quad ...(2.60)$$

The value of $(r_e T_e)$ at any time t (in min) can be expressed as

$$r_e T_e = \alpha_{(q,0)} + 0.25068447t \text{ degree} \quad ...(2.61)$$

where, $\alpha_{(q,0)} = 99.6909833 + 36000.7689T_c + 0.000387087T_c^2$ degree ...(2.62)
(right ascersion)

and T_c is the elapsed time in Julian centuries between $0h$ UT and it is calculated by

$$T_c = \frac{(JD - 2415020)}{36525} \text{ Julian centuries} \quad ...(2.63)$$

where, JD = Julian day

2.2.5. Orbital Elements

Orbital elements are the six quantities required to determine the absolute coordinates of a satellite at time t. Some of the important terms required inorder to understand the orbital elements are given below:

- **Apogee:** The point farthest from the earth.
- **Perigee:** The point of closest approach to earth.
- **Line of absides:** The line joining the perigee and apogee through the center of the earth.
- **Ascending node:** The point where the orbit crosses the equatorial plane going from south to north.
- **Descending node:** The point where the orbit crosses the equatorial plane going from north to south.
- **Line of nodes:** The line joining the ascending and descending nodes through the center of the earth.
- **Inclination:** The angle between orbital plane and the earth's equatorial plane. It is measured at the ascending node from equator to the orbit, going from east to north. The inclination is shown as [i].
- **Prograde orbit:** An orbit in which the satellite moves in the same direction as the earth's rotation. The prograde orbit is also known as the direct orbit. The inclination of a prograde orbit because the earth's rotational velocity provides part of the orbital velocity with a consequent saving in launch energy.
- **Retrograde orbit:** An orbit in which the satellite moves in a direction of satellite motion.
- **Argument of perigee:** The angle from ascending node to perigee, measured in the orbital plane at the earth's center, in the direction of satellite motion. The argument of perigee is as shown in the figure.

The orbital elements are :

- **The semi major axis (a):** It represents half the length of the major axis of the elliptical orbital described by the satellite.
- **The eccentricity (e):** Eccentricity of the ellipse.
- **The mean anomaly (M):** Anomaly is the term for angle. Mean anomaly is the angle which describes the position of the satellite in its orbit relative to the perigee. At perigee, mean anomaly is zero. It increases to 180 degrees at apogee and back to perigee at 360 degrees .
- **The argument of perigee (ω):** It is the angle between the perigee point on the semi-major axis to the semi latus rectum which joins the satellite to the centre of the earth.
- **The inclination (i):** The angle that the orbital plane makes the equatorial plane is called the inclination.
- **The right ascension of the ascending node (r):** It is the angle between the vernal equinox line and the line of nodes measured in the equatorial plane.

2.2.6. Apogee and Perigee Heights

Apogee is that point in the satellite orbit that is the farthest from the center of earth and the apogee distance is shown by r_a.

$$\boxed{r_a = a(1 + e)} \qquad \text{...(2.64)}$$

The satellite velocity is the lowest at the apogee point.

Perigee is the point in the orbit that is nearest to the earth. The perigee distance is shown by 'r_p' and can be written as,

$$r_p = a(1 - e) \quad \text{...(2.65)}$$

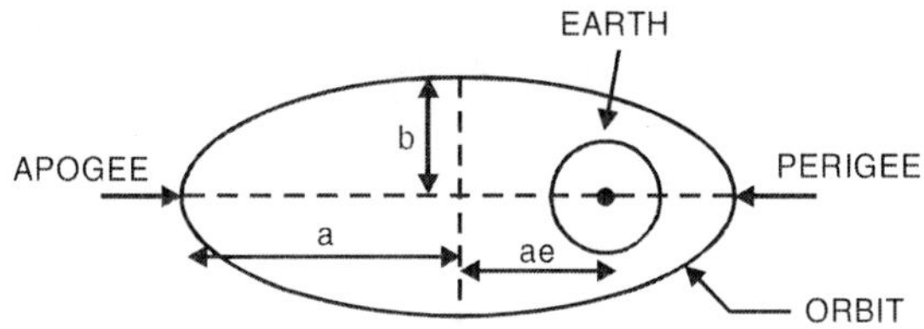

Fig. 2.11.

In order to find out the apogee and perigee heights. The radius of the earth must be subtracted from the radi lengths.

2.3. LOOK ANGLE'S

To maximize transmission and reception, the direction of maximum gain of the earth station antenna (known as bore sight) must point directly at the satellite.

So to align the antenna, two angles must be known. The co-ordinates to which a earth station antenna must be pointed to communicate with a satellite are called the **look angles**. These angles are :

- Azimuth angle (A) : Angle measured from true north.
- Elevation angle (E) : Angle measured up from the local horizontal plane.

These angles can be calculated on the basis of the knowledge latitude (ϕ) and relative longitude (θ) of the earth station.

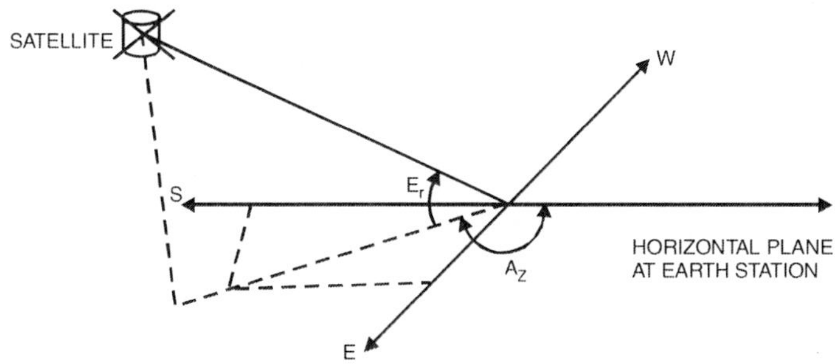

Fig. 2.12.

Earth's constants required in the calculations are :

Mean radius, $R = 6378$ km

Radius of geostationary orbit,

$$\alpha_{gso} = 42164 \text{ km}$$

Other required information are :

- Satellite longitude, ϕ_s

- Earth station longitude, ϕ_E
- Earth station latitude, λ_E

Conventions used in calculations are :

- East longitudes → +ve number
- West longitudes → –ve numbers.

2.3.1. Azimuth Angle

It is defined as the angle by which the antenna pointing at the horizon must be rotated clockwise around its vertical axis from the geographical north to bring the antenna boresight into the vertical plane containing the satellite direction. Its value varies in between 0° and 360°.

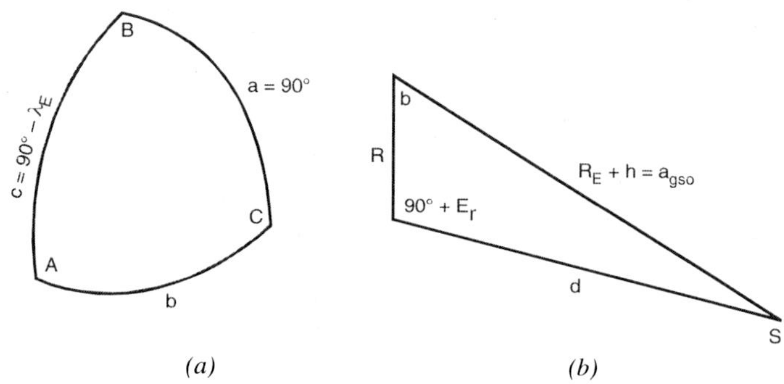

Fig. 2.13.

Property of spherical trigonometry :

All the dimension including the sides are in angular measure.

Angle B = Difference in Longitude between earth station and subsatellite longitude

$$B = \phi_E - \phi_s$$

$$\angle C = 90° - \lambda_E$$

$\angle A$ can be calcualted by Napier's rule

$$\tan A = \frac{-\tan|B|}{\sin \lambda_E}$$

If A is known, azimuth angle can be determined.

Four situations must be considered :

	Condition	Azimuth
1.	$\lambda_E < 0$ and $B < 0$	$A_z = A$
2.	$\lambda_E < 0$ and $B > 0$	$A_z = 360° - A$
3.	$\lambda_E > 0$ and $B < 0$	$A_z = 180° + A$
4.	$\lambda_E > 0$ and $B > 0$	$A_z = 180° - A$

2.3.2. Elevation Angle

It is defined as the angle by which the antenna boresight should be rotated in the vertical plane that contains the satellite direction from the horizontal to the satellite direction.

To find the range and elevation :

Another Napier's can be used.

$$\cos b = \cos \lambda_E \cos B$$

Taking the assumption of the spherical earth and ignoring earth station altitude is quite adequate.

The mean radius is taken as :

$$R = 6371 \text{ km}$$

Applying cosine rule to the plane triangle gives,

$$d = \sqrt{R^2 + a_{gso}^2 - 2Ra_{gso} \cos b}$$

Now **elevation** can be determined from application of the sine rule for plane triangles. It gives :

$$\cos El = \frac{a_{gso}}{d} \sin b$$

Angle of Tilt : It is also known as declination which is used in compass correlation.

$$\delta = 90° - El - \lambda_E$$

or

$$\cos El = \frac{a_{gso}}{d} . \sin \lambda_E$$

Therefore in terms of E_s latitude and distances only, the **angle of tilt** is given by

$$\delta = 90° - a \cos \left(\frac{a_{gso}}{d} \sin \lambda_E \right) - \lambda_E$$

Polarization (ψ) angle is defined as the angle between the polarization plane of a linear polarized wave transmitted by the satellite and the polarization plane of earth station antenna, it may be calculated as

$$\tan \psi = \frac{\sin \theta}{\tan \phi} \quad ...(2.66)$$

2.4. EARTH COVERAGE AND SLANT RANGE

From the Fig. 2.14, the co-ordinates of an earth station may be written as

$$x = R_0 \cos \phi . \sin \theta$$
$$y = R_0 + R_e (1 - \cos \phi . \cos \theta)$$
$$z = R_e \sin \phi$$

The **slant range *R*** from the earth station to the satellite can be given as

$$R^2 = R_0^2 + 2R_e(R_0 + R_e)(1 - \cos\phi \, . \cos\theta) \qquad ...(2.67)$$

for $\frac{R_e}{R_0} \simeq 0.178$, it will be

$$\left(\frac{R}{R_0}\right)^2 = 1 + 0.42\,(1 - \cos\theta \, . \cos\phi) \qquad ...(2.68)$$

From the above equation, it is clear that the maximum value of $\left(\frac{R}{R_0}\right)^2$ is 1.356 and if R^2 is approximately equal to R_0^2 then there will be a error of 1.3 dB in power link. The earth can be covered by a single beam communication satellite and it is desired that it should cover the entire country and this converge should be as maximum as possible. The maximum coverage of the earth by a geosationary satellite is shown in Fig. 2.14.

$$\boxed{\theta_s = 180° - (90° + E + \alpha) = 90° - E - \alpha} \qquad ...(2.69)$$

The maximum geometric coverage is given by the portion of the earth within a cone with the satellite at its apex and tangent to the earth's surface. The coverage angle or the apex angle of this cone is $2\alpha_{max.}$ or

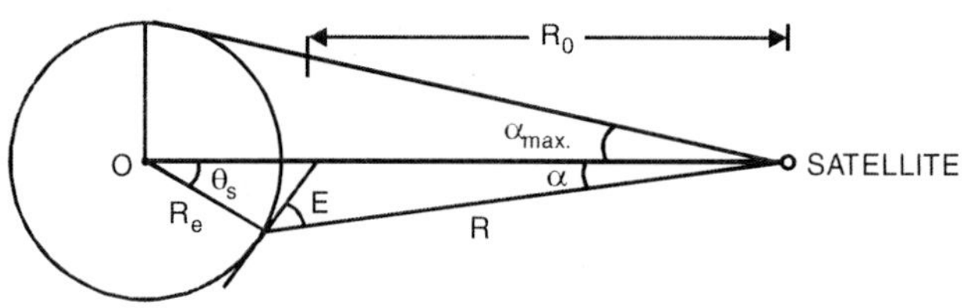

Fig. 2.14. Elevation angle and coverage angle with reference to synchronous satellite.

$$2\, arc \sin\left(\frac{R_e}{R_e + R_0}\right) = 17.4° \qquad ...(2.70)$$

θ_s represents the angular radius of the satellite footprint and is called **central angle.**

For a geostationary orbit θ_s can be calculated by putting $\alpha = \alpha_{max}$ and $E = 0$ in Eq. (2.65)

So, $\theta_s = 81.3°$

But for all practical purpose an earth station requires a minimum elevation angle of 5° above the horizon.

Then, $\theta_s = 76.3°$

So not more than 76.3° of northern or southern latitudes will be covered by the geosynchronous satellite.

Slant range in terms of elevation angle may given as

$$\boxed{R^2 = (R_e + R_0)^2 + R_e^2 - 2R_e(R_e + R_0) \times \sin\left[E + \sin^{-1}\frac{R_e}{R_e + R_0}\cos E\right]} \qquad ...(2.71)$$

For $E_{min} = 5°$

Maximum slant range will be, $R_{max} = 41.127$ km

it gives a satellite **round trip delay of,** $\frac{2d}{c} = 0.274$ s

where, c = Velocity of light

2.5. ORBIT PERTURBATIONS

The type of orbit described for satellite, referred to as **Keplerian orbit.** It is elliptical for the special case of an artificial satellite orbiting the earth. In practice, the satellite and the earth respond to many other influences like :

- assymetry or the earth's gravitational field.
- The gravitational field of the sun and moon.
- Solar radiation pressures.
- Some other factors like atmospheric drag.

But it does affect low orbiting earth satellites below about 1000 km.

2.5.1. Effect of a Non-spherical Earth

The earth is not perfectly spherical, there is an equatorial bulge and a flattering at the poles and this type of shape is known as **oblate spheroid.**

Kepler's third law gives the nominal mean motion as,

$$n_0 = \sqrt{\frac{\mu}{a^3}} \qquad ...(2.72)$$

When the earth oblateness is taken into account, the **mean motion *n*** is modified to,

$$\boxed{n = n_0 \left[\frac{1 + K_1 (1 - 1.5 \sin^2 i)}{a^2 (1 - e^2)^{1.5}} \right]} \qquad ...(2.73)$$

where, K_1 is a constant which evaluates to 66063.1704 km^2.

The orbital period of the earth's oblateness is termed the **anomalistic period.**

The **anomalistic period** is

$$\boxed{P_A = \frac{2\pi}{n} \text{ sec.}} \qquad ...(2.74)$$

where, n = radians per second

Equation (2.73) may be solved for 'a' by finding the root of the following equation.

$$n - \sqrt{\frac{\mu}{a^3}} \left[1 + \frac{K_1 (1 - 1.5 \sin^2 i)}{a^2 (1 - e^2)^{1.5}} \right] = 0 \qquad ...(2.75)$$

The oblateness of the earth also produces two relations of the orbital plane.

- The first is known as **regression of the nodes**, is where the nodes appear to slide along the equator. Line of nodes, which is in the equatorial plane, rotates about the center of earth. Thus r, the right ascension of the ascending node shifts its position.
- The second effect is **rotation of apsides** in the orbital plane.

In addition to the equitorial bulge, the earth is not perfectly circular in the equitorial plane, it has a small eccentricity of the order of 10^{-5}. This is referred to as the **equatorial ellipticity.** This effect setup a gravity gradient which has some effects on satellites in geostationary orbit. The gravity gradient causes the satellite in geostationary orbit to drift to one or two stable points. This effect of equatorial ellipticity is negligible on most other satellite orbits.

2.5.2. Effects of the Sun and Moon

Because of the gravitational attraction of sun and moon there is a change in orbital inclination of a geosynchronous satellite which changes with time. These forces would increase the orbital inclination from an initial 0° at launch to 14.67°, 26.6 years later. But no commercial satellite has a 26 years useful life time. The rate of change varies with the inclination of the moon's orbit, but values of about 0.86° per year are quoted for the 1970-80 time period.

2.6. ORBITAL DETERMINATION

To determine six orbital elements, orbital determination involves making sufficient measurement of the location, or location and velocity of the spacecraft.

- This is an old problem in astronomy and techniques were developed by Gauss, Laplace and others.
- The technique is to calculate an orbit based on the angular position of an object measured at three different times from the same terrestrial location.
- Three angular position measurements are required because there are six unknowns and each measurements provides two equations.
- There may be difficulty because one equation gives azimuth and another equation gives elevation as a function of the six unknown orbital elements.
- Traditional optical astronomy was limited to measure angular position. Radar technique permits measurement of range and range–rate simultaneously with position. These reduces the number of observations required at one location to two.
- Alternatively, accurate time coordination between two or more earth stations permits orbits to determine from simultaneous position and range measurements at two or more sites.

2.7. ORBITAL EFFECTS IN COMMUNICATION SYSTEM PERFORMANCE

2.7.1. Doppler Shift

The frequency of a moving radio transmitter varies with transmitter's velocity to a stationary observer. If,

$$f_T = \text{True transmitter frequency}$$
$$f_R = \text{Received frequency}$$

Then,

$$f_T < f_R = \text{Transmitter is moving towards the receiver}$$
$$f_T > f_R = \text{Transmitter is moving away from the receiver.}$$

Mathematically, the relationship between the transmitted and received frequencies is :

$$\boxed{\frac{f_R - f_T}{f_T} = \frac{\Delta f}{f_T} = \frac{v_T}{v_P}} \quad ...(2.76)$$

where, v_T = Component of the transmit velocity directed toward the receiver.

v_P = Phase velocity of light

If transmitter is moving away from the receiver then v_T is negative. This change in frequency is known as **Doppler shift.**

Doppler shift is more dominated for low orbit satellites and its compensation requires frequency tracking in narrowband receivers. Its effects are negligible for geosynchronous satellites.

2.7.2. Eclipse Effects

Earth and moon affect the working of a satellite because of solar eclipse. So the periodicity and duration of solar eclipse are important. Solar eclipse due to earth lasts for several days so it is more important than the eclipse due to moon on the communication satellite. Summer solstice occurs on 21st June and winter solstice is on 21st December. Similarly spring and autumn equinoxes occur on 21st March and 21st September respectively.

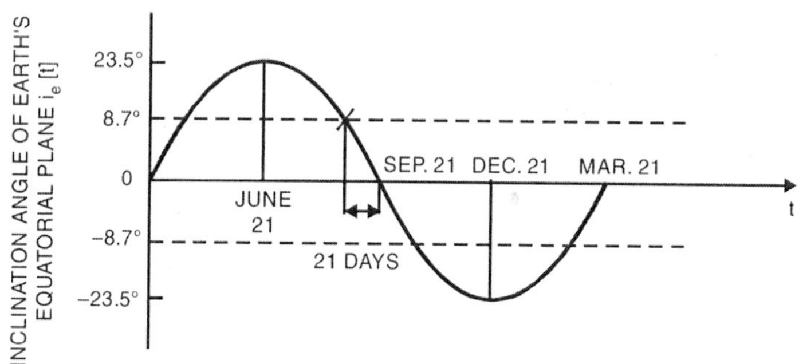

Fig. 2.15. Variation of the earth's equatorial plane inclination angle i_e (*t*).

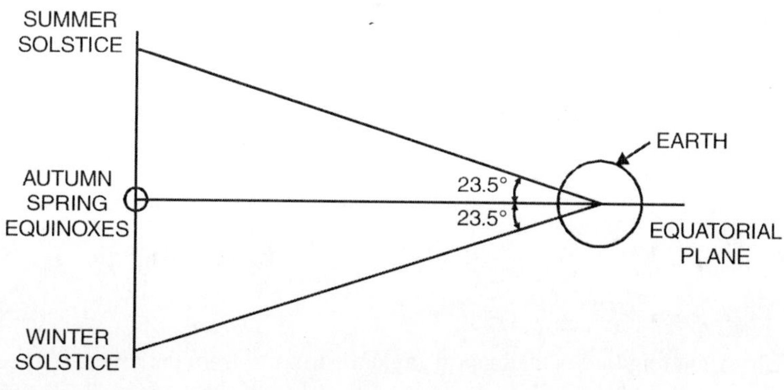

Fig. 2.16. Apparent movement of the sun relative to the earth.

In the Fig. 2.15, the inclination angle of the earth's equatorial plane i_e (t) with respect to the sun's direction is shown. Here the angle i_e (t) is in degree and annual variation is sinusoidal in nature. The apparent movement of the sun relative to the earth is shown in Fig. 2.16.

Mathematically the variation of $i_e(t)$ can be written as

$$\boxed{i_e(t) = 23.5 \sin \frac{2\pi t}{T}} \quad \text{...(2.77)}$$

where, T = Annual period equal to 365 days.

and $$i_e\,(t)]_{max} = 23.5° \quad \text{...(2.78)}$$

i_e (t) is maximum on summer and winter solstices and it is zero at spring and autumn equinoxes.

The satellite is always illuminated at the solstice. it goes the maximum duration of eclipse during equinoxes.

The solar eclipse during equinox is shown in Fig. 2.17. The earth shadow is considered to be a cylinder of constant diameter and **the maximum shadow angle at equinox is given as**

$$\boxed{\phi_{max} = 180° - 2\cos^{-1}\left(\frac{R_e}{R_e + R_0}\right)} \quad \text{...(2.79)}$$

It is same as that of coverage angle. Substituting the value, R_e = 6378 km, R_0 = 35786 km (satellite attitude)

$$\phi_{max} = 17.4°$$

Duration of eclipse whose period is 24 hours is given as

$$t_{max.} = \frac{17.4°}{360°} \times 24 \cong 69.4 \text{ min.}$$

$$\simeq 1.16 \text{ hrs.}$$

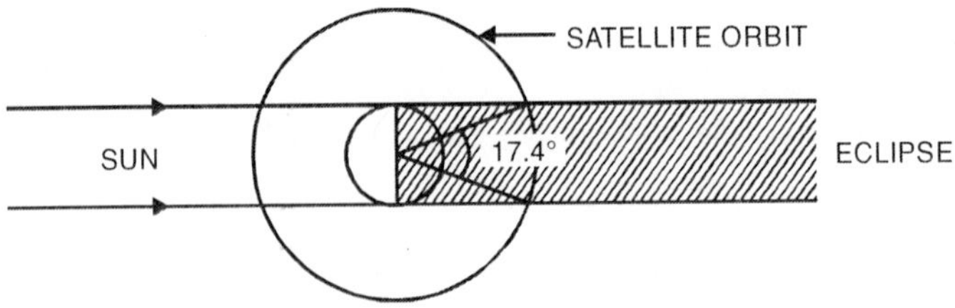

Fig. 2.17. Eclipse due to earth when sun is at equinox.

The relative position of a sun is calculated from the first ellipse day before an equinox and the last day of eclipse after an equinox. In this case, the inclination angle is given as

$$i_e = \frac{1}{2}\phi_{max} = 8.7° \quad \text{...(2.80)}$$

Corresponding to this inclination angle the time 't' from the first day of eclipse to the equinox and the time from the equinox to the last day of eclipse may be calculated as

$$t = \frac{365}{\pi} \sin^{-1}\left(\frac{87}{23.5}\right) \simeq 21 \text{ days} \qquad ...(2.81)$$

So to reach the equinox from the first day of eclipse 21 days are required. Thus in total for 42 days the solar eclipse near equinox affects the satellite working.

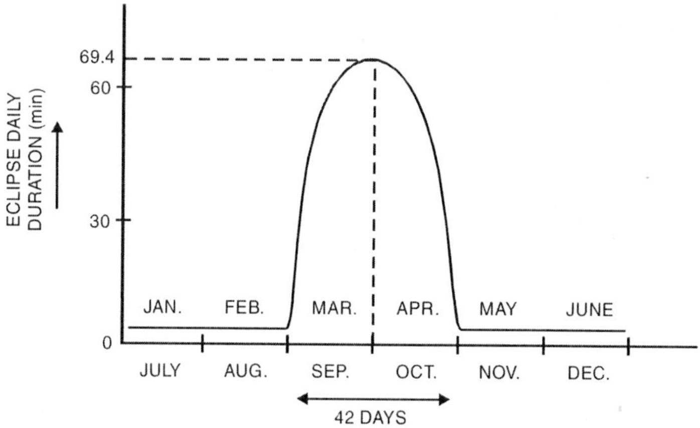

Fig. 2.18. Daily duration of solar eclipse caused by earth.

The solar eclipse caused by moon to the geostationary satellite occurs when the moon passes in front of the sun. It occurs irregulary in duration often it occurs twice within a period of 24 hours and it may range from a few minutes to over two hours period with an average duration of about 40 minutes. In comparison with earth solar eclipse, the number of moon solar eclipse range from zero to four with an average of two per year.

2.7.3. Station Keeping

Station keeping is the process of maintenance of satellite's attitude against different factors that can cause drift with time. Satellites need to adjust their orbit timely because natural forces induces a progressive drift even initially the satellite is placed in correct orbit. The factors that produce drift include :

- Minor gravitational perterbations due to sun and moon.
- Solar radiation pressure.
- The effect of non-spheric earth.

The orbital adjustments are normally carried out by releasing jets of gas or by firing small rockets tied around the body of satellite.

It is already mentioned that a window corresponding to a drift of 75 km on the sphere containing the geostationary satellite orbit is around. It corresponds to an angular shift of ±0.1° in longitude for fixed satellite and also for broadcasting satellite services. This station keeping accuracy has been internationally standardized by CCIR. It is essential that in station keeping orbit correction techniques propellent consumption should be minimum. The strategy usually has certain steps, firstly the direction and the speed of drift of the satellite is determined using a new series of measurement. After that calculations the amplitude and direction of velocity increments are required to modify the orbital parameters.

Compensation of the inclination increments corresponds to North-South (N-S) station keeping and requires a thrust impulse perpendicular to the orbital plane. East-west (E-W) station keeping are obtained by applying thrusts in the orbital plane and it corresponds to ±0.1° E-W accuracy. If the life time of satellite is T_L years, **maximum value of inclination angle** will be

$$\boxed{i_{max.} = \frac{1}{2} i_{av.} \cdot T_L} \qquad ...(2.82)$$

where, i_{av} = Natural annual change of inclination averaged over the life time T_L.

2.7.4. Sun Transit Outage

During the equinox periods, the orbit of the satellite will also pass directly in front of the sun on the sunlit side of the earth. The sun is a hot microwave source with an equivalent temperature of abot 6000 to 10000 K, depending on the time within the 11-year sunspot cycle, at the frequencies used by the communication satellite (4 to 50 GHz). The earth station will therefore receive not only the signal from the satellite but also the noise temperature transmitted by the sun. The added noise temperature will cause the fade margin of the receiver to be exceeded and an outage will occur. This is termed as Sun transit outage.

This effect is predictable and can cause outage for as much as 10 minutes a day for several days and for about 0.2% of an average year. The receiving earth station has no other alternative than to wait for the sun to move out of the main lobe of the antenna.

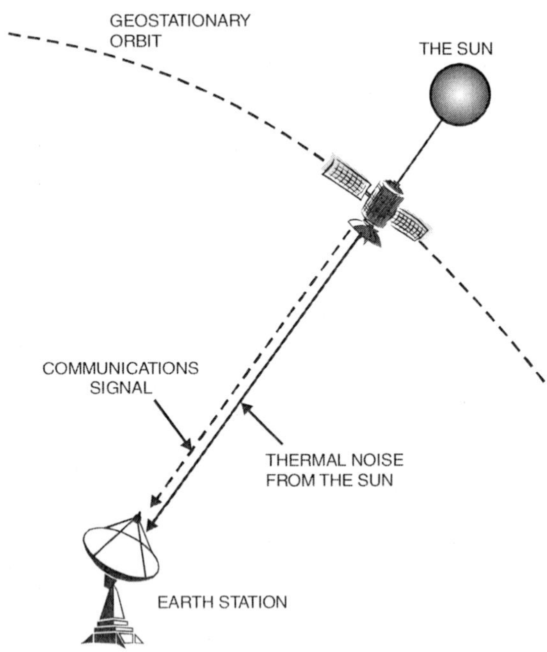

Fig. 2.19

2.8. GEOSTATIONARY AND OTHER ORBITS

Geostationary Orbit

- A geostationary orbit is one in which a satellite moves in the orbit at exactly the same speed as the earth turns and, at the same lattitude specifically zero, the lattitude of the equator.
- These orbits remain in a fixed position w.r.t. a given earth station and have an hour availability time. Therefore, these satellites are used for fixed point communication.

It is an equitorial circular orbit at such a height from earth's surface that a satellite moving from west to east has a velocity so as to orbit the earth once in a time equal to the time taken by earth to complete one rotation about its axis. This time is about 24 hours. In such cases the satellite appears stationary to an observer on earth. The orbital radius of geostationary satellite is 42164.2 km. The movement of satellite is in synchronism with earth's rotation. All geostationary satellites are geosynchronous but vice versa is not true.

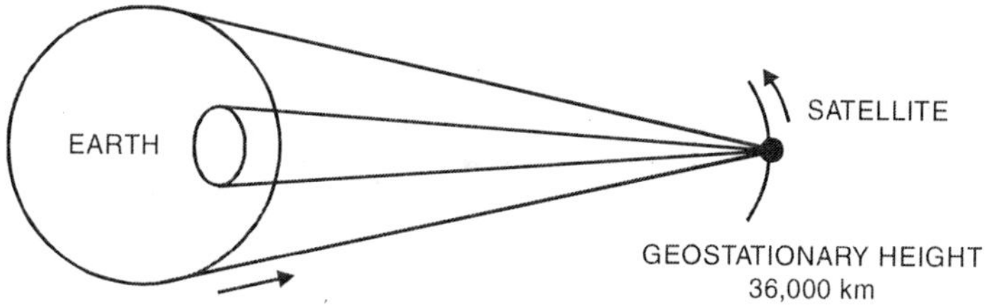

Fig. 2.20. Geostationary orbit periods of revolution equals the earth's rotational period.

The essential requirements for an orbit to be geostationary are :

(1) The satellite should have the angular velocity equal to that of the earth *i.e.,* it should be geosynchronous.

(2) The orbit should have zero degree inclination with the equatorial plane. Any other inclination will not make the satellite staionary for some latitudes, even though it might be geosynchronous.

(3) For zero degree inclination, the orbit must be in the earth's equatorial plane.

(4) The satellite should travel eastward.

(5) As the earth's angular velocity is constant, the orbit has to be circular.

The above criteria can be fulfilled by one orbit only. Hence there is only one geostationary orbit. At present it is the most widely used orbit and is known as **'dark orbit'.**

Polar orbit :

- An orbit which passes over the poles. Polar orbits are relevant for those countries which are located near the poles and it is shown in Fig. 2.21.

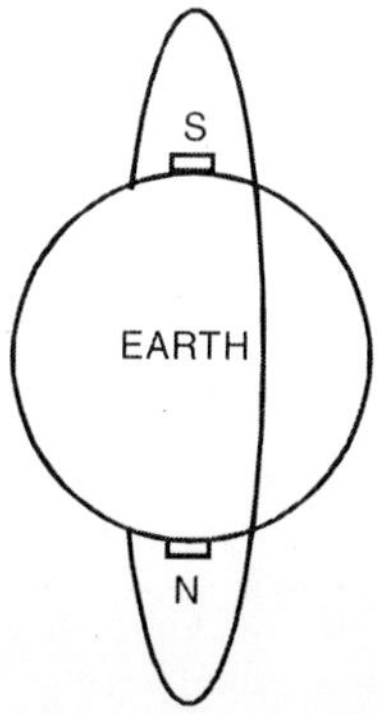

Fig. 2.21. Polar orbit.

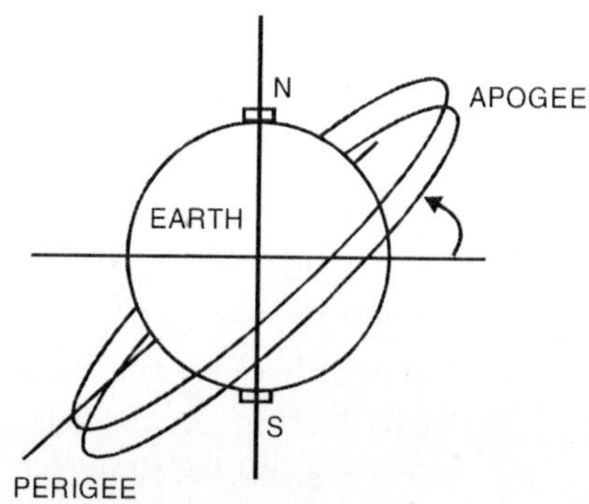

Fig. 2.22. Molniya orbit.

- Polar orbits are used for special applications like navigational satellites.
- Polar orbit has some inclination with respect to the equitorial plane.

Molniya Orbit

It is a highly inclined, highly eccentric, elliptical orbit with an orbital period of about 12 hours. It is popular with russion communication satellites and it serves the purpose of a geosynchronous orbit for high latitude regions. Molniya orbits are capable of providing uninterrupted service if three satellites are placed at different phases and it is shown in Fig. 2.22.

Sun Synchronous Orbit

- The plane of sun synchronous orbit maintain a constant aspect angle with the direction of the sun, the therefore it is known as sun synchronous orbit.
- Earth resource satellites are launched in sun synchronous orbits.

Parking Orbit

Parking orbit is a temporary earth orbit during which the space vehicle is checked out and, its trajectory carefully measured to determine the amount and time of increase in velocity required to send it into a final orbit or into space in the desired direction. (Fig. 2.23)

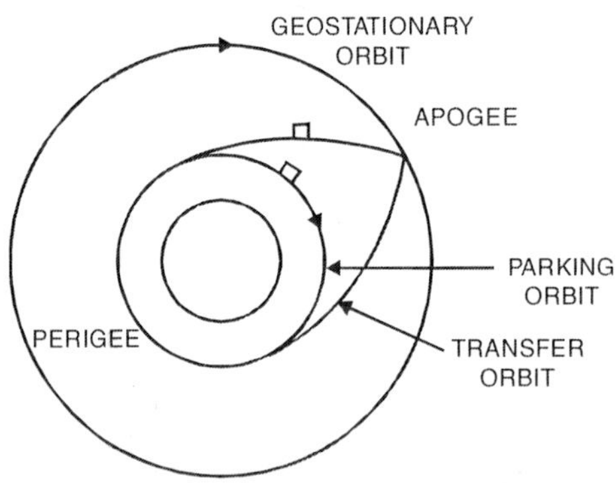

Fig. 2.23. Parking orbit.

Equatorial Orbit

It is a orbit in the plane of equator. It is also known as geostationary orbit.

Inclined Orbit

- Inclined orbits have an inclination between 0° (equatorial orbit) and 90° (polar orbit). These orbits may be determined either by the region on earth or by the lattitude on launch sites.
- These satellites are not sun-synchronous.
- Inclined orbit is used for domestic communication in Russia as its geographical location is near the north pole of the earth.
- The height of inclined orbit is set to cover the required area of interest.

2.9. MECHANISM OF LAUNCHING A SATELLITE

Theoretically a satellite could be placed into geosynchronous orbit in one step but practically it is a two or three step process. The placement of satellite in the geostationary orbit is carried out on the principle of **Hohmann Transfer** and it is shown in Fig. 2.24.

There are **three techniques :**

(*i*) In the first technique, first the satellite is placed in lower circular earth orbit at an altitude of around 300 km. Then a velocity increment is required and it is carried out by various auxiliary propulsion stages. One velocity increment changes the satellites lower circular earth orbit into an elliptical transfer orbit with a perigee of about 300 km and apogee at 42164.2 km (radius of geosynchronous orbit). Then a second velocity increment is required and it finally places the satellite into the desired orbit. **Space transport system (STS)** follows this technique.

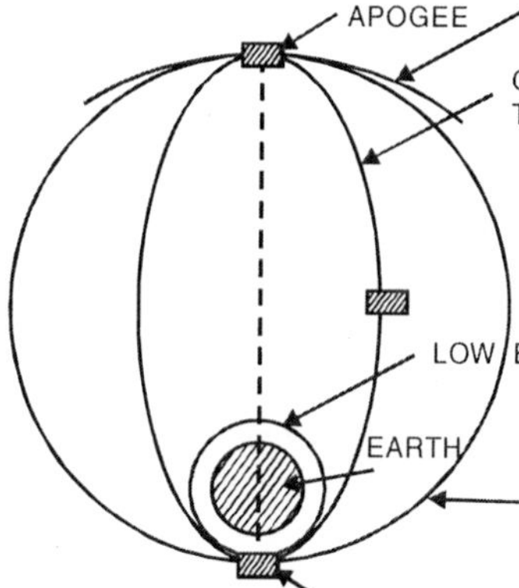

Fig. 2.24. Satellite placement in geostationary orbit.

(*ii*) The second technique has been used by **expandable launch vehicles** such as Ariane, Delta or Atlas-centaur launchers. In this case there is no initial circular orbit and the vehicle provides the necessary velocity at the perigee of the elliptical transfer orbit. Thus here only one velocity increment is required from the satellite at the apogee.

(*iii*) The third technique has been used by **special expendable launch vehicles** such as US Titan IIII C and the USSR Proton launchers. In this case satellites are directly placed into geostationary orbit.

The satellite velocity at the apogee and perigee can be calculated as given below :

$$\boxed{v_s^2 = \frac{2\mu}{r} - \frac{\mu}{a}} \qquad \text{...(2.83)}$$

where, μ = Gravitational constant

a = Semi-major axis

r = Distance from the center of earth to the point where velocity is required.

Solved Examples

Example 2.1. The orbital period of a satellite is 650 min. Determine the semi-major axis of the elliptical orbit.

Solution. Given that,

$$T = 650 \text{ min} = 650 \times 60 \text{ sec.} = 39000 \text{ sec.}$$

$$\mu = GM$$

$$= 6.67 \times 10^{-11} \times 5.98 \times 10^{24} = 39.8 \times 10^{13} \text{ Nm}^2/\text{kg}$$

The orbital time period (T) is given by

$$T = 2\pi\sqrt{\frac{a^3}{\mu}} = 2\pi\sqrt{\frac{a^3}{GM}}$$

$$39000 = 3.14 \times 2\sqrt{\frac{a^3}{39.8 \times 10^{13}}}$$

$$\Rightarrow \quad \sqrt{\frac{a^3}{39.8 \times 10^{13}}} = 6210.191$$

$$a^3 = 1534945636.7398 \times 10^{13}$$

$$a = 535.424 \times 10^4$$

$$\boxed{a = 5350 \text{ km}} \quad \textbf{Ans.}$$

Example 2.2. The apogee and perigee of an elliptical satellite orbits are 3000 km and 200 km. Determine the eccentricity, semi-major axis and semi-minor axis.

Solution. Given that,

$$r_a = 3000 \text{ km}$$

$$r_p = 200 \text{ km}$$

Eccentricity, $$e = \frac{r_a - r_p}{r_a + r_p} = \frac{3000 - 200}{3000 + 200}$$

$$\Rightarrow \quad \boxed{e = 0.875} \quad \textbf{Ans.}$$

Semi-major axis, $$a = \frac{r_a + r_p}{2} = \frac{3000 + 200}{2}$$

$$\Rightarrow \quad \boxed{a = 1600 \text{ km}} \quad \textbf{Ans.}$$

Semi-minor axis, $$b = \frac{r_a - r_p}{2} = \frac{3000 - 200}{2}$$

$$\Rightarrow \quad \boxed{b = 1400 \text{ km}} \quad \textbf{Ans.}$$

Example 2.3. A geostationary satellite moving in an equatorial circular orbit is at a height of 35786 km from the earth's surface. If the radius of the earth is 6378 km. Determine the theoretical maximum coverage angle and maximum slant range.

Solution. Given that,

$$R_0 = 35786 \text{ km}$$

$$R_e = 6378 \text{ km}$$

For theoretical maximum coverage angle,

Elevation angle, $E = 0$

$$\text{Maximum coverage angle} = 2\alpha_{max.}$$

$$= 2\sin^{-1}\left(\frac{R_e}{R_e + R_0}\cos E\right)$$

$$= 2\sin^{-1}\left(\frac{6378}{6378 + 35786}\cos 0\right) = 17.4°$$

If the maximum slant range is R then

$$R^2 = R_e^2 + (R_e + R_0)^2 - 2R_e(R_e + R_0) \times \sin\left[E + \sin^{-1}\frac{R_e}{R_e + R_0}\cos E\right]$$

$$= (6378)^2 + (6378 + 35786)^2 - 2 \times 6378\,(6378 + 35786)\sin(8.7°)$$

$$= 40678884 + 1777802896 - 537843984 \times 0.1512$$

$$R^2 = 1737139041$$

$$\Rightarrow \quad \boxed{R = 41679 \text{ km}} \quad \textbf{Ans.}$$

Example 2.4. A satellite is moving in a circular orbit at a height of 200 km above the surface of earth. Determine its orbital velocity. ($G = 6.67 \times 10^{-11}$ Nm²/kg², $m = 5.98 \times 10^{+24}$ kg)

Solution. The orbital velocity (V) is given by

$$V = \sqrt{\frac{\mu}{(R + H)}}$$

where,

$$\mu = GM$$

$$= 6.67 \times 10^{-11} \times 5.98 \times 10^{24}$$

$$= 39.8 \times 10^{13} \text{ Nm}^2/\text{kg}$$

$$R = \text{Radius of earth} = 6370 \text{ km}$$

$$H = 200 \text{ km}$$

Putting all the values, we get

$$V = \sqrt{\frac{39.8 \times 10^3}{(6370 + 200) \times 10^3}}$$

$$= 10^5\sqrt{\frac{39.8}{6570}} \Rightarrow 10^4\sqrt{0.6057}$$

$$= 10^4 \times 0.778$$

$$\boxed{V = 7.78 \text{ km/sec.}} \quad \textbf{Ans.}$$

Example 2.5. Determine the escape velocity for an object which is launched from the surface of earth to a point where the radius of earth is 6360 km. ($G = 6.67 \times 10^{-11}$ Nm²/kg², $M = 5.98 \times 10^{24}$ kg).

Solution. Escape velocity is given by

$$V_{escape} = \sqrt{\frac{2\mu}{r}}$$

where,
$$\mu = GM = 39.8 \times 10^{13} \text{ Nm}^2/\text{kg}$$
$$r = 6360 \text{ km}$$

Putting these values, we get

$$V_{escape} = \sqrt{\frac{2 \times 39.8 \times 10^{13}}{6360 \times 10^3}} = \sqrt{\frac{79.6}{6360}} \times 10^5 = 11.2 \text{ km/sec.}$$

$$\boxed{V_{escape} = 11.2 \text{ km/sec.}} \quad \textbf{Ans.}$$

Example 26. A satellite is moving in a circular orbit at a height of 150 km above the surface of earth. If the radius of earth is 6360 km, determine the orbital velocity and orbital period of the satellite. ($G = 6.67 \times 10^{-11}$ Nm²/kg, $M = 5.98 \times 10^{24}$ kg).

Solution. Given that, $H = 150$ km, $R = 6360$ km

$$\mu = GM = 6.67 \times 10^{-11} \times 5.98 \times 10^{24} = \mathbf{39.8 \times 10^{13}}$$

Orbital velocity of a satellite is given by

$$V_{orbital} = \sqrt{\frac{\mu}{(R+H)}} = \sqrt{\frac{GM}{(R+H)}}$$

Putting all values we get,

$$= \sqrt{\frac{39.8 \times 10^{13}}{(6360+150) \times 10^3}}$$

$$= \sqrt{\frac{39.8 \times 10^{13}}{6510 \times 10^3}} = 10^4 \times \sqrt{\frac{3980}{6510}}$$

$$\Rightarrow \quad 10^4 \times \sqrt{0.611367} = 10^4 \times 0.782$$

$$\boxed{V_{orbital} = 7.82 \text{ km/sec.}} \quad \textbf{Ans.}$$

Orbital period of a satellites

$$= \frac{2\pi(R+H)}{V_{orbital}}$$

$$= \frac{2 \times 3.14 \times (6360+150)}{7.82} = \frac{6.28 \times 6510}{7.82}$$

$$= 5227.98 \text{ sec.}$$

$$\boxed{T_{orbital} = 87.13 \text{ Min.}} \quad \textbf{Ans.}$$

Example 2.7. A satellite is moving in an elliptical transfer orbit with apogee and perigee at a distance of 35000 km and 500 km. If the radius of earth is 6360 km. Determine the velocity of a satellite at any point in its orbit.

Solution. The velocity (V) at any point along an elliptical orbit is given by

$$V = \sqrt{\mu\left(\frac{2}{r} - \frac{1}{a}\right)}$$

Given that,

$$\mu = GM = 6.67 \times 10^{-11} \times 5.98 \times 10^{24}$$
$$= 39.8 \times 10^{13} \text{ Nm}^2.\text{kg}^2$$
$$a = 35000 \text{ km}$$
$$r = \text{Apogee distance from the center of earth}$$
$$r = 35000 + 6360 = 41360 \text{ km}$$

Putting all these values, we get

$$V = \sqrt{39.8 \times 10^{13}\left(\frac{2}{41360 \times 10^3} - \frac{1}{35000 \times 10^3}\right)}$$

$$= \sqrt{39.8 \times 10^{10}\left(\frac{2}{41360} - \frac{1}{35000}\right)}$$

$$= \sqrt{39.8 \times 10^4 \ (48.3559 - 28.5714)}$$

$$= \sqrt{39.8 \times 10^4 \times 19.7845}$$

$$= 10^2 \times 28.061 = 2.81 \text{ km/sec.}$$

$$\boxed{V = 2.81 \text{ km/sec}}$$ **Ans.**

Example 2.8. Determine the orbit eccentricity of a satellite moving in an elliptical eccentric orbit having the semi-major axis equal to 16000 km. If the difference between the apogee and perigee is 25000 km, and the earth's radius is 6360 km.

Solution.

$$\text{Apogee} = a(1 + e)$$
$$\text{Perigee} = a(1 - e)$$

where a = Semi-major axis = 16000 km

e = Orbital eccentricity

Given that,

$$\text{Apogee} - \text{Perigee} = 25000 \text{ km}$$
$$a(1 + e) - a(1 - e) = 25000 \text{ km}$$
$$2ae = 25000 \text{ km}$$
$$e \times 2 \times 16000 = 25000$$
$$e = \frac{25000}{32000} = \frac{25}{32}$$

$$\boxed{e = 0.78}$$ **Ans.**

Example 2.9. Two satellites are moving in an elliptical eccentric orbit with same perigee but different apogee distances. Satellite 1 is having an orbital period of 5 hours and semimajor axis, 20000 km. While the orbital period of satellite 2 is 2 hours 50 min. Determine the semimajor axis of satellite.

Solution. Orbital period is given by

$$T = 2\pi\sqrt{\frac{a^3}{\mu}}$$

where $\mu = GM = 3.98 \times 10^{13}\ \text{Nm}^2/\text{kg}$

if T_1 and T_2 are the orbital period of satellite-1 and satellite-2 respectively, then,

$$\frac{T_1}{T_2} = \frac{2\pi\sqrt{\frac{a_1^3}{\mu}}}{2\pi\sqrt{\frac{a_2^3}{\mu}}}$$

$$\frac{T_1}{T_2} = \sqrt{\frac{a_1^3}{a_2^3}}$$

Given that,

$$T_1 = 5 \text{ hours} = 300 \text{ min}$$
$$T_2 = 2 \text{ hours, } 50 \text{ min} = 170 \text{ min}$$
$$a_1 = 20000 \text{ km}$$

Substituting these values in above formula, we get

$$\frac{300}{170} = \sqrt{\frac{(20000)^3}{a_2^3}}$$

$$\Rightarrow \quad \frac{(20000)^3}{a_2^3} = 3.114$$

$$\Rightarrow \quad \frac{20000}{a_2} = 1.460 \Rightarrow a_2 = 13698.63 \text{ km}$$

$$\boxed{a_2 = 13698.63 \text{ km}} \quad \textbf{Ans.}$$

Example 2.10. An earth station is located at 30°W longitude and 60° latitude. Determine the earth station azimuth and elevation angle with respect to geostationary satellite located at 50°W longitude. If the orbital radius and earth radius are 42164 km and 6378 km respectively.

Solution. Since the earth is in northern hemisphere and is towards the west of the satellite,

Earth station Azimuth angle, $A = 180 - A'$ is given by,

$$A' = \tan^{-1}\left[\frac{\tan|\theta_S - \theta_L|}{\sin\theta_l}\right]$$

where θ_S = Satellite longitude = 50°

θ_L = Earth station longitude = 30°

θ_l = Earth station latitude = 60°

Putting all these values, we get

$$A' = \tan^{-1}\left[\frac{\tan(50° - 30°)}{\sin 60°}\right]$$

$$= \tan^{-1}\left(\frac{\tan 20°}{\sin 60°}\right) = 22.8°$$

So, Azimuth $(A) = 180 - 22.8°$

$$\boxed{A = 157.2°} \textbf{ Ans.}$$

The earth station elevation (E) is given by

$$E = \tan^{-1}\left[\frac{r - R_e \cos\theta_l \,.\, \cos(\theta_S - \theta_L)}{R \sin \cos^{-1} \cos\theta_l \,.\, \cos(\theta_S - \theta_L)} - \cos^{-1}[\cos\theta_l \,.\, \cos(\theta_S - \theta_L)]\right]$$

where, r = Orbital radius

R = Earth's radius

Putting all these values, we get

$$E = \tan^{-1}\frac{42164 - 6378 \cos 60° \,.\, \cos 20°}{6378 \sin[\cos^{-1}(\cos 60° \,.\, \cos 20°)]} - \cos^{-1}(\cos 60° \,.\, \cos 20°)$$

$$= \tan^{-1}\frac{42164 - 2998}{6378 \sin(\cos^{-1} 0.47)} - \cos^{-1}(0.47)$$

$$= \tan^{-1}\left(\frac{39166}{5631}\right) - 62°$$

$$= 81.8° - 62° = 19.8°$$

$$\boxed{E = 19.8°} \textbf{ Ans.}$$

Example 2.11. A satellite is moving in an elliptical orbit as shown in Fig. 2.25. Determine its semimajor axis.

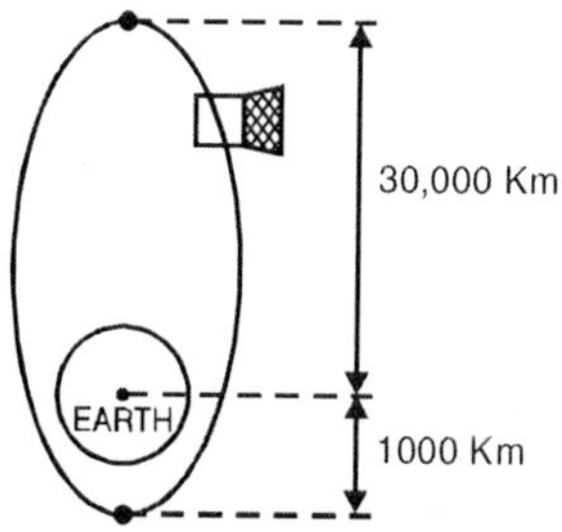

Fig. 2.25.

Solution. Given that,

Apogee = 30,000 km

Perigee = 1000 km

$$\text{Semimajor axis} = \frac{\text{Apogee} + \text{Perigee}}{2}$$

$$= \frac{30{,}000 + 1{,}000}{2} = \frac{31{,}000}{2}$$

$$\boxed{\text{Semimajor axis} = 15500 \text{ km}} \textbf{ Ans.}$$

Example 2.12. Determine the apogee, perigee and orbit eccentricity in a satellite's elliptical eccentric orbit, if the farthest and closest points from earth's surface are 20,000 km and 500 km respectively. The radius of earth is 6370 km.

Solution. Apogee = 20000 + 6370 = 26370 km

Perigee = 500 + 6370 = 6870 km

$$\text{Orbit eccentricity, } e = \frac{\text{Apogee} - \text{Perigee}}{\text{Apogee} + \text{Perigee}} = \frac{26370 - 6870}{26370 + 6870} = \frac{19500}{33240}$$

$$\boxed{e = 0.587}$$

Example 2.13. Determine the average angular velocity of a satellite moving in an elliptical orbit. If the semimajor axis is 42164.8 km and orbital eccentricity is 0.0011.

Given that, $G = 6.67 \times 10^{-11}$

$M = 5.98 \times 10^{28}$

Solution. From the given values, we can calculate μ

$$\begin{aligned}\mu &= GM \\ &= 6.67 \times 10^{-11} \times 5.98 \times 10^{28} \\ &= 39.8 \times 10^{13}\ \text{Nm}^2/\text{kg} \\ &= 39.8 \times 10^{5}\ \text{Km}^2/\text{kg}\end{aligned}$$

Average angular velocity will be

$$\begin{aligned}\eta &= \frac{1}{a}\left(\frac{\mu}{a}\right)^{1/2} = \frac{1}{42164.8 \times 100}\left[\frac{39.8 \times 10^5}{42164.8}\right]^{1/2} \\ &= 7.29 \times 10^{-5}\ \text{rad/s}\end{aligned}$$

$$\boxed{\eta = 7.29 \times 10^{-5}\ \text{rad/sec.}}\ \textbf{Ans.}$$

Example 2.14. A satellite is moving in a circular orbit for the period of 1 day. Calculate the radius of orbit (given that μ = 39.8 × 10^{13} Nm^2/kg).

Solution. The mean motion is given by,

$$n = \frac{2\pi}{T} \qquad \text{where, } T = 1 \text{ day}$$

Thus,

$$n = \frac{2 \times 3.14}{24 \times 60 \times 60}$$

$$n = 7.27 \times 10^{-5}\ \text{rad/sec.}$$

Kepler's third law gives,

$$a = \left(\frac{\mu}{n^2}\right)^{1/3} = \left[\frac{39.8 \times 10^{13}}{(7.27 \times 10^{-5})^2}\right]^{1/3}$$

$$\boxed{a = 42241\ \text{km.}}\ \textbf{Ans.}$$

Example 2.15. Calculate the apogee and perigee heights for the orbital parameters given as

$e = 0.00115$

$a = 7192.3$ km

Assume *a* mean earth radius of 6371 km.

Solution. Given that,

$$R = 6371 \text{ km}$$

Apogee and perigee distance can be calculated as

$$r_a = a(1 + e)$$
$$= 7192.3\ (1 + 0.00115) = 7200.6 \text{ km}$$
$$r_p = a(1 - e)$$
$$= 7192.3\ (1 - 0.00115)$$
$$= 7184.1 \text{ km.}$$

Thus the apogee and perigee heights will be

$$H_a = r_a - R = 7200.6 - 6371$$

$$\boxed{H_a = 829.6 \text{ km.}} \quad \textbf{Ans.}$$

$$H_b = r_b - R = 7184.1 - 6371$$

$$\boxed{H_b = 813.1 \text{ km.}} \quad \textbf{Ans.}$$

Example 2.16. Calculate the time of perigee passage for the NASA elements for 223.8 days, if the values of mean anomaly and mean motions are 246.7° and 14.24 rev/day.

Solution. Given that,

$$M_o = 246.7°$$
$$n = 14.24 \text{ rev/day}$$
$$= 14.23 \times 360 \text{ degree/day}$$
$$t_0 = 223.8 \text{ days}$$

Time of perigee passage can be given as

$$T = 223.8 - \frac{246.7}{14.23 \times 360}$$

$$\boxed{T = 223.8 \text{ days.}} \quad \textbf{Ans.}$$

Example 2.17. The two satellites are moving in different elliptical orbits with the same perigee but different apogee distances. The semimajor axis of the two orbits are 16000 km and 24000 km. Determine the orbital period of satellite 2 if the orbital period of satellite 1 is 600 min.

Solution. Orbital period (T) is given by,

$$T = 2\pi\sqrt{\frac{a^3}{\mu}}$$

If T_1 and T_2 are the orbital periods of satellite 1 and 2 and, a_1 and a_2 are their apogee distances. Then,

$$\frac{T_1}{T_2} = \frac{2\pi\sqrt{\frac{a_1^3}{\mu}}}{2\pi\sqrt{\frac{a_2^3}{\mu}}}$$

$$\left(\frac{T_1}{T_2}\right)^2 = \left(\frac{a_1^3}{a_2^3}\right)$$

$$\frac{T_1}{T_2} = \left(\frac{a_1}{a_2}\right)^{3/2}$$

$$T_2 = 600\left(\frac{24000}{16000}\right)^{3/2}$$

$$T_2 = 110.22 \text{ min}$$

$$\boxed{T_2 = 18.37 \text{ hours.}} \quad \textbf{Ans.}$$

Example 2.18. An earth station is located at 6378 km distance from the center of earth. A satellite is moving at 35786 km above from earth surface, if the minimum possible elevation angle is 5°. Determine the round trip delay for an earth station.

Solution. Given that,

$$R_e = 6378 \text{ km}$$

$$H = 35786 \text{ km.}$$

The slant range can be calculated by

$$d^2 = R_e^2 + (R_e + H)^2 - 2R_e(R_e + H)\sin\left[E + \sin^{-1}\frac{R_e}{R_e + H}\cos E\right]$$

$$= (6378)^2 + (6378 + 35786)^2 - 2 \times 6378 \times 35786$$

$$\times \sin\left[5° + \sin^{-1}\frac{6378}{6378 + 35786}\cos 5°\right]$$

$$= 40678884 + 1777802896 - 126984965 = 1691446815$$

Thus, $d = \sqrt{1691498815}$

$$d = 41127.8 \text{ km.}$$

Round trip time, $\tau = \dfrac{2d}{C} = \dfrac{2 \times 41127.8 \text{ km}}{3 \times 10^5 \text{ km/s}}$

$$= 0.274 \text{ sec.}$$

$$\boxed{\tau = 274 \text{ m sec.}} \quad \textbf{Ans.}$$

Example 2.19. Calculate the velocity of an artificial satellite orbiting the earth in a circular orbit at an altitude of 200 km above the earth's surface.

Solution. Radius of Earth = 6378.140 km

GM of Earth = 3.986005×10^{14} m^3/s^2

Given : $r = (6378.14 + 200) \times 1000 = 6578.140$ m

$$v = \sqrt{\left[\frac{GM}{r}\right]}$$

$$v = \sqrt{\left[\frac{3.986005 \times 10^4}{6578140}\right]}$$

$$\boxed{v = 7784 \text{ m/s.}}$$ **Ans.**

Example 2.20. Calculate the period of revolution for the satellite in Example 19.

Solution. Given, $r = 6578140$ m

$$P^2 = \frac{4 \times \pi^2 \times r^3}{GM}$$

$$P = \sqrt{\left[\frac{4 \times \pi^2 \times r^3}{GM}\right]}$$

$$P = \sqrt{\left[\frac{4 \times \pi^2 \times 6578140^3}{3.986005 \times 10^{14}}\right]}$$

$$\boxed{P = 5310 \text{ s.}}$$ **Ans.**

Example 2.21. Calculate the radius of orbit for a earth satellite in a geosynchronous orbit, where the earth's rotational period is 86164.1 seconds.

Solution. Given, $P = 86164.1$ s

According to above equation

$$P^2 = \frac{4 \times \pi^2 \times r^3}{GM}$$

$$r = \left[\frac{P^2 \times GM}{(4 \times \pi^2)}\right]^{1/3}$$

$$r = \left[\frac{86164.1^2 \times 3.986005 \times 10^{14}}{(4 \times \pi^2)}\right]^{1/3}$$

$$\boxed{r = 42{,}164{,}170 \text{ m.}}$$ **Ans.**

Example 2.22. An artificial earth satellite is in a elliptical orbit which brings it to an altitude of 250 km at perigee and out to an altitude of 500 km at apogee. Calculate the velocity of the satellite at both perigee and apogee.

Solution. Given, $R_p = (6378.14 + 250) \times 1000 = 6628140$ m

$R_a = (6378.14 + 500) \times 1000 = 6878140$ m

Equations (2.16) and (2.17)

$$V_p = \sqrt{\left[\frac{2 \times GM \times R_a}{(R_p \times (R_a + R_p))}\right]}$$

$$V_p = \sqrt{\left[\frac{2 \times 3.986005 \times 10^{14} \times 6878140}{(6628140 \times (6878140 + 6628140))}\right]}$$

$$V_p = 7826 \text{ m/s}$$

$$V_a = \sqrt{\left[\frac{2 \times GM \times R_p}{R_a \times (R_a + R_p)}\right]}$$

$$V_a = \sqrt{\left[\frac{2 \times 3.986005 \times 10^{14} \times 6628140}{(6878140 \times (6878140 + 6628140))}\right]}$$

$$\boxed{V_a = 7542 \text{ m/s.}}$$ **Ans.**

Example 2.23. A satellite in Earth orbit passes through its perigee point at an altitude of 200 km above the earth's surface and at a velocity of 7850 m/s. Calculate the apogee altitude of the satellite.

Solution. Given, $R_p = (6378.14 + 200) \times 1000 = 6578140$ m

$V_p = 7850$ m/s

$$R_a = \frac{R_p}{\left[\dfrac{2 \times GM}{(R_p \times V_p^2)} - 1\right]}$$

$$R_a = \frac{6578140}{\left[\dfrac{2 \times 3.986005 \times 10^{14}}{(6578140 \times 7850^2)} - 1\right]}$$

$$R_a = 6805140 \text{ m.}$$

$$\boxed{\text{Altitude @ apogee} = \frac{6805140}{1000} - 6378.14 = 427.0 \text{ Km.}}$$ **Ans.**

Example 2.24. Calculate the eccentricity of the orbit for the satellite in example 2.23.

Solution. Given, $R_p = 6578140$ m

$V_p = 7850$ m/s

$$e = \frac{R_p \times V_p^2}{GM} - 1$$

$$e = \frac{6578140 - 7850^2}{3986005 \times 10^{14}} - 1$$

$$\boxed{e = 0.01696.}$$ **Ans.**

Example 2.25. A satellite in earth orbit has a semi-major axis of 6700 km and an eccentricity of 0.01. Calculate the satellite's altitude at both perigee and apogee.

Solution. Given,

$$a = 6700 \text{ km}$$
$$e = 0.01$$
$$R_p = a \times (1 - e)$$
$$R_p = 6700 \times (1 - 0.01)$$
$$R_p = 6633 \text{ km}$$
$$\boxed{\text{Altitude @ perigee} = 6633 - 6378.14 = 254.9 \text{ km}} \quad \textbf{Ans.}$$
$$R_a = a \times (1 + e)$$
$$R_a = 6700 \times (1 + 0.01)$$
$$R_a = 6767 \text{ km}$$
$$\boxed{\text{Altitude @ apogee} = 6767 - 6378.14 = 388.9 \text{ km}} \quad \textbf{Ans.}$$

Example 2.26. A satellite is in an orbit with a semi-major axis of 7500 km and an eccentricity of 0.1. Calculate the time it takes to move from a position 30 degrees past perigee to 90 degrees past perigee.

Solution. Given,

$$a = 7500 \times 1000 = 7500000 \text{ m}$$
$$e = 0.1$$
$$t_0 = 0$$
$$V_0 = 30 \text{ deg} \times \frac{\pi}{180} = 0.52360 \text{ radians}$$
$$V = 90 \text{ deg} \times \frac{\pi}{180} = 1.57080 \text{ radians}$$
$$\cos E = \frac{(e + \cos V)}{(1 + e \cos V)}$$
$$E_0 = \text{Across}\left[\frac{(0.1 + \cos(0.52360))}{(1 + 0.1 \times \cos(0.52360))}\right]$$
$$E_0 = 0.47557 \text{ radians}$$
$$E = \text{Across}\left[\frac{(0.1 + \cos(1.57080))}{(1 + 0.1 \times \cos(1.57080))}\right]$$
$$E = 1.47063 \text{ radians}$$
$$M = E - e \times \sin E$$
$$M_0 = 0.47557 - 0.1 \times \sin(0.47557)$$
$$M_0 = 0.42978 \text{ radians}$$
$$M = 1.47063 - 0.1 \times \sin(1.47063)$$
$$M = 1.37113 \text{ radians}$$
$$n = \sqrt{\left[\frac{GM}{a^3}\right]}$$

$$n = \sqrt{\left[\frac{3.986005 \times 10^{14}}{7500000^3}\right]}$$

$$n = 0.00097202 \text{ rad/s}$$

$$M - M_0 = n \times (t - t_0)$$

$$t = \frac{t_0 + (M - M_0)}{n}$$

$$t = \frac{0 + (1.37113 - 0.42978)}{0.00097202}$$

$$\boxed{t = 968.4 \text{ s}} \quad \textbf{Ans.}$$

Example 2.27. The satellite in Example 2.26 has a true anomaly of 90 degrees. What will be the satellite's position, *i.e.,* it's true anomaly, 20 minutes later ?

Solution. Given,

$$a = 7500000 \text{ m}$$
$$e = 0.1$$
$$t_0 = 0$$
$$t = 20 \times 60 = 1200 \text{ s}$$

$$V_0 = 90 \times \frac{\pi}{180} = 1.57080 \text{ rad}$$

From Example 26,

$$M_0 = 1.37113 \text{ rad}$$
$$n = 0.00097202 \text{ rad/s}$$
$$M - M_0 = n \times (t - t_0)$$
$$M = M_0 + n \times (t - t_0)$$
$$M = 1.37113 + 0.00097202 \times (1200 - 0)$$

$$\boxed{M = 2.53755}$$

Method #1,Low Accuracy :

$$v \sim M + 2 \times e \times \sin M + 1.25 \times e_2 \times \sin 2M$$
$$v \sim 2.53755 + 2 \times 0.1 \times \sin (2.53755) + 1.25 \times 0.12 \times \sin (2 \times 2.53755)$$
$$v \sim 2.63946 = 151.2 \text{ degrees}$$

Method #2, High Accuracy :

$$M = E - e \times \sin E$$
$$2.53755 = E - 0.1 \times \sin E$$

By iteration, $E = 2.58996$ radians

Equation (2.34),

$$\cos E = \frac{(e + \cos V)}{(1 + e \cos V)}$$

Rearranging variables gives,

$$\cos V = \frac{(\cos E - e)}{(1 - e \cos E)}$$

$$V = \text{Across}\left[\frac{(\cos (2.58996) - 0.1)}{(1 - 0.1 \times \cos (2.58996))}\right]$$

$$\boxed{V = 2.64034 = 151.3 \text{ degrees}} \quad \textbf{Ans.}$$

Example 2.28. For the satellite in Examples 2.26 and 2.27, calculate the length of its position vector, its flight-path angle, and its velocity when the satellite's true anomaly is 225 degrees.

Solution. Given, $a = 7500000$ m

$e = 0.1$

$V = 225$ degrees

Length of position vector is given by

$$r = \frac{a \times (1 - e^2)}{(1 + e \times \cos V)}$$

$$r = \frac{7500000 \times (1 - 0.1^2)}{(1 + 0.1 \times \cos (225))}$$

$$\boxed{r = 7989977 \text{ m}} \quad \textbf{Ans.}$$

$$\phi = \text{Arctant}\left[\frac{e \times \sin V}{(1 + e \times \cos V)}\right]$$

$$\phi = \text{Arctant}\left[\frac{0.1 \times \sin (225)}{(1 + 0.1 \times \cos (225))}\right]$$

$$\boxed{\phi = -4.351 \text{ degrees}} \quad \textbf{Ans.}$$

$$v = \sqrt{\left[\frac{GM \times a (1 - e^2)}{(r \times \cos \phi)}\right]}$$

$$v = \sqrt{\left[\frac{3.986005 \times 10^{14} \times 7500000 \times (1 - 0.1^2)}{(7989977 \times \cos (-4.351))}\right]}$$

$$\boxed{v = 6828 \text{ m/s.}} \quad \textbf{Ans.}$$

Example 2.29. Calculate the velocity charge required to transfer a satellite from a circular 600 km orbit with an inclination of 28 degrees to an orbit of equal size with an inclination of 20 degrees.

Solution. Given, $r = (6378.14 + 600) \times 1000 = 6978140$ m

$q = 28 - 20 = 8$ degrees

$$V_i = \sqrt{\left[\frac{GM}{r}\right]}$$

$$V_i = \sqrt{\left[\frac{3.986005 \times 10^{14}}{6978140}\right]}$$

$$V_i = 7558 \text{ m/s}$$

$$D_V = 2 \times V_i \times \sin\left(\frac{\theta}{2}\right)$$

$$D_V = 2 \times 7558 \times \sin\left(\frac{8}{2}\right)$$

$$\boxed{D_V = 1054 \text{ m/s}} \quad \textbf{Ans.}$$

Example 2.30. Calculate the escape velocity of a spacecraft launched from an Earth orbit with an altitude of 300 km.

Solution. Given, $r = (6378.14 + 300) \times 1000 = 6678140$ m

Escape velocity of spacecraft launched from earth orbit is given by

$$V_{ESC} = \sqrt{\left[\frac{2 \times GM}{r}\right]}$$

$$V_{ESC} = \sqrt{\left[\frac{2 \times 3.986005 \times 10^{14}}{6678140}\right]}$$

$$\boxed{V_{ESC} = 10926 \text{ m/s}} \quad \textbf{Ans.}$$

Summary

1. The orbit of a satellite is geostationary and it has to be maintained but regarding the orbit two main problems are there launching and putting the satellite in orbit and second is to maintain it.
2. Satellites move around the earth and three Kepler's laws are applicable to them.
3. Newton's gravitational law of force also controls the satellite orbit. The satellite position is described by a second order linear vector differential equation. The solution of this equation involves six constants, which are called the orbital elements.
4. The satellite orbit is elliptical in nature with the earth at one focus. The furthest point in the satellite orbit from the centre of earth is known as apogee and the nearest point from the center of earth is known as perigee.
5. Look angles are the coordinates to which an earth station antenna must be pointed to communicate with the satellite. These are azimuth and elevation angle and these may be calculated from the orbital radius.
6. The argument of perigee (w) is the angle measured along the orbit from the ascending node to perigee and it is required to locate the orbital co-ordinate system with respect to equitorial co-ordinate system.

7. Because of solar eclipse earth and moon affect the working of a satellite and the two major parameters of solar eclipse are periodicity and duration.
8. The process of maintenance of satellites altitude against the factors that can cause drift with time is known as station keeping.
9. The max. coverage angle of the earth by a geostationary satellite is 17.4°.
10. At the present time, satellites may be launched by expandable launch vehicles or by the space shuttle. In both methods satellite is placed a low inclined elliptical orbit called a transfer orbit before it is enjected into synchronous orbit.
11. Vertical equinox is a point on the ecliptic where the first day of spring begins.
12. A point where the orbit crosses the equitorial plane going from south to north is known as the ascending node while the point where the orbit crosses the equitorial plane going from north to south is known as desending node.
13. A satellite, which is orbiting the earth, is defined by six orbital elements. Sometimes they are all referred as the Keplerian element set.
14. The sub-satellite point is the location on the surface of the earth that lies directly between the center of the earth and the satellite. It may be defined as projection of the satellite on the earth's surface along the local vertical.
15. Basically there are three common types of satellite earch orbits : geostationary earth orbits (GEO), low earth orbits (LEO) and medium earth orbits (MEO).
16. Satellites can be classified according to their orbital altitude from earth. Accordingly, satellites have been classified in three categories : low earth orbit satellites, medium earth orbit satellites, geosynchronous satellites.
17. A circular geosynchronous orbit in the plane of the earth's equator has a radius of approximately 42,164 km, approximately 35,790 km above mean sea level.
18. Low earth orbit has the advantages of short propagation delays, low transmission power and, low price for satellite and equipment.
19. A network of atleast 6 LEO satellites is required to cover a region continuously. This is the disadvantage of low earth orbit.
20. The coverage spot of medium earth orbit is greater than a LEO, but still less than a GEO.
21. Geosynchronous orbit has the advantages of large area coverage, synchronous satellite rotation to earth and less system complexity.
22. In addition to the classification based on their orbital altitudes, satellites can be grouped into two classes are : bent-pipe satellites, on board processing satellites.
23. Azimuth angle is defined as the angle by which the antenna pointing to the horizon must be rotated clockwise around its vertical axis from the geographical north to bring the antenna boresight into the vertical plane containing the satellite direction.
24. Elevation angle is the angle by which the antenna boresight should be rotated in the vertical plane that contains the satellite direction from the horizontal to the satellite direction.
25. The azimuth angle can be calculated by

$$r = \text{arc } t_g \left(\frac{t_g . \theta}{\sin \phi} \right)$$

26. Elevation angle is defined as

$$E = \frac{\text{Arc } t_g \left[\cos \theta . \cos \phi - \dfrac{R_e}{R_e + R_0} \right]}{\left[1 - (\cos \theta . \cos \phi)^2 \right]^{1/2}}$$

27. For minimum elevation angle *i.e.* 5°, the slant range will be maximum and its value is 41.127 km.

28. If the life time of satellite is T_L years, maximum value of inclination angle will be

$$i_{max.} = \frac{1}{2} i_{av} \cdot T_L$$

29. Ascending node is a point where the orbit crosses the equatorial plane going from south to north.

30. Descending node is a point where the orbit crosses the equatorial plane going from north to south.

31. Line of ascending node is defined which connects the ascending node and center of the earth.

32. Space shuttle is a manned and is launched into a circular orbit. This is the initial low orbit and is approximately 300 km above the earth's surface. Then it is placed into an elliptical orbit and this orbit is known as transfer orbit.

33. ELV is an unmanned launch vehicle. It is used to place the satellite directly in an elliptical orbit. The highest point in this elliptical orbit is known as apogee and the lowest point is the perigee.

34. ITU and CCIR prescribe the parking slots for individual countries and this helps in avoiding the uncessary losses or traffic capacity.

35. The equation of orbit is given by

$$r_0 = \frac{P}{1 + e \cos \theta_0}$$

36. The average angular velocity of a satellite is given by

$$\eta = \frac{1}{a}\left(\frac{\mu}{a}\right)^{1/2}$$

37. For a circular orbit the satellite velocity is,

$$v = \sqrt{\frac{\mu}{a}}$$

Review Questions

1. Explain the Kepler's law of planetary motion and how are they applicable to the geostationary satellite.
2. What is orbit ? Derive an expression for the equation of satellite orbit.
3. What is meant by look angles ? Explain them with reference to a geostationary satellite and earth station. **[UPTU 2005-06]**
4. Explain briefly the orbital parameters required to determine a satellite's orbit.
5. Explain the concept of earth coverage and slant range for geostationary satellite. What are the maximum values of these parameter.
6. Explain the concept and significance of station keeping. What are N-S and E-W station keeping.
7. Explain how a satellite is placed into geostationary orbit from earth ?
8. What is meant by orbit perturbations ? Explain in brief.
9. Differentiate between geosynchronous and geostationary orbit.
10. Briefly discuss the various types of orbits.

11. List the advantages and disadvantages of geostationary orbit.

12. Briefly discuss, how does the solar eclipse affect the working of a communication satellite ? Write the month and the duration when the eclipse effects are maximum.

13. What do you mean by ascending and descending nodes ? In what units would these be measured ?

14. Explain how a satellite is placed into a geostationary orbit ? **[UPTU 2003-04]**

15. What do you mean by coverage angle and slant range ? **[UPTU 2004-05]**

16. Mention the duration and month when the eclipse effects are maximum.

17. What do you understand by sun-transit outage ?

18. What do you understand by line of apsides and line of nodes ?

19. A satellite is moving in a circular orbit at a height of 322 km. Determine

(*a*) Orbital angular velocity in radians per second.

(*b*) Orbital period in minutes.

(*c*) Orbital linear velocity.

20. The semimajor axis of the orbit of an earth orbiting satellite is 9500 km. Determine the mean anomaly 10 min. after passage of perigee.

21. The satellite orbit has an eccentricity of 0.002 and a semimajor axis of 10,000 km. Determine the values of

(*a*) The latus rectum.

(*b*) Minor axis

(*c*) Distance between foci

22. An earth station is located at latitude 35° North and longitude 100° West. Calculate the antenna look angle for a satellite at 67° W.

23. A satellite orbit has an eccentricity of 0.2 and a semi-major axis of 10,000 km. Find the values of the semi-minor axis and the distance between foci.

24. Determine the ratio of the orbital period of the two satellites moving in an elliptical orbit with a semi-major axis equal to 2400 km and 1800 km.

25. The semi-major axis and semi-minor axis of an elliptical satellite orbit are 28000 km and 18000 km respectively. Determine the apogee and perigee distances.

26. Determine the orbital eccentricity, semi-major axis and semi-minor axis of an eccentric elliptical orbit of a satellite with an apogee and perigee points at a distance of 50,000 km and 8,000 m.

27. Determine the escape velocity and the velocity at that point on the orbit for an object to be launched from the surface of a earth from a point where earth's radius is 6360 km.

28. A satellite is orbiting in the equitorial plane with a period from perigee to perigee of 12 hrs. Calculate the semi-major axis if the eccentricity is 0.002 and earth's radius is 6378 km.

29. Explain what is mean by apogee and perigee heights. The cosmos 1675 satellite has an apogee height of 39,342 km and a perigee height of 613 km. Determine the semi-major axis and eccentricity of its orbit. Assume a mean earth's radius is 6371 km.

30. The farthest and closest points in a satellite elliptical eccentir orbit from earths surface are 30,000 km and 150 km respectively. Determine the orbit eccentricity. The radius of earth is 6370 km. **[UPTU 2003-04]**

27. Calculate the orbital velocity of a satellite moving in a circular orbit at a height 200 km above the surface of earth ($G = 6.67 \times 10^{-11}$ N-m^2/kg^2, $M = 5.98 \times 10^{24}$ kg, R_e = 6370 km) **[UPTU 2003-04]**

28. Explain the effect of solar eclipse on the performance of an geostationary satellite. In what way it is related to fixing the parking place of a satellite. **[UPTU 2007-08]**

29. A low orbit satellite has an 8-hours prograde orbit. How long duration each orbit will an *Es* be able to communicate with it above an elevation angle of 15% ? **[UPTU 2007-08]**

30. A satellite is orbiting in a geosynchronous orbit of radius 41500 im. Find the velocity and time of orbit. What will be the change in velocity if the radius reduces to 36000 km ? Assume g_0 = 398600.5 Km^{3}s^2. **[UPTU 2007-08]**

31. What are the torques that affect the position of a geostationary satellite ? **[UPTU 2007-08]**

32. Explain the following w.r.t. satellites :

(*a*) ascending and descending modes

(*b*) satellite axes

(*c*) look angles

(*d*) inclination perigee and apogee. **[UPTU 2007-08]**

Questions Type Questions

1. Which of the following law governs the movement of satellites in earth orbits

(*a*) Newtons law of mechanics (*b*) Laws of quantum mechanics

(*c*) Galilean laws (*d*) Kepler's laws

2. Which of the communications satellites is in a highly eccentric, inclined orbit

(*a*) Molniya series (*b*) Raduga

(*c*) Ekran (*d*) Gorizont.

3. A geostationary satellite

(*a*) Is motionless in space

(*b*) Appears to be stationary over earth's magnetic pole

(*c*) Orbits the earth with 24 hours period

(*d*) Is located at a height of 36000 km to ensure global coverage.

4. Time (t) taken by a satellite to complete one revolution around the earth (of radius R) in a circular orbit at a distance of (h) from earth's surface is given by

(*a*) $t = \dfrac{2\pi(R+h)}{v}$ (*b*) $t = \dfrac{2\pi(R-h)}{v}$

(*c*) $t = \dfrac{2\pi h}{v}$ (*d*) $t = \dfrac{2\pi R}{v}$

5. Taking orbital period of a satellite to be approximately 24 hrs, the height of three satellites above earth's surface is (Assume a circular orbit)

(*a*) Indeterminate from given data (*b*) 35,800 km

(*c*) 40,000 km (*d*) 36,400 km

6. The velocity with which a satellite orbits round the earth

(*a*) Depends upon the force with which it is launched

(*b*) Given by $v = R\sqrt{\dfrac{R}{g} + h}$

(*c*) Given by $v = R\sqrt{\frac{g}{R} + h}$

(*d*) Depends upon the nature of the orbit

7. Geostationary satellites are generally placed in

(*a*) Equatorial orbit (*b*) Polar orbit
(*c*) Inclined orbit (*d*) None of these

8. Geostationary satellites follow

(*a*) Circular path (*b*) Elliptical path
(*c*) Inclined elliptical path (*d*) None of these

9. The eccentricity of an elliptical eccentric orbit with apogee and perigee distances can be calculated as

(*a*) $e = \frac{r_p}{r_a}$ (*b*) $e = \frac{r_a - r_p}{r_a + r_p}$

(*c*) $e = \frac{r_a + r_p}{r_a - r_p}$ (*d*) $e = \frac{r_a}{r_p}$

10. The orbital velocity of the satellite

(*a*) is directly proportional to its distance from earth's surface
(*b*) is inversely proportional to square root of its distance from earth's center
(*c*) depends upon the thrust with which it is launched
(*d*) is continuously changing as the satellite revolves.

11. For orbital velocity between the first orbital velocity (v_1) and the second orbital velocity (v_2), the satellite orbit would

(*a*) be circular (*b*) be elliptical
(*c*) have zero eccentricity (*d*) be eccentric
(*e*) both (*a*) and (*c*) (*f*) both (*b*) and (*d*)

12. The earth coverage angle for a geostationary satellite at an altitude of 42000 km from center of earth for a 0° earth station elevation would be assume earth's radius = 6370 km)

(*a*) $2 \tan^{-1} 0.15$ (*b*) $\cos^{-1} 0.15$
(*c*) $2 \sin^{-1} 0.15$ (*d*) $\sin^{-1} 0.15$

13. Clarke orbit is another name for a

(*a*) Sun synchronous orbit (*b*) Moving a orbit
(*c*) Geostationary orbit (*d*) Low earth orbit

14. In a circular geostationary orbit in the equatorial plane

(*a*) The apogee equals the perigee (*b*) The apogee is twice the perigee
(*c*) The perigee is twice the apogee (*d*) None of these

15. For an eccentric elliptical orbit with eccentricity (*e*) and semimajor axis (*a*), the perigee point distance from the center of earth having a radius (*R*) is given by

(*a*) $r_p = a(1 - e) / R$ (*b*) $r_p = a(1 - e)$
(*c*) $r_p = a(1 + e)$ (*d*) $r_p = aR/(1 - e)$

ANSWERS

1. (*d*)	**2.** (*a*)	**3.** (*d*)	**4.** (*a*)	**5.** (*a*)
6. (*d*)	**7.** (*a*)	**8.** (*a*)	**9.** (*b*)	**10.** (*b*)
11. (*f*)	**12.** (*c*)	**13.** (*c*)	**14.** (*a*)	**15.** (*b*)

❑❑❑

3

CHAPTER

SATELLITE SUBSYSTEMS

INSIDE THIS CHAPTER

3.1. INTRODUCTION

Geostationary communication satellites are very complex and, very expensive to purchase and launch. It is very difficult to maintain a communication system 35,800 km out in space. The life of communication satellite is 5 to 10 years. The cost of a satellite is several million dollar and it is because of the need of dedicated earth station for the monitoring and control of the satellite. The revenue of these costs is obtained by selling the communication capacity of the satellite to the users.

It can be done by two ways :

- Way of leasing circuits of transponder.
- Charging for circuit use as in International Telephone Service.

Since its cost is very high, the operator of the system try to recover the initial and operating cost within the expected life time of the spacecraft. The designer must provide a satellite that can survive the hostile environment of outer space for its long life. In this chapter we will discuss the communication subsystems required on a spacecraft in details.

3.2. SATELLITE SUBSYSTEMS

Figure 3.1 shows a typical block diagram of a satellite subsystem.

A complete satellite consists of several subsystems, but most important of these are :

1. Power supply system
2. Attitude and orbit control system
3. Telemetry, tracking and command system
4. Communication subsystem

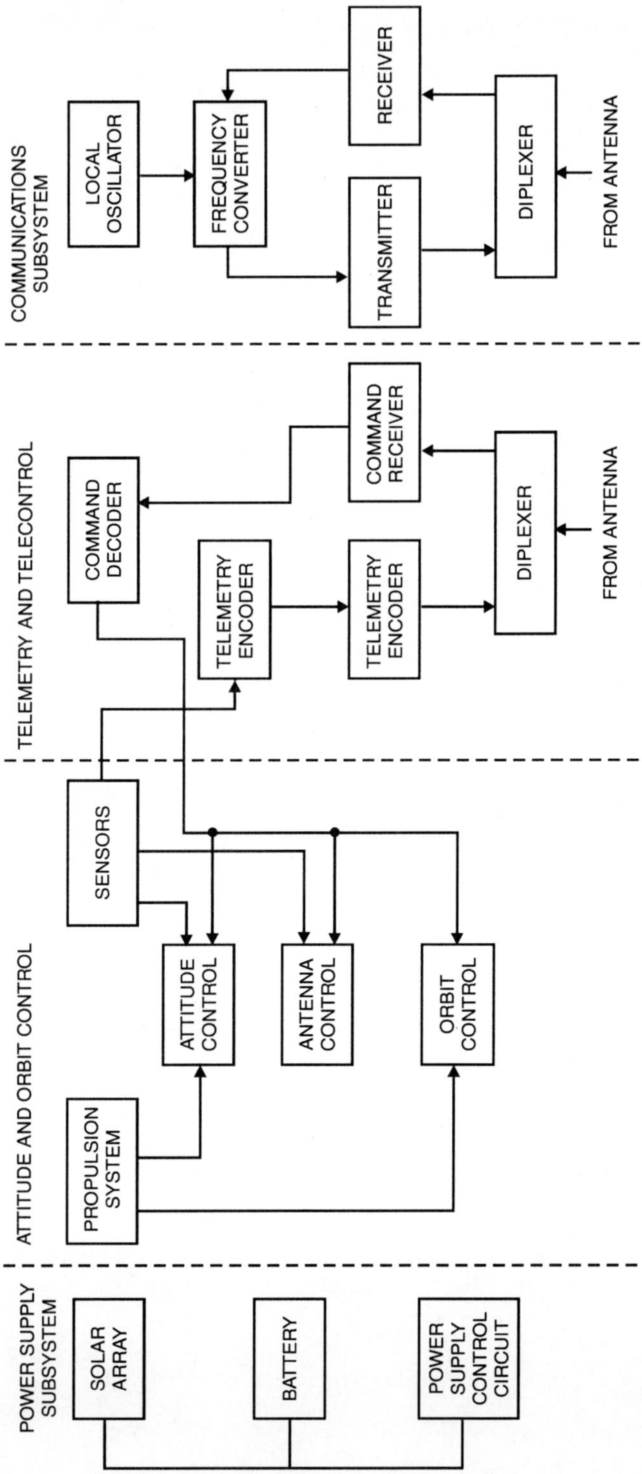

Fig. 3.1. Block diagram of a satellite showing various subsystems.

3.3. POWER SUPPLY SYSTEM

In communication satellite the electric power supply system provides electrical energy required to run the spacecraft, when the satellite is in stationary orbit. This sub-system is further divided into three parts :

(*a*) Solar array

(*b*) Battery

(*c*) Power supply control circuit

In construction solar cell is a metallurgical junction between two dissimilar semiconductor materials. **Solar cells** are photovoltaic cells that converts solar radiation into electricity and then it is converted to the required voltage level in a power supply. The conversion efficiency of solar cell is around 12% to 15% and typical satellite solar cells generate 1 to 2 kW of power.

The following are the **possible structure of that can be used for the fabrication of solar cells :**

- P-N homojunction
- P-N heterojunction
- Tandem or two junctions having a p-i-n structure.
- Metal semiconductor (Schottky) junction.
- Metal insulator semiconductor (MIS) junction.

Out of all these, silicon homo P-n junction solar cell is the most common and widely used because of :

- abundent availability of silicon
- high stability
- low cost
- simple to fabricate

Thousands of minute solar cells are placed on the surface of the satellite. If its surface area is inadequate, then separate projections are used. These so-called **solar panels** convert sun energy into electrical energy. Under the effect of radiation the efficiency is decreases by about 30-35% in seven years. During the time of ejection of satellite into transfer orbit, the electro mechanical accumulators provide the electric power supply. The power is mainly used by the communication system, especially in its transmitter and also by all other systems on the spacecraft. Later use term is house keeping, since these subsystems supports the communication system.

3.4. ATTITUDE AND ORBIT CONTROL SYSTEM (AOCS)

Attitude and orbit control is used to control the orbit of the satellite, besides helping to maintain stabilization and its position. The control can be effected from the satellite itself and from ground. This subsystem consists of four major parts :

1. Sensors
2. Propulsion system
3. Attitude control
4. Orbit control

Forces Acting on a Satellite

There are a number of forces working on a satellite which tend to change attitude and orbit of a satellite beyond the permissible limit. **These forces** are :

(*a*) **Asymmetry of the earth's gravitational field :** Earth is not a true sphere. It is bulging at the equator by about 65 km at the longitudes of 165°E and 15°W. This causes an acceleration and hence for accurate positioning, the satellite must be accelerated in the opposite direction by firing the rocket motors called **thrusters** at appropriate intervals. Earth is flattered at the poles by about 20 km but this has little effect on a geostationary satellite.

(*b*) **Gravitation due to sun, planets and moon :** They set up rotational moments if the satellite is not perfectly balanced. Moon being the nearest heavy mass, has maximum effect.

(*c*) **Pressure due to solar radiation :** It can also change orientation (spin-axis) of the satellite.

(*d*) **Magnetic field of the earth :** It can exert forces on the satellite if a net magnetic moment is present, thus affecting its velocity and orientation.

Need of Attitude and Orbit Control

Change in the spin axis of the satellite will cause pointing error between the satellites antenna and the earth station's antenna. This will reduce the signal at the receiving side and will degrade C/N ratio. Hence the need of maintenance of correct attitude. As the extra forces can change velocity of the satellite and therefore angle of inclination of the orbital plane with respect to the equitorial plane, the satellite will become non-geostationary. This will require steering of the earth station antenna to get the best signal from the satellite. Thus correct attitude and correct orbit are essential requirements for the optimized performance of the satellite links.

3.4.1. Sensors

For attitude control, two types of sensors are used in the satellite :

1. Earth sensor
2. Sun sensor

3.4.1.1. Earth Sensor

It is a passive infrared device, operating in 14-16 μm wavelength. It senses the infrared rays coming from around the horizon. There is a sharp temperature difference between the space and the earth's horizon, as space is cool and earth is warm. Two earth sensors are used, positioned 5° north and 5° south of the spin axis. When the spin axis of the satellite is correctly maintained the output of north and south sensors are in phase, otherwise they are out of phase. The phase difference pulses are sent to the earth station and they measure **earth aspect angle.**

3.4.1.2. Sun Sensors

It has a fan shaped field of view. It operates in the visible spectrum and uses a photocell for detecting solar radiations. There are two solar sensors, one parallel to the spin axis and the other canted 35°. Pulses from the sun sensors are sent to the earth station to determine **solar aspect angle.**

Data from the earth and sun sensors is analyzed and orientation of the satellite is accurately determined by the computers at the earth's station. Commands are generated and sent to the satellite to fire the rocket motor for correcting the axis.

3.4.2. Propulsion System

It is the reaction control system carried by the satellite in the geostationary orbit so as to generate forces on it whenever required. The reaction control system has a supply of fuel and it helps the satellite to move, to its assigned position in orbit, to maintain it in that position and to maintain the direction of spin axis attitude control in the case of forces that perturb it. Usually a propulsion system consists of 3 units :

1. Low thrust actuators : Devoted to attitude and orbit corrections that provide an annual velocity increment of the order of 50 m/s.

2. High thrust motor : Which provides the velocity increment required for the geostationary orbit injection at the transfer orbit apogee.

3. With space shuttle launched satellites : It provides the velocity increment required to inject the satellite into the transfer orbit.

Out of these three propulsion units the low thrust actuators are of much importance for geostationary orbit because it is responsible for keeping the satellite in its orbit with its perfect attitude till its life end. The low thrust actuators can be either **chemical** ones or the **electrical** ones.

The chemical thrusters have a thrust level of between 0.5 N and few 10000 N. In chemical thrusters gases are generated at high temperatures by chemical reaction of propellents which may be either solid or liquid. The gases are then accelerated by a nozzle.

The electric thrusters produce a thrust only in between 2 and 10 mN. They provide thrust by accelerating ionized mass in an electromagnetic or electrostatic field. In communication satellites chemical thrusters are used.

3.4.3. Attitude Control System

The attitude control of a spacecraft is necessary so the antennas must be pointed correctly at the earth. Gravitational forces from the sun, moon and planets will setup rotational moments when the spacecraft is not perfectly balanced. These are carried out by control loops based on error detection and decision systems as shown in Fig. 3.2.

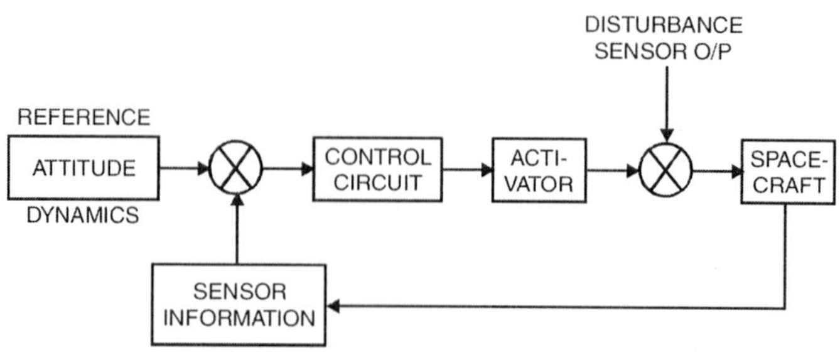

Fig. 3.2. Control scheme.

They use the telemetry, tracking and command (TT&C) subsystems for information exchange between satellite and earth. The attitude control may be :

- Passive
- Active

In **passive attitude control** the required attitude corresponds to a position of a satellite, while for **active attitude control** the satellite is unstable or sufficiently stable within the desired attitude configurations. Mainly four operations are required for this :

1. Detection of the satellite attitude
2. Comparison with the reference axis
3. Determination of the corrective torque.
4. Correction of the altitude by actuators mounted on the satellite.

Any spacecraft is governed using three axis :

- Yaw
- Roll
- Pitch.

In order to control the attitude in space, the satellite has to be properly oriented using momentum wheels and thruster motors in these three axis. The two major methods used are :

- Spin stabilization.
- Three axis body stabilization.

3.4.3.1. Spin Stabilization

It is the most commonly employed method where the entire spacecraft is rotated at 30 to 100 rotations per minute. This spin provides a powerful gyroscopic action to maintain the spin axis in the correct direction. Such satellites consist of cylindrical drum covered by solar cells and the rocket motors. The transponder is mounted on the top of the drum. It is driven by an electric motor in the appropriate direction to that of the drum, so that the antennas remain pointing towards the earth. This opposite motion is called **despun**. The despun section is kept stationary by counter rotation provided is by small gas jets mounted on the periphery of the drum. Figure 3.3 shows the spin stabilization in the geostationary orbit. Spin stabilized satellites are mainly the communication satellites.

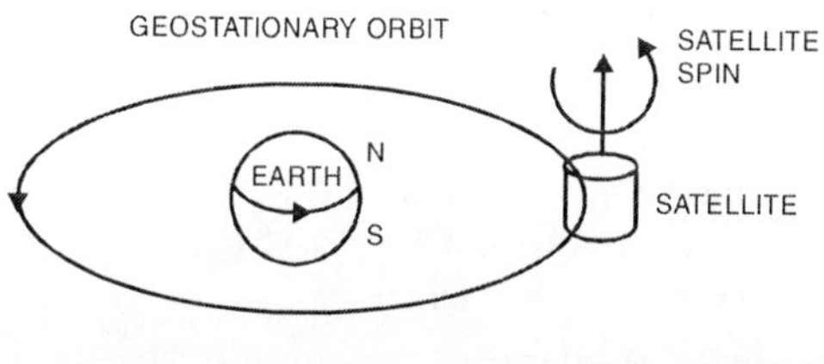

(a) Spin stabilization in the geostationary orbit

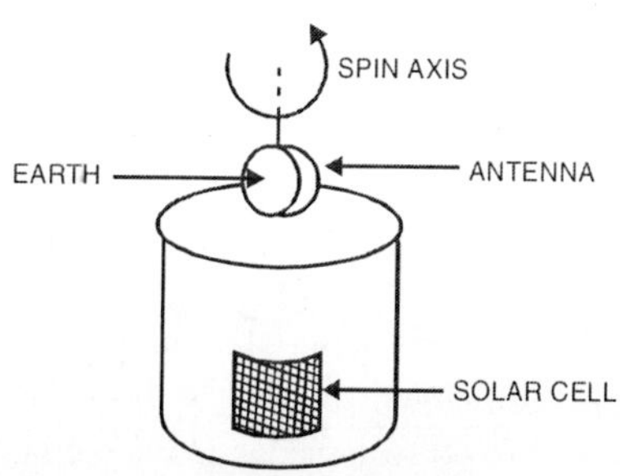

(b) Spin stabilized satellite

Fig. 3.3.

3.4.3.2. Three Axis Body Stabilization

On the other hand, a satellite can rotate about the three axis termed as yaw, roll and pitch axis. When a satellite is stabilized about these axis, it is called three axis body stabilization.

In this method stability is achieved by mounting three momentum wheels on three mutually perpendicular axis as shown in Fig. 3.4.

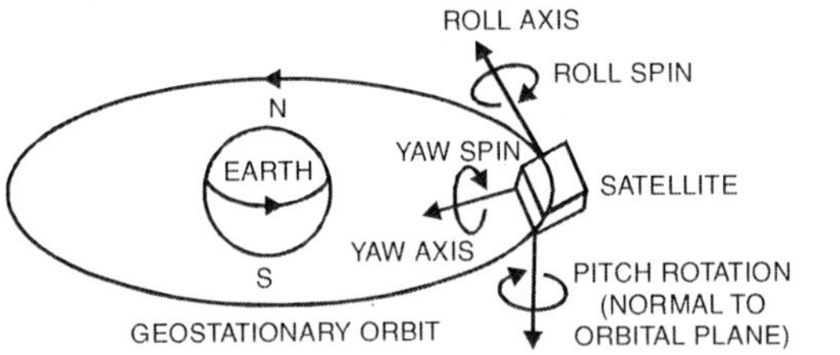

(a) *Three axis stabilization in the geostationary orbit.*

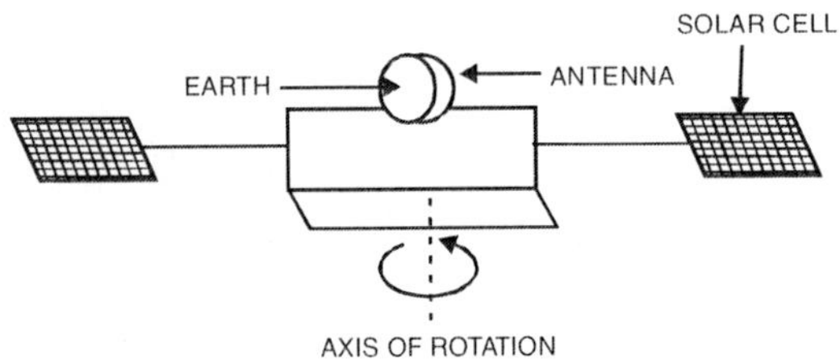

(b) *Three axis stabilized satellite.*

Fig. 3.4.

A momentum wheel is a high speed wheel driven by a motor. It is kept in a sealed evaluated chamber. Increase in its speed increases the angular momentum. Change in the attitude are transmitted to the earth station by telemetring the data from the sensors. The data is analysed and commands are sent to the satellite to increase or decrease the speed of the momentum wheels as per requirement to correct the attitude about its three aixs : role axis (the orbital plane), pitch axis (normal to the orbital plane) and yaw axis (the local vertical plane facing the earth station). Antennas are mounted on the satellites surface facing the earth.

3.4.4. Orbit Control System

A GEO satellite is subjected to several forces that tend to accelerate it away from its required orbit. It is the function of the orbit control system to return it to its correct orbit.

For orbit control, sensors are used in the satellite to measure linear acceleration (momentum wheels cannot be used for this control because they cause rotational torques), changes in velocity, sensed by the velocity sensors, are transmitted to the earth's controlling station. These changes are analysed and, appropriate commands are generated and sent to the satellite for correcting the velocity.

To correct change in the inclination angle (i), change of velocity at right angles to the orbital plane is needed. This correction requires more fuel than for any other correction. A typical satellite weighing 1000 kg may need 30 kg of fuel to maintain inclination within ±0.1°. This puts a penality by reduction in the communication payload and therefore reduction in satellite's capacity. Hence in most of the satellite systems, inclination control is not used. Instead, the satellite is launched with an initial inclination of about 3°. It will be reduced by 0.85° per year due to forces working on the satellite. Thus after three or four years, the orbit would be in the equatorial plane. With earth stations having steerable antennas, drift in inclination can be managed by proper tracking and saving in fuel can be used to carry more transponders.

3.5. TELEMETRY, TRACKING AND COMMAND (TT&C)

Data received from the satellite about status of attitude, orbit and other parameters, is processed at the ground station. Telemetry, tracking and command subsystem is a part of satellite management task and it involves an earth station. **The main functions of the TT&C subsystems are :**

1. Measurement of angle and range for the localization of the satellite.
2. Transmission of housekeeping information.
3. Status of satellite to the ground control station.
4. Receiving command signals for station keeping operations of the onboard equipments.

Figure 3.5 shows a typical arrangement of a telemetry, tracking and command system.

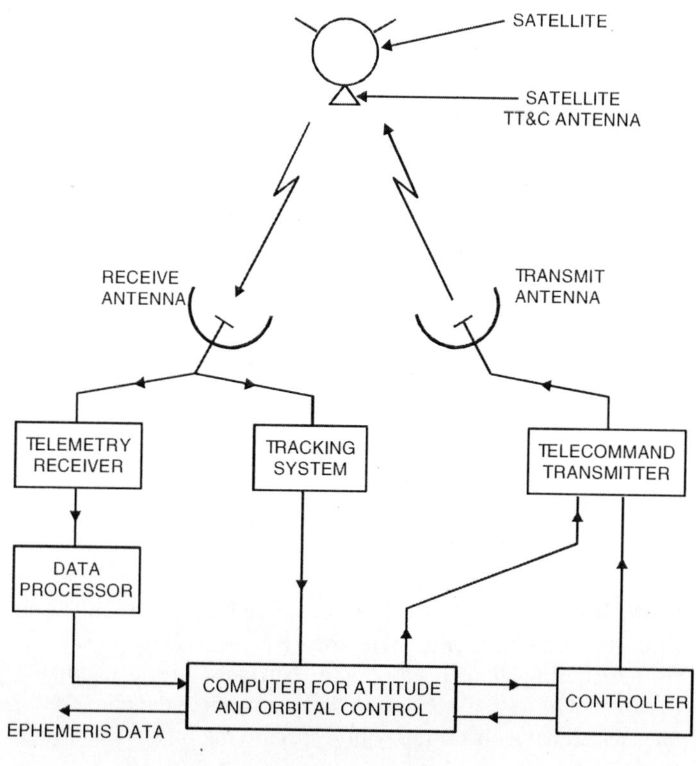

Fig. 3.5. General diagram of telemetry, tracking and command system.

3.5.1. Telemetry

Telemetry system collects the data from many sensors and sends this data to the controlling earth station. **The sensors are mounted on the satellite and they monitor :**

- The pressure in fuel tank.
- Voltage and currents in the power conditioning unit.
- Current drawn by each subsystem.
- Critical voltage and currents in the communication electronics.
- Temperature of other subsystems.
- Position of switches and attitude.

Typically 100 sensors are required to monitor these data. When the satellite is in transfer orbit, the telemeter transmitter is connected to a TWTA in the satellite repeater. The telemetry data is digitized in nature and transmitted as frequency or phase shift keying (FSK or PSK) of a low power telemetry carrier using time division techniques.

Figure 3.6 shows the block diagram of telemetry system.

To maintain a high carrier to noise ratio and narrow bandwidth, a low data rate is used to allow the receiver at the earth station. The entire TDM frame can take several seconds to transmits thousands of bits of data. The controlling earth station uses a computer to monitor, store and decode the telemetry data so that the status of any system or sensor on the spacecraft can be determined. Alarms can also be used if any parameter goes outside the allowable limits.

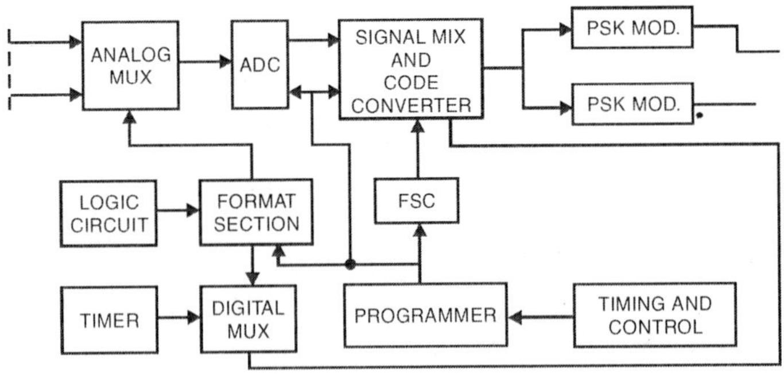

Fig. 3.6. Telemetry system.

3.5.2. Tracking or Ranging

It determines the slant range for accurate determination of satellite orbit. It can be done in two ways :

1. By transmitting a command carrier modulated by multiple tones from the earth station to the satellite. The carrier is received by the command receiver in the satellite, demodulated and then goes to the telemetry or beacon transmitter for retransmission to the earth, where the phase difference of the received signal is measured with respect to the transmitted signal.

2. Velocity and acceleration sensors can be used in the satellite to sense the change in the orbit. This data can be sent to the earth station through telemetry, where commands are generated and sent to the satellite for orbit correction.

3.5.3. Command Subsystem

For common operation the satellite contains a receiver which works only for command signal transmitted from TTC earth stations. By this command subsystem controls the satellite operation by receiving and decoding command from the TTC earth station. The communication antenna is used for command and ranging when the satellite is in synchronous orbit and, during the satellites position in transfer orbit the omnidirectional antenna is used for command and ranging.

During the enjection of satellite into geostationary orbit a back up system is used which provides :

- The control of apogee boost motors.
- Attitude control systems.
- Orbit control thrusters.

This back up system works with omnidirectional antenna at either UHF or S-band (2-4 GHz) and sufficient margin is required in the signal to noise ratio (S/N) at the satellite receiver. In case of failure of main TT&C system, the back up system canbe used to keep the satellite on station. It is also used to enject the spacecraft from geostationary orbit and to switch off all transmitters, when it reaches the end of its useful life.

The block diagram of spacecraft command system is shown in Fig. 3.7.

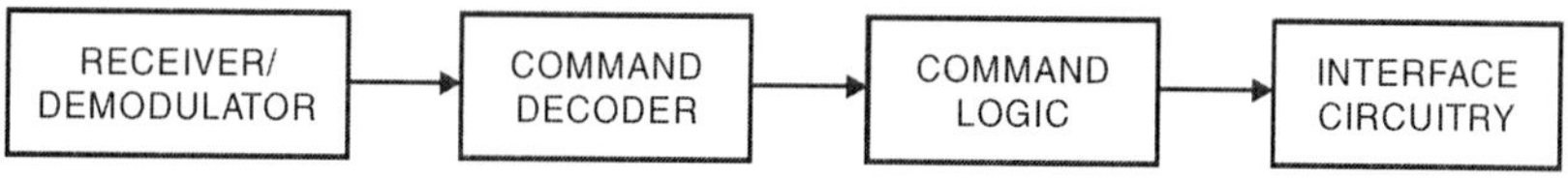

Fig. 3.7. Block diagram of spacecraft command system.

- Decoder reproduce command messages and produce lock/enable and clock signals.
- Command logic validates the command.
  1. Default is to reject if any uncertainty of validity
  2. Drives appropriate interface circuitry.
- Command decoder detects PCM encoding and outputs binary stream in non-return to zero format. It activates downstream command subsystem components.
- The function of Interface circuitry are latching relays with teletales, pulse commands, level commands and, serial and parallel data commands.

3.5.4. TT&C Interface

The different subsystems and their requirements are listed in Table 3.1 :

Table 3.1.

S. No.	Subsystem	Requirements
1.	Attitude determination and control	● Antenna pointing
2.	Command and data handling	● Command telemetry data rates ● Clock, bit sync, timing requirements ● Two way communication requirements ● Autonomous fault detection and recovery ● Command and telemetry electrical interface
3.	Electrical power subsystems	● Distribution requirements
4.	Thermal/structural	● Heat sinks for TWTAs ● Heat dissipation for all active boxes ● Location of TT&C subsystem ● Ceal field of view and movements for all antennas
5.	Payload	● Storing mission data ● RF interface requirements ● Special requirements for modulation and coding

3.6. COMMUNICATION SUBSYSTEM

The communication subsystem is the major part of the operation of a spacecraft. The communication system provides the only link between an operational vehicle and the ground control station. Examples of communicated data can include scientific research, telemetry, payload results, or relayed information from other spacecraft. This subsystem plays a crucial role in the proper functioning of the space system. Its purpose is to describe the major elements of a communication subsystem and to provide the equations for preliminary sizing of the major components.

3.6.1. Communication Subsystem Characterization

Broadly it can be defined as the system which allows data transfer to and from extravehicular sources. The primary functions of a communication system are mention in Table 3.2.

Table 3.2.

Functions	Specific Functions
● Receive signals ❑ from earth ❑ from another satellite	● Carrier tracking ❑ 2 way coherent communication ❑ 2 way non-coherent communication ❑ 1 way communication ● Command reception and detection ❑ Acquire and track uplink carrier

Functions	Specific Functions
	❑ Demodulate carrier and subcarrier ❑ Derive bit timing and detect data bits ❑ Resolve data base ambiguity if it exists ❑ Forward command data, clock, and in lock indicator to the subsystem for command and data handling. ● Telemetry modulation and transmission ❑ Receive telemetry data streams from the command and data handling subsystem or data storage subsystem ❑ Modulate downlink subcarrier and carrier with mission or science telemetry ❑ Transmit composite signals to earth or relay satellite. ● Ranging ❑ Detect and retransmits ranging pseudo-random code or ranging tone signals ❑ Retransmits either phase coherently or non-coherently ● Subsystem operations ❑ Receiver commands from the sub-system for command and data handling ❑ Provide health and status telemetry to the C&DH subsystem ❑ Perform antenna pointing for any antenna requiring beam steering ❑ Perform mission sequence operations per stored software sequence ❑ Autonomously select omni antenna when spacecraft attitude is lost. ❑ Autonomously detect falts and recover communicates using stored software sequence.

In general there are various types of communication subsystems that describe the operational systems in current spacecraft. Three **major categories** of communications systems are used when characterizing a mission. These are :

1. Near earth communication
2. Long range data relay
3. Telecommunication.

3.6.2. Near Earth Communications

Many satellites in low earth orbit utilize the communications system to maintain contact with ground control centers for the purpose of telemetry, tracking and control (TT&C) and mission data transmissions. Figure 3.8 shows some of the near earth applications of the communications subsystems on vehicles in various stages of orbital insertion and operation.

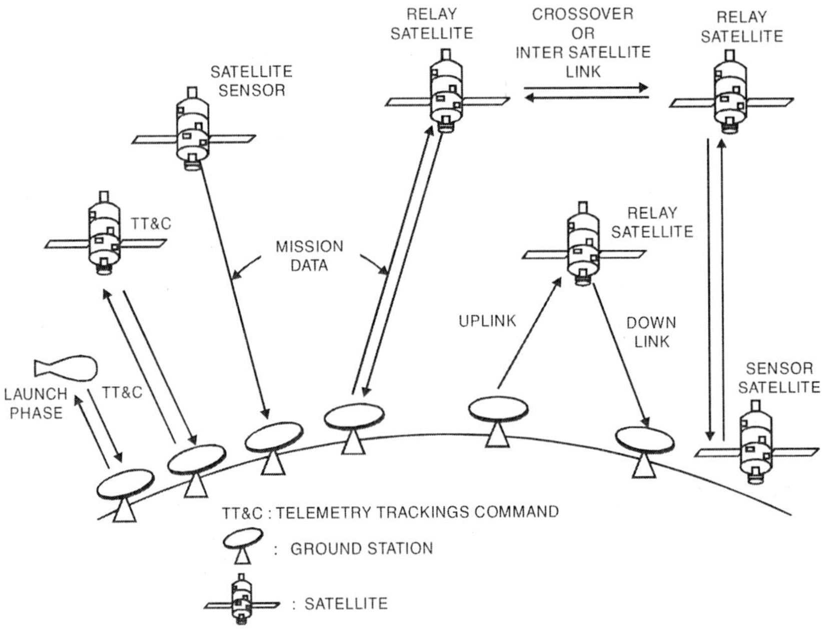

Fig. 3.8. Near earth communication architecture.

These missions typically have scientific observation or reconnaissance related payloads that are the major design drivers of the spacecraft. These system require more design resources and thus often force constraints on the communication system. In a number of cases, the design limitations lead the vehicle to rely on relay satellites to communicate with the ground. This process reduces the percentage of total mass and power that a communications subsystem requires as opposes to directly communicating with a ground station.

3.6.3. Long Range Data Relay

Long range spacecraft systems have particular communications requirements that keeps more emphasis on the design and resources of the communication subsystem. In these long range vehicles the communicated data is crucial for mission success, where the data is typically scientific in nature and can include video images and other data intensive forms of communication. The power that the communications system requires is a major design driver.

Power saving techniques can include low frequency transmissions, non-continuous communications and data encryption. Additionally, the antenna's selection is a key design concern in the context of accurate transmission at adequate data rates under obvious power constraints. In spite of these efforts, the communication subsystem requires more of the design resources to ensure successful data transmission over the long interplanetry distances. Examples of long range spacecraft are Pioneed, Voyager, Magellan and Galileo.

3.6.4. Telecommunications

The space system in which communications plays an important role is known as telecommunications satellite. In these spacecraft, the communicatons payload drives the design of the other subsystems. These vehicles can be viewed as orbiting communications subsystems that possess the other subsystems as suppose for their mission. The communications system utilizes a significant degree of the design resources for these spacecraft. Some examples of telecommunications spacecraft are TDRS, comstar, telestar, syncon and ACTS.

Table 3.3 gives the details of the subsystem requirements and some available options to satisfy them.

Table 3.3. Communication subsystem requirements.

S. No.	Requirement	Option	Comment
1	Data Rates ● Command ● Health and status telemetry ● Mission/ science	● 1000 bps typical, 8 to 64 bps deep space ● 2400 bps is common Low = 10-1000 bps Medium = 1-100 kbps High = 1000 kbps to Gbps	● 8-2000 bps for command ● 40-10,000 bps for health and status Mission dependent for mission/ science Some experiments Real-time imaging
2.	Data Volume	Record data and transmit during longer windows	Data rate × transmission duration ● Shorter duration increases data rate ● May require data compression techniques
3.	Data Storage	Tape recorders 75 × 10^9 bits Solid state recorders 128 × 10^6 bits Bubble memory 128 × 10^6 bits	Policy may dictat all data to be stored that is not immed, transmitted Mission may require that data be stored them played back later
4.	Frequency	Used existing assigned frequencies and channels Used systems that are compatible to the existing system	Policy set by FCC, ITU and NTIA Refer to the frequency atmospheric absorption charts
5.	Bandwidths	Use C.E. Channon's theorem to calculate channel capacity.	Primary driver is data rate Secondary driver modulation scheme
6.	S/C power	Use larger antennas/ higher efficiency amplifiers Reconsider data requirements	S/C power may limit size of communications system transmitter.
7.	S/C mass	Use TWTAs for high rf power output too reduce antenna size, reconsider data requirements	S/C communications system mass allocation may limit size or antennas

S. No.	Requirement	Option	Comment
8.	Beamwidth	Various antenna types, beam shapes, and beamwidths have different values	Ground coverage area rqmts or the radiation footprint on the ground Antenna gain null requirements Antenna pointing error
9.	EIRP (Effective Isotropic Radiated Power)	For a constant EIRP : As antenna size (gain) is increased, the transmitter power requirement decreases	EIRP = transmitted power × antenna gain- From end losses Min EIRP required EIRP = space loss + atmospheric loss + antenna pointing loss – receiver antenna gain – receiver sensitivity
10.	G/T [Receiver antenna gain (dB)/receiver system noise temp (K)]	Various values of communication system temperatures and G/Ts	G/T is the sensitivity of the receiving station and common figure of merit for an existing satellite link, a ground station can only very its antenna gain and sys noise temp to improve the sys signal-to-noise ratio

3.7. COMMUNICATION SUBSYSTEM COMPONENTS

In general a communication subsystem is comprised of similar components regardless of the mission for which it is designed. Some of these components include the systems antenna, transponder, amplifier, modulating/demodulating equipment, noise filters and encryption/decryption equipment.

3.7.1. Transponder

The transponder is the component of the subsystem that receives the signal and shifts its frequency for transmission. It is the path of each channel from the receiving antenna to the transmitting antenna. Its basic function is to isolate the neighbouring RF channels and to translate, and amplify the frequencies for retransmission. Typically uplink and downlink frequencies are separate to minimize interference between transmitted and received signals. Usually a lower frequency is chosen for downlink since it requires a lower output power of the satellite.

3.7.1.1. Satellite Transponder Communication Link

A transponder is a broadband RF channel used to amplify one or more carriers on the downlink side of a geostationary communication satellite. It is a part of the microwave repeater and antenna system that is housed onboard the operating satellite. Examples of these satellites include AMC 4 and Telestar 5, located at 101° and 97° west longitude, respectively. These satellites and most of their wholes in the geostationary orbit have bent pipe repeaters using C and Ku bands bent pipe repeater is simply one that receives all signals in the uplink beam, beam translates them to the downlink band and separates them into individual transponders of fixed bandwidth. Figure 3.9 shows its basic concept.

Each transponder is amplified by :

- Travelling wave tube amplifier (TWTA)
- Solid state power amplifiers (SSPA)

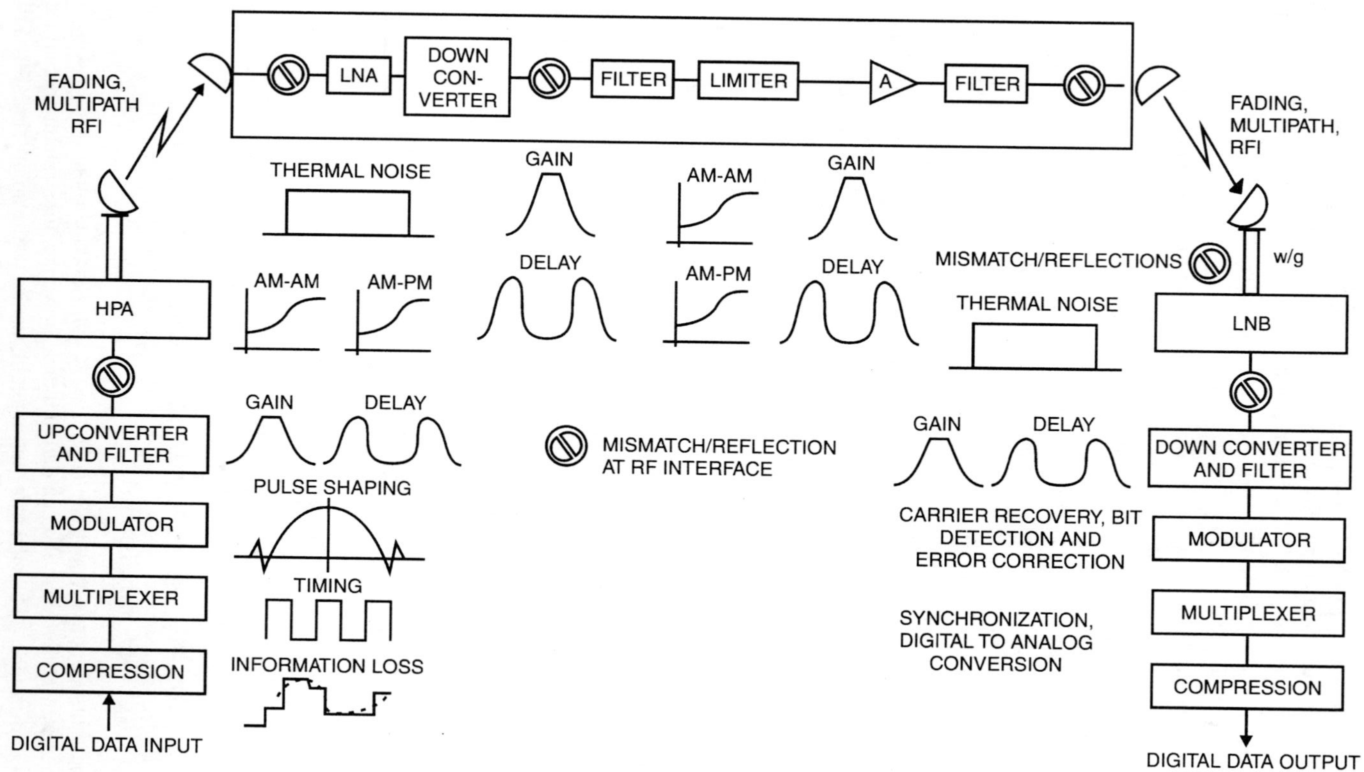

Fig. 3.9. Element of basic satellite link showing ground stations and the satellite transponder.

Satellites of this type are very popular for transmitting TV channels to broadcast stations, cable TV systems and directly to the home. Other applications include very small aperture terminal (VSAT), data communication networks, international high bit rate pipes and rural telephony. Integration of these information types is becoming popular as satellite transponders can deliver data rates in the range of 50 to 150 Mbps. Achieving these high data rates requires careful consideration of the design and performance of the repeater.

The most significant impairments to digital transmission come about in the filtering, which constrains bandwidth and introduces delay distortion, and the power amplification, which produces AM/AM and AM/PM conversion. For maximum power output with highest efficiency, it should be operated at its saturation point. However many services are sensitive and susceptible to AM/AM and AM/PM conversion, for which back off is necessary. With such an operating point, intermodulation distortion can be held to an acceptable level, however, back off also reduces downlink power. The transponder itself is simply a repeater. It takes the signal from the uplink at a frequency f_1, amplifies it and sends it back on a second frequency f_2. Figure 3.10 shows a typical frequency plan with 24 channel transponder.

The uplink frequency is at 6 GHz and the downlink frequency is at 4 GHz. The 24 channels are separated by 40 MHz and have a 36 MHz useful bandwidth. The guard band of 4 MHz assures that the transponders do not interact with each other.

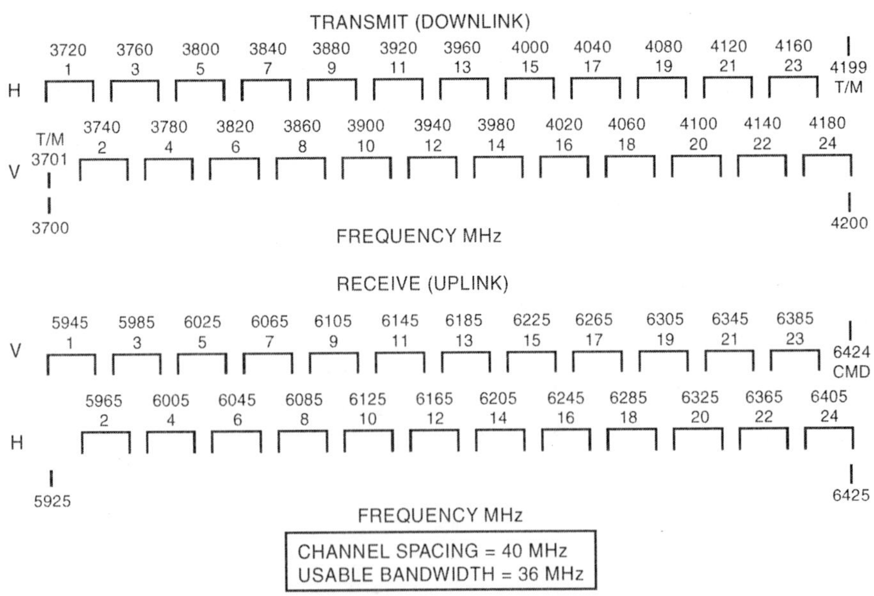

Fig. 3.10. Basic 24 channel C-band transponder frequency plan.

3.7.1.2. System Impairments

The transponder is a central element in the end to end communication link. This provides a simplified system block diagram that shows the impairments that affect the system performance. Thus the transmitting earth station on the uplink side will cause its share of distortion as with the receiving earth station on the downlink side. Some of this distortion is uncorrelated, which means that its contribution can be added more or less algebraically.

1. Filter gain and phase distortion : These are common elements in a communication link. Most filters are designed in the frequency domain in terms of their type and order. This information is linked to the filter poles and hence the frequency response. In a time domain simulator such as system view by Elanix the time impulse response is derived from the frequency response, and the filtering action is convolution operation. In a frequency domain type implementation, the data is processed in blocks, which leads to signal discontinuities at the block transitions.

2. Thermal Noise : Thermal noise is the most common impairment in a wireless communication system. There are three general sources

1. The noise that enters the antenna with the signal, called **antenna noise**.
2. The noise generated due to **ohmic absorption** in the various passive hardware components.
3. Noise produced in **amplifiers** through thermal action within semiconductors.

3. Signal Fading : Uplink and downlink transmissions can experience various forms of fading as signal passes through the transponder and ionosphere. Sometimes this is called **multipath fade**. Basically the signal from the transmitter to the receiver can bounce off various objects or bend due to variations is refractive index and can combine destructively at the antenna. The net received signal experiences time varying fading.

4. Radio frequency interference from internal and external sources.

5. Because of TWT AM/AM and AM/PM conversion

6. Pre-amplifier and mixer non-linearities.

3.7.1.3. Quasilinear Repeaters

Quasilinear repeaters describe the tendency of the satellite transponder amplifiers to exhibits a nonlinear response close to maximum power output and a more linear response at lower power levels. This repeater receives, separates and amplifies its assigned uplink carrier, translates the frequency to the downlink band and amplifies the signal for retransmission on the downlink. This transponder converts from uplink to downlink band in one step and is thus sometimes known as a **single conversion type**.

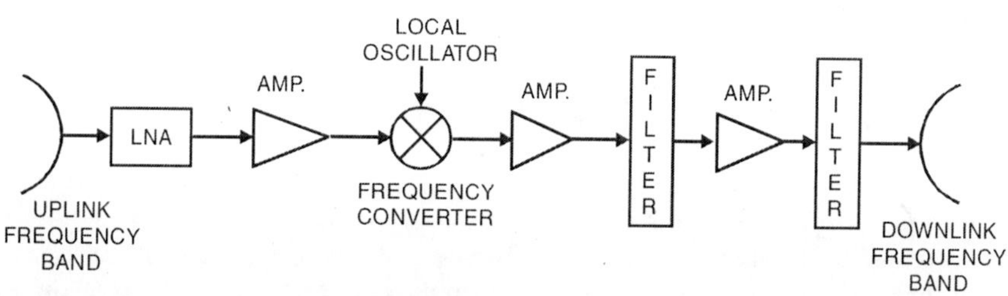

Fig. 3.11. Quasilinear repeater transponder.

Figure 3.11 shows the block diagram of a quasilinear repeater. Travelling wave tube amplifiers are usually used in the construction of final high power output stage of the transponder. In the earlier stages of the transponder, after the signal have been received, the transponder must include:

- A filter to eliminate energy outside the operating band.
- A low noise amplifier (LNA) to increase signal power.
- A broadband frequency converter to shift the operating band from downlink to uplink.

Another set of filters are used to separate the operating bandwidth into individual transponder channel bands. High power amplifiers are then used to amplify the band signal for each channel. Finally out of band products of the amplifier nonlinearity are eliminated after passing through a band pass filter. The outputs of several channels HPAs are then combined in an output mixers and fed to a common antenna system for transmissions.

3.7.1.4. Regenerative Repeaters

The regenerative repeaters gives improved performance over the quasilinear repeaters but it is more complex. Figure 3.12 shows a block diagram of regenerative repeater.

It performs the transmitting and receiving functions in the same manner as the quasilinear repeater but the regenerative repeater contains a demodulator in each transmission link that demodulates the uplink signal to the digital base band signal and a modulator that remodulates the signal on a downlink carrier. The demodulated signal is received and restored to standard form. This approach effectively isolates the uplink performance from the downlink performance, preventing the accumulation of noise and distortion over the two links.

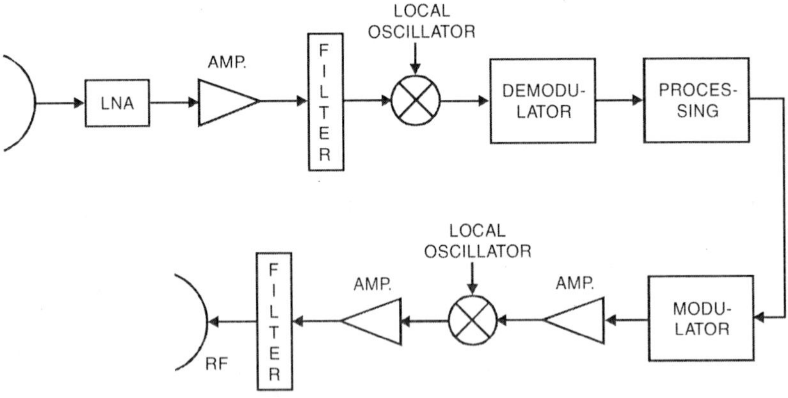

Fig. 3.12. Regenerative repeater.

A considerable increase in the communication capacity of the satellite can be achieved by combining on board processing with switched beam technology. A **switched beam satellite** generates a narrow transmit beam for each earth station with which it communicates and then transmits sequentially to each one using time division multiplexing the signals. The narrow beam has to cover only one earth station, allowing the satellite transmit antenna to have a very high gain compared to a zone coverage antenna. A narrow scanning beam or a combination of fixed and scanning beams can also be used. The high gain antennas used in swtiched beam systems raise the EIRP of the satellite transmitter and thus increase the capacity of the downlink. Because of the low transmit power available from the satellite, the downlink usually limits the capacity of a transponder.

3.7.1.5. Other Transponder Uses

1. Frequency reuse and Transponder Gain Adjustment : Modern communication satellites often provide the capability to double the use of available satellite bandwidth by providing dual polarized transmit and receive antennas with two sets of transponders, one set for operation with each polarization. The gain of a transponder chain can be adjusted by ground command over a wide range of values. There is the flexibility in the optimization of the earth station transmitter power and, costs over a much wide range of earth station sizes and capacities than is possible with a fixed gain transponder. This technique can double the capacity of the satellite system and is effective as long as the polarization isolation exceeds about 30 dB.

2. Cross band Operation : In some satellite applications, a transponder must interconnect with stations operating in two different frequency bands. Example : Maritime satellite service operating in C-band. Transmissions from a ship at sea to shore are received at the satellite in the 1.6 GHz band uplink and transmitted to shore in the 4 GHz band. Transmissions from shore to a ship at sea are received in 1.5 GHz band and transmitted to the ship in 1.5 GHz band. In such a case, separate transponders are always provided for each direct transmissions.

3.7.1.6. Travelling Wave Tube Amplifiers

A travelling wave tube is a sophisticated vacuum tube which uses a narrow electron beam guided by a shaped magnetic field. Amplification is achieved by transferring energy from this beam to a microwave signals as it enters at one end and exists at the other. A TWT requires a fairly complex power supply to convert the dc voltage from the spacecraft electrical system because of high voltages needed to generate and control the beam. So a tube itself is called a TWT, while the tube with its associated power supply is called a TWTA.

3.8. SATELLITE ANTENNA

An antenna is simply a means for radiating or receiving the radio waves. The antenna converts electronic carrier signals to polarized electromagnetic fields and vice versa. A transmitting antenna is composed of a feed assembly that illuminates an aperture reflecting surface from which the electromagnetic field radiates. A receiving antenna has an aperture or surface focussing an incident radiation field to a collecting feed, producing an electromagnetic signal proportional to the incident radiation.

3.8.1. Antenna Parameters

The **most important parameters to consider in antenna design are its gain, beamwidth and sidelobes.**

(a) Antenna Gain

The gain of an antenna refers to the energy radiated or received by a particular antenna compared to the energy radiated of received by some reference antenna. The reference antenna is chosen as an isotropic radiator which radiates energy in all directions from a point source. Thus the gain of an isotropic radiator is unity.

Some antennas concentrate energy radiated into a smaller region or beamwidth, which increases the signal amplitude within the beamwidth. This concentration of energy in one direction provides the gain of an antenna as compared to an isotropic radiator. A **reflector** is used to focus or concentrate the energy from a lightbulb in a flashlight into a light beam. The gain is usually defined in the direction of maximum signal strength. The maximum gain would occur at the center of the signal. The following expression provides the gain relationship :

$$G = \frac{\eta (4\pi A)}{\lambda^2} \quad \text{...(3.1)}$$

where,
G = Antenna gain (dB)
A = Antenna aperture (cm^2)
η = Antenna efficiency (0.55 to 0.70)
λ = Signal wavelength

Larger the antenna, higher the gain and narrower the beamwidth. Typically high gain antennas are used to support high data rates with low transmitter power.

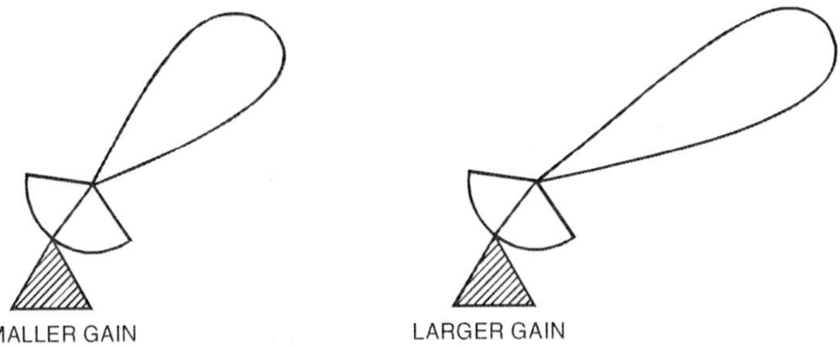

Fig. 3.13.

(b) Antenna Efficiency

Antenna efficiency is also an important factor and it includes the following effects :

1. **Surface accuracy :** A measure of the RF smoothness of the reflecting surface of the antenna.
2. **Profile accuracy :** A measure of how closely the shape of the dish approximates a true parabolic shape.
3. **Physical size :** For larger dish sizes (say above 2.4 m) it becomes impractical to manufacture and more importantly to transport dishes as a single object. So they tend to be built in petals. The shape and surface accuracy become dependent on supporting streets. This tends to mean that it becomes harder to achieve high efficiency from larger dishes.
4. **Aperture blockage :** The effect where the LNC or subreflector reduces the amount of signal that reaches the parabolic collecting area of the dish.
5. **Losses between feed and LNC.**
6. **Mismatch effect :** This is a measure of how much of the energy reflected from the parabolic surface of the dish is used by the feed.

(c) Focal Length to Diameter Ratio

Another consideration of antenna design is the focal length to diameter ratio (f/D). It provides a measure of the depth or shallowness of the dish cross section. A shallow dish has the feed a long way from the surface of the dish ($f/D = 0.4 - 0.5$) and will tend to be more efficient as the wavefronts are more likely to arrive in phase at the feed. Deep dishes (f/D) in the range (0.25 to 0.35) occupy less space but tend to have poorer efficiency. They also require specially designed feeds.

(d) Aperture Diameter

It is a critical factor and it determines the size of the antenna needed for the mission. This value is directly related to the beamwidth. Different antenna types have different beamwidth/aperture diameter relationship. Table 3.4 displays the various antenna configurations and the corresponding formulas.

Table 3.4. Antenna types for satellite systems.

S. No.	Antenna type	Parabolic reflector	Helix	Horn	Biconical Horn
1.		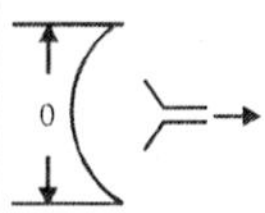	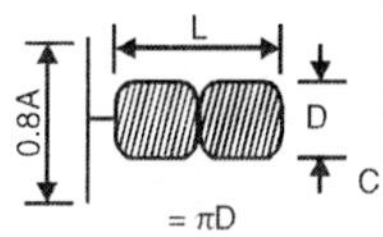	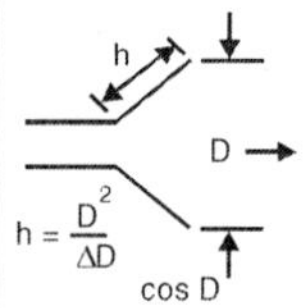	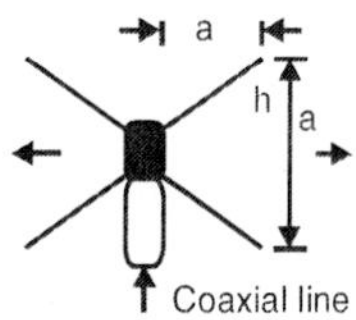
2.	Beam shape	Conical	Conical	Conical	Toroidal
3.	Typical Max	15-65	5-20	5-20	0-5
	Gain (dB)				
4.	Peek Gain (dB)	$17.8 + 20 \log D$ $+ 20 \log f$ ($\eta = 0.55$)	$10.3 + 10 \log \left(\frac{C^2 L}{\lambda^3}\right)$ $0.8 \le C/\lambda \le 1.2$ ($\eta = 0.70$)	$20 \log \left(\frac{C}{\lambda}\right) - 2.8$ ($\eta = 0.52$)	$5 \log \left(\frac{h}{\lambda}\right) + 3.5$ $R > 2\lambda$ $a = \sqrt{ah\lambda}$
5.	Half-Power Band width (deg)	$\frac{21}{fD}$	$\frac{52}{\sqrt{C^2 L/\lambda^3}}$	$\approx \frac{225}{(C/\lambda)}$	Typically 40° × 360° for gain ≥ –1 dB 70° × 360° for gain ≥ –3 dB
6.	Peak Gain & Dimensions of 18° Beam at 400 MHz	G = 19.1 dB D = 2.9 m	G = 19.4 dB D : 0.19 m 0.24 m or L : 9.8 m 6.2 m	G = 19.1 dB D = 3 m h = 4 m	...

3.8.2. Types of Antenna

Basically four main types of antennas are used on spacecraft. These are :

1. Wire antennas : monopoles and dipoles
2. Horn antennas

3. Array antennas
4. Reflector antennas.

The gain of an antenna is a measure of the antenna's ability to direct energy in one direction rather than all around. An antenna pattern is a plot of the field strength when the antenna is driven by the transmitter. Figure 3.14 shows typical satellite antenna coverage zones. The pattern is specified by 3 dB beamwidth.

3.8.2.1. Wire Antennas

- They are mainly used at VHF and UHF to provide communications for telemetry, tracking and command systems.
- These antennas provide omnidirectional coverage that are used primary during launch and orbit insertion, when the main antennas have not been deployed or property positioned.
- They are of different shapes like straight wire (dipole), monopole etc.

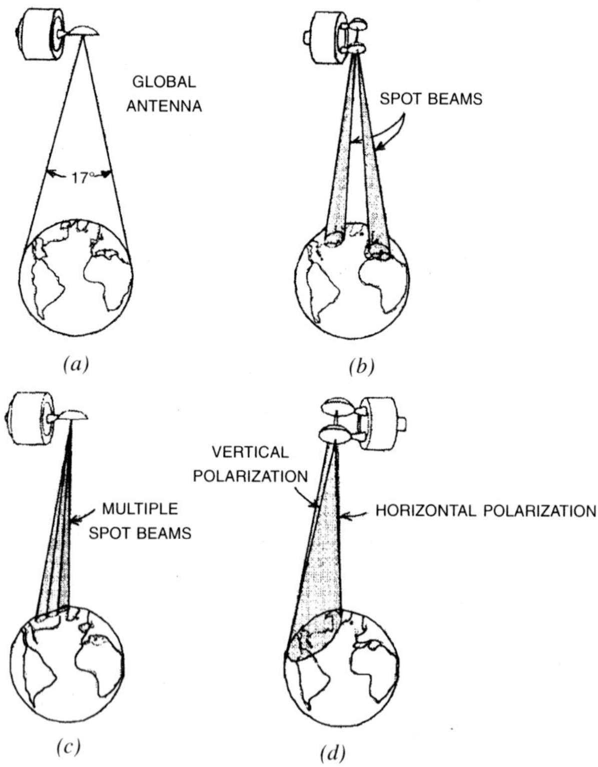

Beamwidth	*Earth converage diameter*
10°	3921 miles
5.7°	2235
2.8°	1117
1.0°	392
5.7°	223

Fig. 3.14. Satellite antenna coverage zones.

3.8.2.2. Horn Antenna

- These types of antennas are commonly used at the microwave frequencies.
- A horn antenna is used for the transmission and reception of microwave signals. It derives its name from the characteristics flared appearance.
- The flared portion can be square, rectangular or conical. The maximum radiation and response corresponds with the axis of the horn.
- In this respect, the antenna resembles on acoustic horn. It is usually fed with a waveguide.

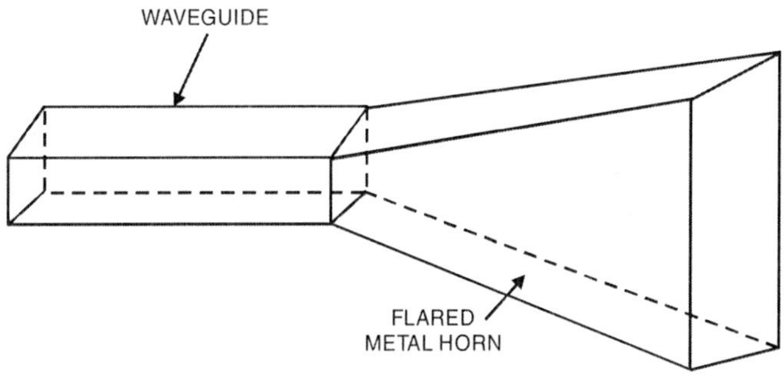

Fig. 3.15. Horn antenna.

- In order to function properly, a horn antenna must be a certain minimum size relative to the wavelength of the incoming or outgoing electromagnetic field. If the horn is too small or the wavelength is too large the antenna will not work efficiently.
- Horn antennas are **commonly used as the active device in a dish antenna.** The horn is pointed towards the center of the dish reflector. The use of a horn rather than a dipole antenna or any other type of antenna, at the focal point of the dish minimizes loss of energy around the edges of the dish reflector.
- It also minimizes the response of the antenna to unwanted signals not in the favoured direction of the dish.
- Horns are also used as feeds for reflectors either singly or clusters. Horns and reflectors are examples of aperture antennas that launch a wave into free space from a waveguide.
- It is difficult to obtain gains much greater than 23 dB or beamwidths narrower than about 10° with horn antennas. For higher gains or narrower beamwidths a reflector antenna or array must be used.

3.8.2.3. Antenna Array

- For satellite communication phased array antenna is of more interest.
- They are based on a principle that a pointable focused beam can be formed by simultaneously phase shifting an electromagnetic carrier feild in *N* separate branched, and radiating each shifted component as the output on an individual radiator (dipole, horn etc.). The resulting radiated pattern will reinforce in some direction and interfere in others, the net result being a combined beam that points

in a direction determined by the phase shifts used. By properly selecting the phase shifts, the directivity of the beam can be oriented in a given direction. Hence a phased array can theoretically produce beams in arbitrary off axis directions.

- The array gain increases with the square of the number of array elements and thus high gain is achieved with large arrays.
- When a reflector is placed behind the dipole array, radiation occurs in one direction and is shown in Fig. 3.16.

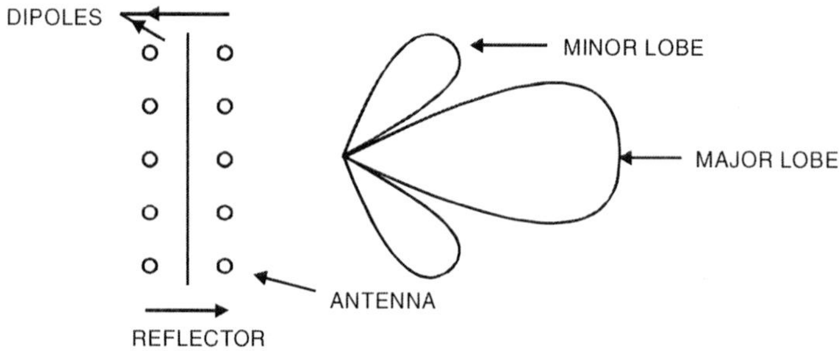

Fig. 3.16. Field pattern of an antenna array.

Basically two types of antennas array are there : **parasitic** and **driven.** The antenna shown in Fig. 3.16 is example of parasitic array. Reflector is not directly connected to the energy source. Driven arrays, in which all the radiating elements are connected to the energy source, have smaller losses than parasitic arrays while retaining some of the narrow beam characteristics.

- Parasitic arrays are used primarily as antennas in fire control radars and other installations, such as microwave communication systems and they require very accurate (narrow) beams.

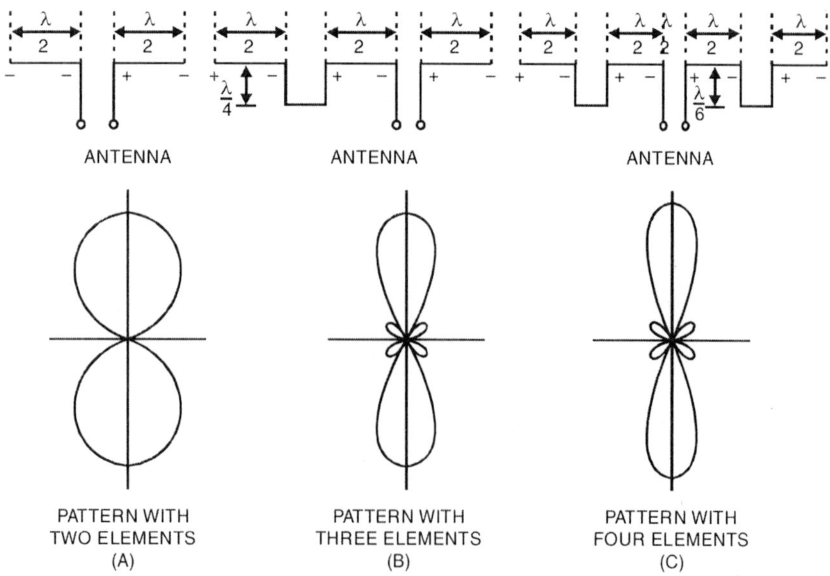

Fig. 3.17. Horizontal array field patterns.

- Driven arrays are used primarily as search radar antennas because extremely narrow beam are less critical than low losses. Driven arrays are usually made up of a number of half wave dipoles positioned and phased so that the desired directional pattern will be achieved.

Figure 3.17 shows a simple antenna array consisting of two horizontially mounted elements, each a half wavelength long and fed in phase. The resulting radiation pattern is in a direction at right angles to the plane containing the antenna conductor.

3.8.2.4. Reflector Antennas

- A spherical wavefront (one in which the energy spreads out in all directions) spreads out as it travels away from the antenna and produces a pattern that is not directional.
- A wavefront that exists in only one plane does not spread because all the wavefront moves forward in the same direction. For an antenna to be highly directive, it must change the normally spherical wavefront into a plane wavefront.
- Many highly directive microwave antenna produces a plane wavefront by using a reflector to focus the radiated energy. **The parabolic reflector is most often used for high directivity.**

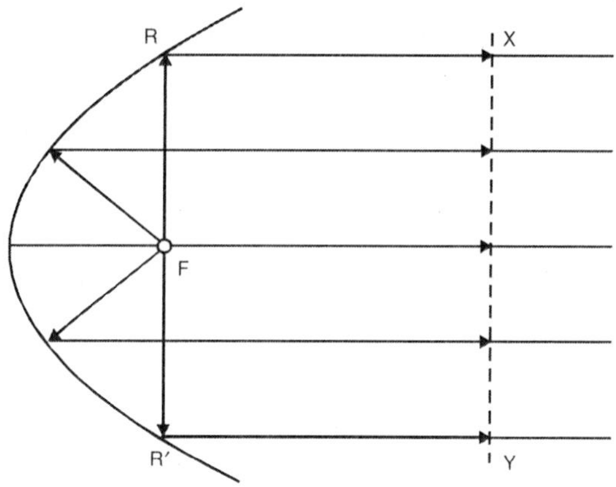

Fig. 3.18. Parabolic reflector antenna.

- Microwaves travel in straight lines as do light rays. They can also be focussed and reflected just as light array can, illustrated by the antenna shown in Fig. 3.18.
- A microwave source is placed at focal point F. The field leaves this antenna as a spherical wavefront. As each part of the wavefront reaches the reflecting surface, it is phase shifted 180°. Each part is then sent outward at an angle that results in all parts of the field travelling in parallel paths. Because of the spherical shape of a parabolic surface, all paths are from F to the reflectors and back to the line *XY* are the same length. Therefore, when the parts of the field are reflected from the parabolic surface, they travel to line XY in the same time.
- A dipole is used as the source of transmission, energy will be radiated from the antenna into space as well as toward the reflector.

- Energy which is not directed toward the paraboloid, has a wide beam characteristic which will destroy the narrow pattern of the parabolic reflector. However, a hemisphere shield may be used to direct most of the radiation toward the parabolic surface and thus prevent the destruction of the narrow pattern.
- Direct radiation into space is eliminated, the beam is made sharper, and more power is concentrated in the beam.
- Without the shield, some of the radiated field would leave the radiator directly. Since this part of the field that would leave the radiator would not be reflected, it would not become a part of the main beam and could serve no useful purpose.
- Figure 3.19 shows the radiation pattern of a paraboloid reflector contains a major lobe and several minor lobes.

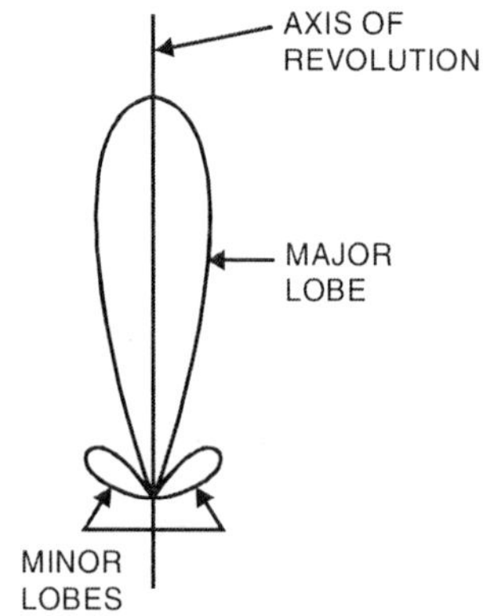

Fig. 3.19. Radiation pattern of parabolic reflector antenna.

3.8.3. Spacecraft Antennas in Practice

Usually the antennas of a communication satellite are a limiting element in the complete system. In an ideal spacecraft, there would be one antenna beam for each earth station. It is completely isolated from all other beams for transmit and receive. Therefore, if the two earth stations are 300 Km apart on the earth's surface and the satellite is in geostationary orbit, their angular separation at the spacecraft is 0.5°.

For $\theta_{3\text{ dB}}$ to be 0.5°, D/λ must be 150, which requires on aperture diameter of 11.3 m at 4 GHz. Although antennas of this large size have been flown on spacecraft are not yet in regular shape.

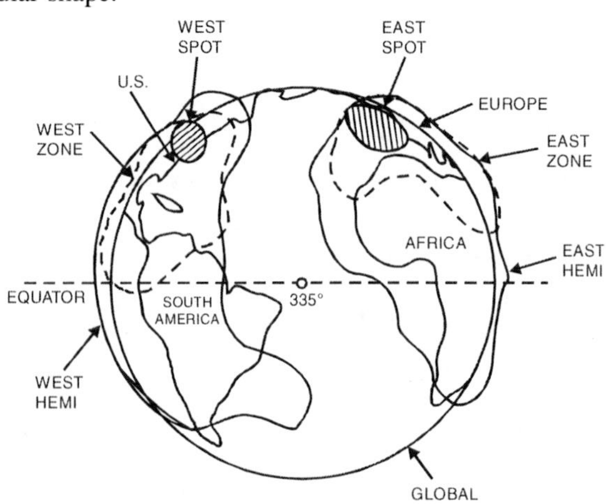

Fig. 3.20. Beam coverage patterns for INTELSAT V at 335°.

To provide a separate beam for each earth station, it would also require one antenna feed per earth station if a multiple feed antenna with a single reflector were used. In many satellites, a compromise between one beam per station and one beam for all stations has been used by using zone coverage beams and orthogonal polarization within the same beam to provide more channels per satellite. INTELSAT V Spacecraft has four reflector antennas. Each reflector is illuminated by a complex feed that provides the required beam shape to permit communication between earth stations within a given coverage zone. Figure 3.20 shows the coverage zones provided by INTELSAT V.

The largest reflector on the spacecraft transmits a 4 GHz and produces the **"peanut"** shaped pattern for the zone beams, which are designated to concentrate the transmitted energy onto densely populated areas such as North America and Western Europe, where much telecommunication traffic is generated. The smaller antennas are used to provide hemisphere transmit and receive beams and the 14/11 GHz spot beams. There are also four horn antennas that provides global beam coverage.

The antennas on intelsat V are highly complex and provide seven independent beams, with four fold frequency reuse in the 6/4 GHz band. This gives an available bandwidth of 1470 MHz with a further 780 MHz provided by the 14/11 GHz system. Most domestic satellites do not have such complex antenna system but never the less use orthogonal polarization frequency reuse to double the effective bandwidth at 6/4 GHz.

3.8.4. Frequency Reuse Antennas

Mainly there are two types of frequency reuse

(1) Spatial beam separation

(2) Orthogonal polarization.

In spatial beam frequency reuse, the same frequency bands are used for transmission and reception to geographically separated regions of the earth surface. The positioning of each satellite in the intersat global system has been selected to provide an east and a west zone by locating each spacecraft over an ocean. The angular separation of the east and west zones is about 10° and the coverage zones are approximately 5° wide. By generating shaped beams using large clustered feeds, it is possible to reduce the sidelobes from one beam to an acceptable level in the other zone. For FM transmission, it is necessary to keep interference between beams down to about –25 dB. With digital transmission systems, higher interference level can be tolerated, perhaps upto –17 dB.

In orthogonal polarization frequency reuse, each beam is generated in two polarizations. Example : the INTELSAT 6/4 GHz system uplink at 6 GHz on INTELSAT V uses left hand circular polarization (LHCP) for each of the hemi beams. The two zone beams each use right hand circular polarization (RHCP) and overlay the hemibeams. This places a very stringent requirement on the antenna, which must maintain an isolation between the two polarizations of atleast 25 dB throughout the coverage zones of both beams. A single antenna is used for the transmit beams at 4 GHz and a second antenna for reception at 6 GHz.

Summary

1. Spacecraft that carry communication relay provides a stable platform in geostationary orbit for a period of 5 to 10 years.
2. The attitude and orbit control system keeps the satellite in the correct orbit and on station, and pointing in the correct directions.

3. The telemetry, tracking and command system allows an earth station to control the subsystems on the spacecraft and monitor their operations.
4. The power system provides the electrical energy needed to run the spacecraft and the communication system.
5. Solar cells generate the power, power condition unit controls its distribution and batteries provide power during launch and eclipse.
6. Uplink and downlink antennas are highly directive, so parabolic reflectors with horn antenna at the focus used mainly. They provide very high gain also.
7. The available bandwidth of the transponder in 614 GHz band is 500 MHz but it can be increased upto 1000 MHz by using orthogonal polarizations.
8. The two major elements of a communication subsystems are transponder and antenna. The spacecraft antenna generates beam with carefully designed pattern to cover specific parts of the earth's surface.
9. Modern communication satellites are having the capability of reuse of available satellite bandwidth by providing dual polarized transmit and receive antennas with two sets of transponder.
10. The cost of satellite is several million dollar so the revenue of its cost is obtained by selling the communication capacity of the satellite to the users.
11. The conversion efficiency of solar cell is about 12 to 15% and typical satellite solar cell generate 1 to 2 KW of power.
12. Number of forces working on a satellite tend to change the attitude and orbit of a satellite, and they are : assymmetry of the earth's gravitational field, gravitation due to sun and planets, pressure due to solar radiation and magnetic field of the earth.
13. Earth sensor is a passive infrared device, operating in 14-16 μm wavelength and it senses the infrared rays.
14. Sun sensors uses a photocell and has a fan shaped field of view. It operates in the visible spectrum.
15. The entire spacecraft is rotated at 30 to 100 rotations per minute and this spin provides a powerful glyoscopic action to maintain the spin axis in correct direction.
16. Tracking determines the slant range for accurate determination of satellite orbit. It is also known as ranging.
17. Basically four types of antennas can be used on the spacecraft—wire antenna, horn antenna, array antennas, reflector antennas.
18. Attitude and orbit control system (AOCS) consists of locket motors that are used to move the satellite back to the correct orbit.
19. The tracking system is located at this earth station and provides information about the range, elevation and azimuth angles of the satellite.
20. To provide a gyroscopic action, the entire body of the spacecraft can be rotated at 30 to 100 rpm, which maintains the spin axis in the same direction. Such spacecrafts are known as spinners.
21. In three axis stabilization, with three momentum wheels, rotation of an spacecraft about each axis can be commanded from earth by increasing or decreasing the appropriate momentum wheel speed.
22. A satellite attitude may be altered along one or more of 3 axis. They are
    - Yaw axis—Directed towards the center of earth.
    - Pitch axis—Normal to the orbital plane.
    - Roll axis—Tangent to the orbit.
23. There are two ways to make and keep satellite stable in orbit. They are
    - Spin stabilization.
    - 3-axis body stabilization.
24. Beam width is defined as an angular width of the major lobe between points where radiated power have fallen to half of its maximum value.

25. The command system is used to make changes in attitude and corrections to the orbit and to control the communication system.
26. Eclipses occur twice per year, around the spring and fall equinoxes, when the earth shadow passes across the spacecraft.
27. Signals *i.e.* known as carriers transmitted by an earth station are received at the satellite by either a zone beam or a spot beam antenna.
28. Zone beams can receive from transmitters anywhere within the coverage zone, whereas spot beams have limited coverage.
29. Intermodulation distortion is likely to occur whenever we use travelling wave tube amplifier as wideband amplifier for more than one signal and drive the amplifier into saturation.
30. The transponder is the component of the subsystem that receives the signal and shifts its frequency for transmission.
31. Basically two types of repeaters are used in communication subsystems
    - Quasilinear repeaters
    - Regerative repeaters
32. The gain of an antenna is defined as

$$G = \frac{\eta 4\pi A}{\lambda^2}$$

33. Wire antennas are mainly used at VHF and UHF to provide communications for telemetry, tracking and command systems.
34. A horn antenna is used for the transmission and reception of microwave signals. It derives its name from the characteristics flared appearance.
35. A horn antenna must be a certain minimum size relative to the wavelength of the incoming or outgoing electromagnetic field so that it can function properly.
36. For satellite communication phased array antenna is most important and they are of two types :
    - *Parasitic :* Parasitic arrays are used primarily as antennas in fire control radars and other installations such as microwave communication systems.
    - *Driven :* Driven arrays are used mainly as search radar antennas. They are usually made up of a number of half wave dipoles positioned and phased so that the desired directional pattern will be achieved.

Review Questions

1. Discusss the main advantages and applications of satellites. Draw the block diagram of a satellite showing various subsystems.
2. Describe the telemetry, tracking and command facilities of a satellite communication system. Do they are a part of space segment or ground segment ? **[UPTU 2008-09]**
3. What is the attitude and orbit control subsystem ? Explain as to how these perform their functions.
4. Briefly describe the three axis method of satellite stabilization. **[UPTU 2007-08]**
5. What is meant by frequency reuse ? Explain it.
6. Explain the working of a transponder with the help of a block diagram. Why does the conversion in a transponder use 'RF to RF' conversion instead of 'RF to IF to RF' ?
7. What is the propulsion subsystem ? Explain its constituent and their function ?
8. Which kind of antennas are used for TTC signal transmission during transfer orbit and an orbit ?

9. What is the multiple beam technique ?
10. Explain why some satellites employ cylindrical solar arrays whereas other employ solar rail arrays for the production of primary power. State the typical power output to be expected from each type. Why is it necessary to carry batteries in addition to solar cell array. **[UPTU 2007-08]**
11. What is a despun antenna ? How is despunning achieved ? Explain. **[UPTU 2007-08]**
12. What kinds of antenna systems are being used in satelite communication subsystems ?
13. What is the redundancy and why it is necessary for the satellite subsystem mainly in repeaters ?
14. What do you mean by sensor ? Explain its types also.
15. What is the need of attitude and orbit control ? Explain it.
16. Draw the block diagram of spacecraft common system. Explain their functions also.
17. How many types of communication systems are there ? Give their applications also.
18. Draw the block diagram of a regenerative repeaters and explain the function of each block.
19. What is the function of transponder. Draw and discuss its block diagram. **[UPTU 2007-08]**
20. What is spot beams ? How can it be generated ?
21. What do you understand by frequency reuse ? Briefly explain its types also.
22. What do you mean by satellite stabilization ? Explain the importance of stabilization.
23. Discuss the various types of antennas used on spacecraft. **[UPTU 2008-09]**
24. What do you understand by tracking and point with reference to earth station ? **[UPTU 2003-04; 2007-08]**
25. Describe the following terms :
    (*a*) Attitude control (*b*) Despun
    (*c*) Reliability (*d*) Threster
    (*e*) Spot beam (*f*) Stabilization.

Objective Type Questions

1. Satellite transponder a
   (*a*) Receiver and a transmitter designed to relay microwave transmissions from one point on earth to another.
   (*b*) Device that echoes the radiation without change from one point to another.
   (*c*) Device that transforms the message sent from one location on earth to a different code for transmission to another location.
   (*d*) None of these.
2. Satellite may reuse the same frequency in the same area by
   (*a*) Having many small antennas (*b*) Overlapping radiation zones
   (*c*) TDMA (*d*) FDMA
   (*e*) Any of the above
3. The satellite in which the antennas are mounted on a despun plateform is the
   (*a*) Geostationary satellite (*b*) Sun synchronous satellite
   (*c*) Spin stabilised satellite (*d*) 3-axis body stabilised satellite
4. In a spin stabilized geostationary satellite, the spin axis is
   (*a*) Perpendicular to the orbital plane
   (*b*) In the plane of the orbit
   (*c*) Inclined at 45° to the orbital plane
   (*d*) None of these.

5. The available bandwidth of a transponder in 6/4 GHz band is
   (*a*) 500 MHz (*b*) 600 MHz
   (*c*) 550 MHz (*d*) 1000 MHz
6. Commonly used antenna in spacecraft subsystem is
   (*a*) Parabolic reflector (*b*) Yagi-uda
   (*c*) Log-periodic (*d*) Array antenna
7. Telemetry data are transmitted as a frequency or phase shift keying using
   (*a*) Time division technique (*b*) Frequency division technique
   (*c*) Any of above two (*d*) None of these.
8. Satellite transponder can deliver data rates in the range of
   (*a*) 50–150 Mbps (*b*) 20–50 Mbps
   (*c*) 150-250 Mbps (*d*) 50–150 Kbps
9. A body stabilized satellite
   (*a*) Uses solar panels whose cells are continuously oriented towards the sun
   (*b*) Uses solar cells mounted on a cylindrical body that continually rotates so that about 40% of the cells receive solar radiation at a given time.
   (*c*) Uses a spinning satellite to maintain orientation in phase
   (*d*) Uses a spinning movements what to maintain orientation in space
   (*e*) Both (*a*) and (*d*)
   (*f*) Both (*a*) and (*b*)
10. The satellite in which the solar cells are mounted on panels external to the space craft body, allowing them to be oriented towards the sun is the
    (*a*) Communication satellite
    (*b*) Direct broad cast satellite
    (*c*) 3-axis body stabilized satellite
    (*d*) Spin stabilized satellite
11. The conversion efficiency of solar cell is about
    (*a*) 12% to 15% (*b*) 15% to 20%
    (*c*) 12% to 25% (*d*) None of these
12. Propulsion subsystem unit consists of
    (*a*) Low thrust actuators (*b*) High thrust motor
    (*c*) With space shuttle launch satellite
    (*d*) All of these (*e*) None of these.
13. The type of antenna commonly used in fire control radars
    (*a*) Parasitic arrays (*b*) Driven array
    (*c*) Horn antenna (*d*) Yagi-uda antenna

ANSWERS

1. (*a*)	**2.** (*a*)	**3.** (*d*)	**4.** (*b*)	**5.** (*a*)
6. (*d*)	**7.** (*b*)	**8.** (*a*)	**9.** (*e*)	**10.** (*c*)
11. (*a*)	**12.** (*d*)	**13.** (*a*)		

❑❑❑

4

CHAPTER

SATELLITE CHANNEL AND LINK DESIGN

INSIDE THIS CHAPTER

4.1. INTRODUCTION

The design of satellite communication is very complex. In order to obtain the maximum performance some **dominating factors related to the design** are :

1. Weight of the satellite
2. DC power generated on board
3. Frequency band allocation for satellite communication
4. Multiple access techniques

Technically the **radio frequency range of satellite communication** is determined by

1. absorption, scattering and refraction effects in the atmosphere.
2. galactic noise and thermal noise because of the atmosphere.
3. Appropriate satellite antenna gain.

Satellite link gives the estimate power that a satellite will receive from the transmitting earth station and power received by the receiving earth station from the satellite repeater. The factor effecting the link design are discussed earlier. The major factor in the link design is the frequency of uplink and downlink (6/4 GHz).

The **satellite link may be :**

- Point to point
- Point to multipoint

Present days the **communication satellite uses microwave bands beacause of the following advantages :**

1. Microwave antennas have very high directivity. The directive gain as high as 50 dB can be easily achieved by using parabolic reflectors of feasible dimensions.

2. External noise from atmosphere are insignificant above 2 GHz. This causes high carrier to noise ratio at the input.
3. Frequency modulation used with microwaves gives specific improvements in signal to noise ratio in the receiver.

The weight of satellite is limited by the cost of launching a spacecraft into geostationary orbit. This cost is extremely high and is \$ 100,000 per kg per launch with final positioning. Since the size and weight of satellite is limited so it limits the number of solar cells and antenna size. These limitations in size produces a situation in which the spacecraft has a limit RF output power and it can not be concentrated on a very small areas of the earth. For a typical communication satellite the resulting flux density at the earth's surface will be –127 dB W/m^2. A direct broadcast satellite using a 200 W transmitter and directional beam can achieve 100 dB W/m^2.

There are a large number of earth stations sharing one satellite and providing many interconnecting paths. Some of these may be through one transponder, with the result that the available communication capacity of the transponder must be shared among several earth stations. This sharing technique is called **"multiple access"** and may be achieved by sharing the transponder bandwidth in seperate **frequency slots (FDMA)** or the transponder availability in discrete **time slots (TDMA)**. One more technique **code division multiple access** or spread spectrum, shares the transponder by allowing coded signals to overlap in time and frequency. A communication system must be designed to achieve some minimum performance standards within the limitations of transmitter power and RF bandwidth. The most important factor in information channel is signal to noise ratio (S/N) and it depends on carrier to noise (C/N) ratio of the RF or IF signal in the receiver, type of modulation and the baseband channel bandwidths in the receiver.

4.2. BASIC TRANSMISSION THEORY

The calculation of power received by an earth station from a satellite transmitter is very important in satellite communication link design. This calculation is quite similar to that of radio link.

Consider a transmitting source to be located in free space and supposed to radiate a power P_T watt uniformly in all directions. Such a source is called isotropic and ideally it is not possible because it could not create transverse polarized electromagnetic waves. Normally we use directive antenna to limit our transmitted power to be radiated primarily in one direction and its directivity is represented by a finite beam of width θ.

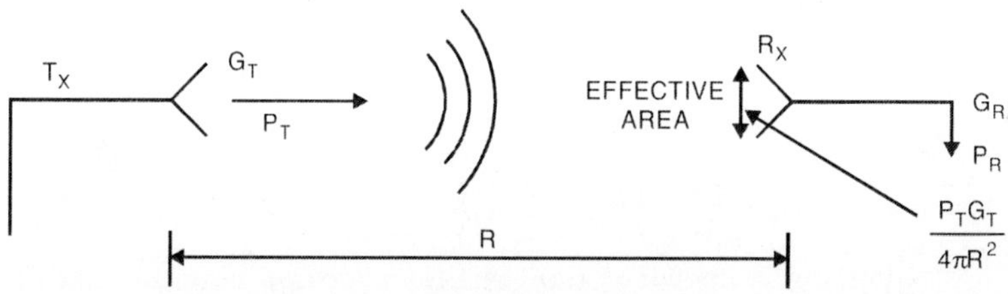

Fig. 4.1. Link budget calculation.

The distance between a transmitter and receiver is R. For a transmitter with P_T watt driving a lossless antenna with gain G_T, the flux density in the direction of antenna boresight at distance R metre is

$$F = \frac{P_T G_T}{4\pi R^2} \text{ W/m}^2 \qquad ...(4.1)$$

The product $P_T G_T$ is called **effective isotropic radiated power (EIRP)**. It gives a measure of power flux. If the aperture area of ideal receiving antenna is A_R m^2, as shown in given Fig. 4.2, we would correct received power $P_R W$ given by

$$P_R = F.A_R \qquad ...(4.2)$$

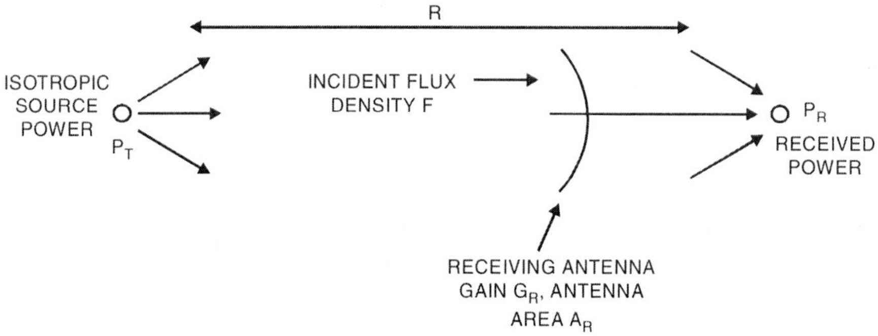

Fig. 4.2. Power received by antenna with area A_R.

In practical antenna with effective area A_R m^2, some of the energy incident on the aperture is reflected away from the antenna, and some is absorbed by lossy components. This reduced efficiency is described by an effective aperature A_e, where

$$A_e = \eta \cdot A_R \qquad ...(4.3)$$

η = aperture efficiency of the antenna and it is because of all the losses between the incident wavefront and the antenna output port.

For **paraboloidal reflector antennas** η is typically 50 to 75% in range and for horn antennas it approaches 90%. So the power received by an real antenna with physical receiving area A_R and effective aperture area A_e m^2 is

$$P_R = \frac{P_T G_T A_e}{4\pi R^2} \qquad ...(4.4)$$

The receiving antenna gain G_R is related to its effective area A_R by the relationship

$$G_R = \frac{4\pi A_e}{\lambda^2} \qquad ...(4.5)$$

Substituting Eq. (4.5) into Eq. (4.4) we get,

$$\boxed{P_R = P_T G_T G_R \left(\frac{\lambda}{4\pi R}\right)^2} \qquad ...(4.6)$$

This expression is known as the **Friis transmittion equation** and it is very important in calculation of power received in any radio link.

The term $\left(\frac{4\pi R}{\lambda}\right)^2$ is known as **path loss** or **free space loss (L_{FS}).**

Collecting the various factors we can write

$$\text{Power Received} = \frac{\text{EIRP} \times \text{Receiving Antenna Gain}}{\text{Path Loss}} \quad ...(4.7)$$

In **terms of decidel**, we have

$$P_R = (\text{EIRP} + G_R - L_{FS})\text{dB W} \quad ...(4.8)$$

where, $EIRP = 10 \log_{10} (P_T G_T)$ dB W

$$G_R = 10 \log_{10}\left(\frac{4\pi A_e}{\lambda^2}\right) \text{dB}$$

$$L_{FS} = \text{Path loss} = 20 \log_{10}\left(\frac{4\pi R}{\lambda}\right) \text{dB}$$

Equation (4.8) represent an ideal case, in which there are no additional losses in the link. In real sense there would be a variety of losses so,

$$L = L_{FS} * L_A \quad ...(4.9)$$

where, L_A = Additional losses.

The additional losses be written as

$$L_A = L_{ta} * L_{ra} * A_{AG} * A_{rain} * L_{point} * L_{POL} \quad ...(4.10)$$

where, L_{ta} = Losses associated with transmitting antenna.

L_{ra} = Losses associated with receiving antenna.

A_{AG} = Attenuation by the atmosphere and ionosphere.

A_{rain} = Attenuation due to precipitations and clouds.

L_{POL} = Losses caused by polarisation mismatch between the transmitting and the receiving antenna.

L_{point} = Losses caused by antenna depointing.

So the Eq. (4.10) can be written as

$$P_R = EIRP + G_R - L_{FS} - L_{ta} - L_{ra} - A_{AG} - A_{rain} - L_{POL} - L_{point} \quad ...(4.11)$$

The power received P_R from eqns. (4.4) and (4.6) is known as **carrier power (*C*)** because most satellite links use either frequency modulation for analog transmission or phase modulations for digital systems. In both modulation systems when data are modulated onto carrier, the amplifier of the carrier is not changed, so carrier power C is always equal to received power P_R.

4.3. SYSTEM NOISE TEMPERATURE, C/N AND G/T RATIO

4.3.1. Noise Temperature

Noise temperature is very important parameter since it provides a way of determining thermal noise generated by the devices in the receiver system. The most important source of noise in the receiver is the thermal noise in its preamplifier. The noise power P_N in the receiver is given by

$$\boxed{P_N = KT_nB} \quad ...(4.12)$$

where, K = Boltzmann constant (1.38×10^{-23})

T_n = Receiver noise temperature

B = Bandwidth.

P_N is the **available noise power** and will be delivered only to a device that is impedence matched to the source.

Term, KT_n = Noise spectral density (watts per hertz)

The density is constant for all radio frequencies upto 300 GHz. To maintain the best possible carrier to noise ratio *i.e.* quality of communication, noise at the receiver must be reduced. This can be done by keeping the B.W. in the receiver, usually in the IF amplifier stages, to be large enough to allow the signal to pass unrestricted, while keeping the noise power to the lowest possible value. The performance of the receiver system is determined by system noise temperature T_S. T_S is also called **effective input noise temperature** of the receiver and is defined as the noise temperature of a noise source located at the input of a noiseless receiver which would produced the same contribution to the receiver output noise as the internal noise of actual system itself.

In the receiver before demodulator a RF amplifier and IF amplifier exists. So if the overall gain of the receiver is G and bandwidth is B, the **noise power at the demodulator input** is

$$\boxed{P_n = KT_S\, B\, G} \quad ...(4.13)$$

If P_R is the signal power at the input to the receiver then the signal power at the demodulator input will be P_RG.

Hence the **carrier to noise ratio** at the demodulated input is given by

$$\boxed{\frac{C}{N} = \frac{P_R\, G}{KT_S\, BG} = \frac{P_R}{KT_S B}} \quad ...(4.14)$$

Several sources of noise in the receiver is replaced by single system noise temperature T_S.

4.3.2. Calculation of System Noise Temperature

Figure 4.3 shows a communication receiver

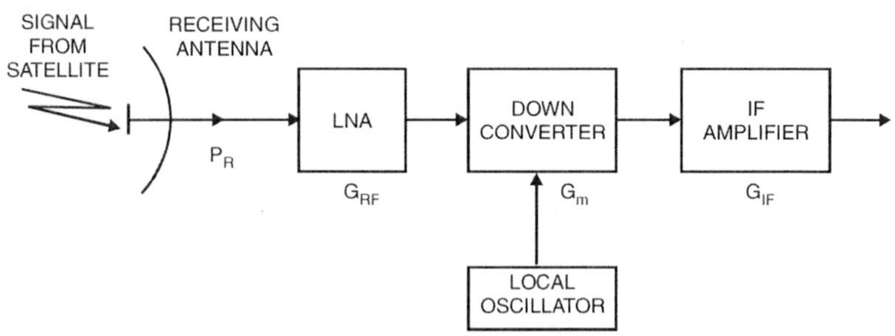

Fig. 4.3. Earth station receiver.

The equiivalent circuit used to represent this receiver is shown in Fig. 4.4 in which each device is replaced by single noise source with temperature, T_S.

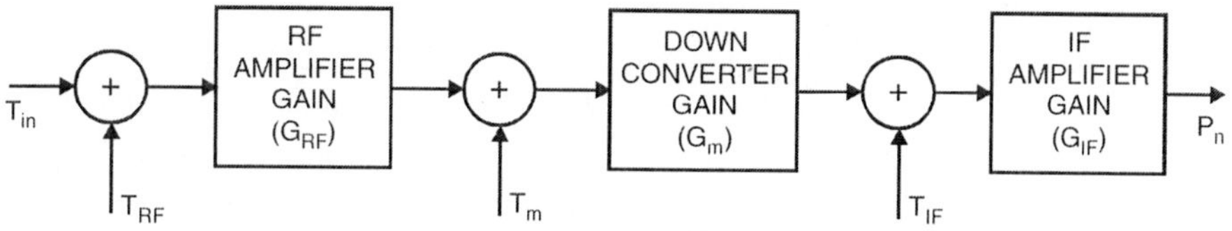

Fig. 4.4. Equivalent circuit of receiver.

Let the gain of RF amplifier, down converter and IF amplifiers are G_{RF}, G_m, G_{IF} and their respective equivalent noise temperatures are T_{RF}, T_m and T_{IF}.

If T_{in} is the input noise temperature then the **total noise power at the output of the IF amplifier** is given by

$$P_n = G_{IF} KT_{IF}B + G_{IF} G_m KT_m B + G_{IF} G_m G_{RF} K_B (T_{RF} + T_{in}) \quad ...(4.15)$$

$$= G_{IF} G_m G_{RF} \left[\frac{KT_{IF} B}{G_m G_{IF}} + \frac{KT_m B}{G_{RF}} + KB (T_{RF} + T_{in}) \right]$$

$$\boxed{P_n = G_{IF} G_m G_{RF} KB \left[T_{RF} + T_{in} + \frac{T_m}{G_{RF}} + \frac{T_{IF}}{G_m G_{RF}} \right]} \quad ...(4.16)$$

With the system noise temperature T_S, the noise power, P_n at the output of IF amplifier is given by

$$P_n = G_{IF} G_m G_{RF} KT_S B \quad ...(4.17)$$

from Eq. (4.16) and Eq. (4.17) we get,

$$KT_S B = KB \left[T_{RF} + T_{in} + \frac{T_m}{G_{RF}} + \frac{T_{IF}}{G_m G_{RF}} \right]$$

or

$$\boxed{T_S = \left[T_{RF} + T_{in} + \frac{T_m}{G_{RF}} + \frac{T_{IF}}{G_m G_{RF}} \right]} \quad ...(4.18)$$

Since the gain of each stage is added in the receiver so the succeeding stages contribute less and less noise.

4.3.3. Noise Figure and Noise Temperature

Noise figure is used to specify the noise generated within a device. The operational noise figure is defined as the ratio of input signal to noise power to the output signal to noise power. It is given by

$$\boxed{\text{Noise Figure } (NF) = \frac{(S/N)_{in}}{(S/N)_{out}}} \quad ...(4.19)$$

The relationship between a noise temperature and noise figure is given by

$$\boxed{T_d = T_0 (NF - 1)} \quad ...(4.20)$$

where, T_0 = reference temperature used to calculate the standard NF.

Usually noise figure is given in decibles.

4.3.4. Antenna Noise Temperature

Antenna noise temperature is a result of other noise sources including galactic and other satellites.

The antenna noise is due to energy, which is fed to the antenna by unwanted radiation sources such as stars, galaxies and other communication signals. The atmosphere itself behaves as a resistive medium, which supplies noise power to the antenna. The output noise power from the antenna,

$$N_A = KT_AB \quad ...(4.21)$$

It will depend on the positions, temperatures and emisivities of the noise sources and the gain, and polar radiation pattern of the antenna.

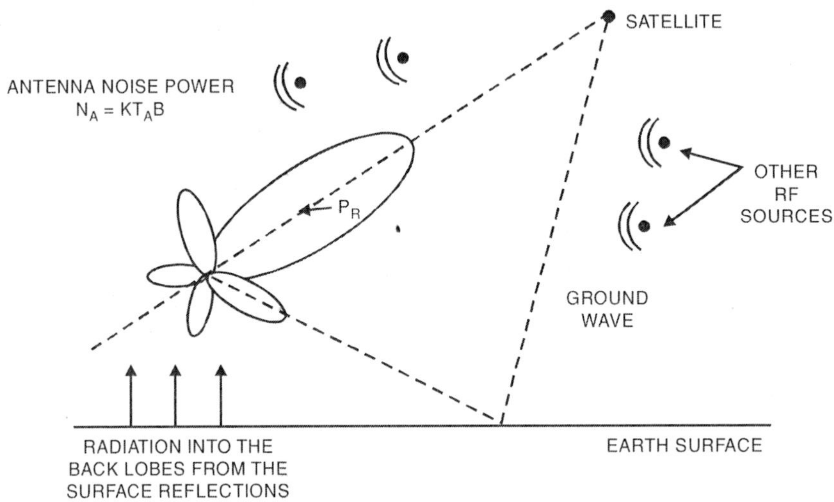

Fig. 4.5.

Antenna Pointing to the Sky : In this case the output noise power from the antenna has two components which are represented by the sky temperature T_{sky} and the earth temperature T_{earth}. T_{sky} is due to noise originating in the atmosphere. It varies with frequency and the elevation angle E of the antenna. The sky temperature is higher for $E = 0°$ (antenna pointing to the horizontal) because of the longer path of the radiation through the atmosphere. Elevation angles of less than 10° are usually avoided.

Figures 4.6 and 4.7 show T_{sky} for different frequency ranges.

For $E \geq 10°$ and $f \leq 15$ GHz. $T_{sky} \leq 40$ K. T_{earth} arises from radiation which feeds into the antenna via the back lobes of the antenna radiation pattern.

For a large (≃ 5 m) cassegraine antenna

$$T_{earth} \approx 10 \text{ K}$$

For a small (≈ 0.5 m) antenna

$$T_{earth} \approx 100 \text{ K}$$

If an antenna points towards the sun and noise effective temperature is about 10,000 K, the situation should be avoided.

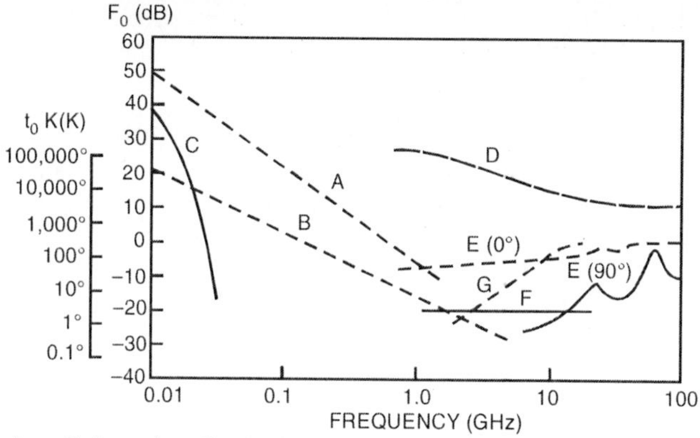

A : *Estiamted median business area man-made noise*

B : *Galactic noise*

C : *Atmosphere C noise, value exceeded 0.5% of time*

D : *Quiet sun (½ degree beamwidth directed at sun)*

E : *Sky noise due to oxygen and water vapour (very narrow beam antenna); upper curve, 0° elevation angle; lower curve 90° elevation angle*

F : *Black body (cosmic background), 2.7 K*

G : *Heavy rain (50 min/n over 5 km)*

Fig. 4.6. Antenna sky noise temperature as a function of frequency and antenna angle.

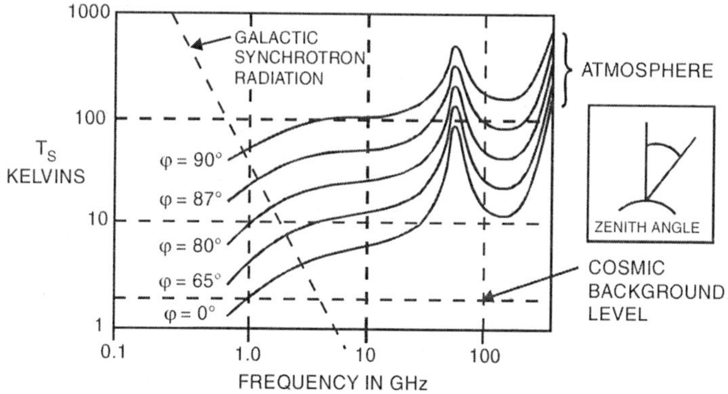

Fig. 4.7. Sky noise for clear air and 7.5 g/m^3 of water vapour concentration (φ is the elevation angle).

Antenna pointing to the earth : Usually the beam width is less than or equal to the angle subtended by the earth so that the earth fills the beam. Then the noise temperature of the antenna is about 290 K which is the physical temperature of the earth.

4.3.5. Noise Density

It is a power presented at 1 kHz bandwidth.

It is defined by

$$N_0 = \frac{N}{B}(\text{W/Hz}) = KT \qquad \text{...(4.22)}$$

Since ($N = KTB$)

$$N_0\ (\text{dBW/Hz}) = 10 \log N - 10 \log B$$

$$= 10 \log K + 10 \log T \qquad ...(4.23)$$

Remains constant for all frequency upto 300 GHz as long as T remains constant.

$$\frac{C}{KT} = \text{A ratio often used to state the quality of a satellite channel}$$

$$= \frac{\text{Received carrier power}}{\text{Noise density}} \qquad ...(4.24)$$

$$= \frac{P_r}{N_0} = \frac{P_r}{N/B} = \frac{P_r}{KT}$$

The carrier to Noise ratio is related to the quantity of information that can be transmitted in unit bandwidth as shown by Shanon's law

$$\boxed{W = B \log_2 \left(\frac{C}{KT} + 1 \right)} \qquad ...(4.25)$$

where, W = Information carrying capacity of the link

B = Usable bandwidth

4.3.6. G/T Ratio for Earth Station

From the Eq. (4.14) the link equation can be rewritten as

$$\frac{C}{N} = \frac{P_T G_T G_R}{K T_S B} \left[\frac{\lambda}{4\pi R} \right]^2$$

$$\boxed{\frac{C}{N} = \frac{P_T G_T}{KB} \left[\frac{\lambda}{4\pi R} \right]^2 \frac{G_R}{T_S}} \qquad ...(4.26)$$

Thus, $$\left(\frac{C}{N} \right) \propto \left(\frac{G_R}{T_S} \right)$$

and for a given satellite λ and R will be constant. $\frac{G_R}{T_S}$ ratio is widely used to specify the **quality** of an earth station and is known as ***G/T* ratio** or the **figure of merit.**

If $N_0 = \frac{N}{B}$ which is also called the **noise power spectral density** $\left(\frac{\text{W}}{\text{Hz}} \right)$. The carrier power to noise power spectral density ratio is given by

$$\boxed{\left(\frac{C}{N_0} \right)_{\text{dBHz}} = 10 \log P_T G_T - 20 \log \frac{4\pi R}{\lambda} + 10 \log \frac{G_R}{T_S} - 10 \log \text{K}} \qquad ...(4.27)$$

A standard earth station used in Intelsat Network is required to have a G/T ratio of 40.7 dB K^{-1}, at 4.0 GHz and 5° elevation angle. It is necessary to specify the frequency and elevation angle because G_R varies as f^2 across the frequency band and T_S depends on sky temperature, which increases as the elevation angle is reduced below 10°. At higher frequencies say 11 GHz, the variation in G_R and T_S are quite small because noise contribution of LNA is higher and narrow functional bandwidth employed. Numerical value of G_R is smaller than T_S because satellite antenna may have negative G_R/T_S which is below 0 dB K^{-1}.

4.4. UPLINK DESIGN

The design of uplink is easier than down link because an accurate specified carrier power density must be presented at satellite transponder and high power transmitters at earth stations can be designed. In comparison with the receiving equipment, the cost of transmitter is high. The generation of high power microwave carrier is very expensive and it require considerable care to control spurious emission and group delay effect. So the transmission in satellite communication is point to multipoint and a high cost transmitter earth station provides services to many low cost receive only station.

The satellite transponder is a quasilinear amplifier and the received carrier level determines the output level. where a travelling wave tube is used as the output high power amplifier in the transponder. As an example, for an INTELSAT-IV A satellite transponder operated in the FDMA mode, an input carrier density of –73.7 dB W/m^2 to –67.5 dB W/m^2 depending on the transponder gain setting and it can be controlled via the TT&C system. The uplink (C/N) for the given flux density for an Intelsat IV – A satellite is in the range of 20.9 to 29.0 dB.

The transponder input stage is often a GaAs FET amplifier and the front end G/T of the satellite is –17.6 dB K^{-1} on global beam and –11.6 dB K^{-1} on spot beam. Since during propogation the input noise temperature does not change significantly so noise contribution from the uplink is generally constant.

Non-linear effects in the uplink are included by including back off terms. In this case equation of C/N_0 for uplink is given by

$$\left(\frac{C}{N_0}\right)_{\text{uplink}} = 10 \log P_T G_T - 20 \log \frac{4\pi R}{\lambda} + 10 \log \frac{G}{T} - 10 \log K - BO_i \quad \text{...(4.28)}$$

where, BO_i = input back off in dB and it is of the order of 3 to 7.

4.5. DOWNLINK DESIGN

There are two major points that must be remember in the design of downlink of any satellite communication system.

(1) It should carry maximum number of channels at a minimum launching cost and minimum maintenance cost.

(2) With a given (S/N) ratio, for a given percentage of time (typically 99.9%), it gives guarantee continuity of the link.

The **first objective** requires a compromise between antenna cost, receiver cost, tracking accuracy, station keeping, modulation and multiple access technique. The **second objective** requires a minimum (C/N) at the receiver input for 99.9% of the time and will almost certainly require a modulation scheme or signal processing technique that gives a (S/N) improvement over the receiver (C/N).

Usually the modulation is FM in analog system and PSK in digital system but companded SSB is also used.

By using frequency modulation (FM) and frequency division multiple access, the transponder cannot be run at its maximum output power because of its non-linear characteristics. Intermodulation arises because of the non-linearity of the transponder output amplifier and these intermodulation product increases the overall noise level and thus reduces the (C/N) of the wanted signals at the receiving earth stations. We cannot increase the (C/N) by decreasing the bandwidth in order to decrease the noise power unless we are prepared to reduce the total number of channels transmitted through the transponder. This is the disadvantage of FDMA system because this makes the cost of low capacity earth station is very high as the cost of high capacity stations.

An alternative multiple access technique is TDMA that allows large and small earth stations to share the same transponder with increased efficiency. In systems using frequency modulation, the improvement in (S/N) is obtained by using wide deviation FM.

$\frac{C}{N_0}$ for downlink is given by

$$\boxed{\left(\frac{C}{N_0}\right)_{\text{down link}} = 10 \log P_T G_T - 20 \log \frac{4\pi R}{\lambda} + 10 \log \frac{G}{T} - 10 \log K - BO_0} \quad \text{...(4.29)}$$

where, BO_0 = Output back off in dB

Input and output back off relationship for a satellite travelling wave tube amplifier is given by,

$$[BO]_i = [BO]_0 + 5 \text{ dB} \quad \text{...(4.30)}$$

4.6. EFFECTS OF RAIN

Upto this point, calculations have been made for clear sky conditions. In the *C* band, more especially the Ku band, rainfall is the most significant cause of signal fading. Rainfall results in attenuation of radio waves by scattering and by absorption of energy from the wave, and it increases with increase in frequency.

For 99% of the time, attenuation will be equal to or less than 0.2 dB; for 99.5% of time, it will be equal to or less than 0.3 dB and for 99.9% of time, it will be equal to or less than 1.3 dB.

4.6.1. Uplink Rain Fade Margin

Rainfall results in attenuation of the signal and an increase in noise temperature. It decreases the (C/N_0) at the satellite in two ways.

One is the **increase in noise** and this is because the satellite antenna is pointed toward a "hot" earth, this added to the satellite receiver noise temperature and tends to mask any additional noise included by rain attenuation. But it is not a major factor for uplink.

Other is the **uplink power control** for rain fades that is uplink carrier power at the satellite must be held within close limits for certain modes of operation. Power output from the satellite may be monitored by a central control station or in some cases by each earth station. To compensate rain fading power output from any given earth station may be increased.

4.6.2. Downlink Rain Fade Margin

Rainfall introduces attenuation by absorption and scattering of signal energy and the absorptive attenuation introduces noise. Let $[A]$ dB represents the rain attenuation caused by absorption the corresponding power loss ratio is

$$A = 10^{[A]/10} \quad \text{...(4.31)}$$

The **effective noise temperature** of the rain is

$$\boxed{T_{\text{rain}} = T_a\left[1 - \left(\frac{1}{A}\right)\right]} \quad \text{...(4.32)}$$

The total sky noise temperature is the clear sky temperature T_{CS} plus the rain temperature,

$$\boxed{T_{\text{sky}} = T_{CS} + T_{\text{rain}}} \quad \text{...(4.33)}$$

where, T_{CS} = Noise temperature of clear sky

The downlink $\left(\frac{C}{N_0}\right)$ power ratio is related to the clear sky value by

$$\boxed{\left(\frac{N_0}{C}\right)_{\text{rain}} = \left(\frac{N_0}{C}\right)_{CS}\left[A + (A-1)\left(\frac{T_a}{T_{S,CS}}\right)\right]} \quad \text{...(4.34)}$$

where, $T_{S,\,CS}$ = System noise temperature at clear sky condition.

For satisfactory reception of the signal, a minimum value of (C/N) is required. In the case of frequency modulation, a min value is set by the threshold level of the FM detector and a **threshold margin** is normally allowed.

4.7. SATELLITE SYSTEMS USING SMALL EARTH STATIONS

In certain applications we want that satellite should carry only one or two telephone or data channels or a direct broadcast TV signal, therefore we prefer to use small, low cost earth stations. These earth station costs a few hundred or a few thousand dollars.

To allow us to use a smaller receiving antenna, two parameters need to be adjusted at the satellite :

- Satellite transmitted power
- Satellite antenna beam width
- In domestic satellite systems, narrow beams are used for transmitting from the space craft to provide coverage to required limited area.

- Since the dimensions of the antenna mounted on space craft is limited, the coverage zone cannot be made arbitrary small.
- When observed from geostationary orbit, the earth's disk subtend an angle of 17°. It can be illuminated with a microwave horn having an aperture a few wavelengths in diameter.
- At 4 GHz to obtain a 4° spot beam, a dish 1.4 m in diameter is required.

As the frequency is increased, the diameter of the spacecraft antenna in wavelength is increased for a given dish diameter and makes it feasible to use more directive beams. If switched or multiple beam system is used, the single transmit (or receive) beam must cover the whole region that the domestic satellite serves.

The gain of the earth station at the edge of the coverage zone is typically 3 dB or lower, or 29 dB. The received carrier level and (C/N) can be easily calculated for an earth station with a 6 m antenna, using a transponder with an output power of 5 W at 4 GHz and transmitting a single carrier.

Ignoring all losses and using a receiving antenna gain of 45.6 dB (60% efficiency) and a system noise temperature of 120 K.

$$P_r = 7 + 29 - 196 + 45.6$$

$$\boxed{P_r = -114.4 \text{ dBW}} \quad \text{...(4.35)}$$

The noise power at the input to a low noise receiver with BW 36 MHz and system noise temperature 120 K is :

$$N = KTB = -228.6 + 20.8 + 75.6 \text{ dBW}$$

$$\boxed{N = -132.2 \text{ dBW}} \quad \text{...(4.36)}$$

Therefore, for this system, the (C/N) is 17.8 dB. This is same 6.8 dB above an FM threshold of 11 dB and provides an adequate margin for an operational system.

The diameter of receiving antenna can be reduced if an FM demodulator with a lower threshold is used. For high quality demodulator, the typical threshold figure is 11 dB. Threshold is recognized by an increase in noise at the FM demodulator output caused by noise spikes in the IF input to the demodulator.

A 3-m dish was the smallest earth station that could be used for reception of TV transmission from the first generation of U.S. domestic satellites.

4.7.1. Direct Broadcast Television

Planned broadcasting directly to home TV receivers takes place in the Ku-band *i.e.* 12 GHz. This service is known as direct broadcast satellite (DBS) service. There is some variation in the frequency bands assigned to different geographic regions. In DBS the reception of TV signals from satellite is direct and is meant for the individual reception and community reception.

The idea of direct broadcasting of TV from a satellite was introduced around 1986 with the existing communication satellites. The major feature of direct broadcast satellite in that it allows earth station to be mounted in domestic area. The dish size is of the order of 0.4 to 0.75 metre.

DBS Frequencies Allocated to India

Table 4.1 shows the DBS frequencies allocated to India with their orbit location and EIRP.

Table 4.1.

Channel no.	Frequency	Orbit location	EIRP
1.	11727.48	56°	63.1
		68°	63.3
2.	11746.66	56°	63.9
		68°	63.3
3.	11765.84	56°	63.9
		84°	63.3
4.	11785.02	56°	63.6
		68°	63.7
5.	11804.2	56°	63.1
		68°	63.4
6.	11823.38	56°	63.6
		68°	64.0
7.	11842.56	56°	63.4
		68°	63.5
		68°	63.5
8.	11861.74	56°	63.7
		68°	63.7
9.	11880.92	56°	63.2
		68°	63.5
10.	11900.10	56°	63.0
		68°	64.0
11.	11919.28	56°	63.5
		68°	63.5
12.	11938.46	56°	63.8
		68°	63.8
13.	11957.64	56°	63.3
		68°	63.5
14.	11976.82	56°	63.7
		68°	64.1
15.	11996.00	56°	63.5
		68°	63.6
16.	12015.18	56°	64.8
		68°	65.5
17.	12034.36	56°	64.3
		68°	63.6
18.	12053.54	56°	63.8
		68°	63.8
19.	12072.72	56°	64.3
		68°	63.8
20.	12091.90	56°	63.8
		68°	63.8
21.	12111.08	56°	64.4
		68°	63.7
22.	12130.26	56°	63.8
		68°	63.8
23.	12149.44	56°	64.4
		68°	63.7
24.	12168.62	56°	63.9
		68°	63.9

For DBS TV transmission the down link frequency is about 12 GHz and uplink frequency is about 14.5–14.8 GHz. There are several advantages to use Ku band for DBS service :

(1) Reflector diameter at the receiver may be less than 1 metre.

(2) Reduction in the effective area of the antenna (Ae) will be fully compensated by more gain (G). Hence there would be no adverse effect on the overall gain due to the reduction in the size of antenna.

(3) There is more path loss at Ku band and higher bands, but this is compensated by higher EIRP for DBS system.

(4) Bandwidth of 500 MHz can accommodate 32 TV channels, each of which is 24 MHz wide. The overlapping of channel may appears as, $32 \times 24 = 768$ MHz, but this is made possible by using orthogonal polarizations.

The block diagram of a DBS system is shown in Fig. 4.8.

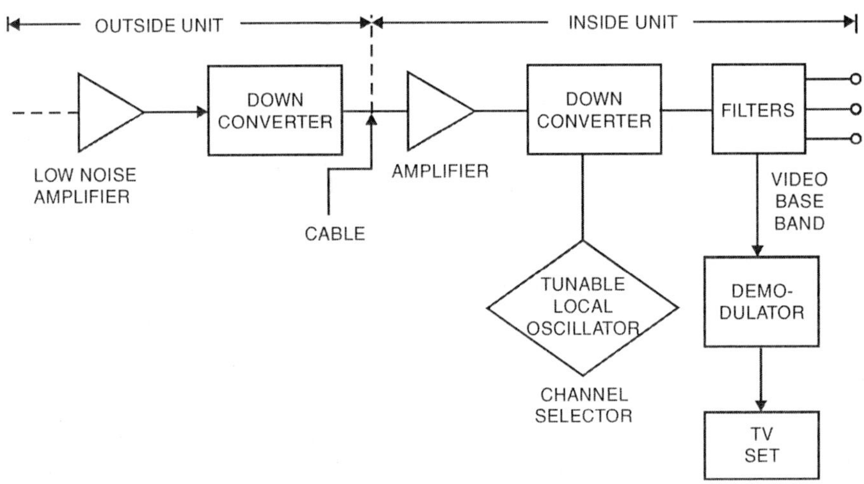

Fig. 4.8. Block diagram of DBS system.

In the block diagram of DBS system, there are two units, outside unit and inside unit. The various components used in both of the units are shown in Fig. 4.8 and their description are given as

Outside Unit : It consists of a receiving antenna feeding directly into a low noise amplifier/converter combination. The receiving antenna is a parabolic reflector with a horn as a active element. The horn can be directly in front of the reflector or it may use an offset feed. The reflector diameter may be 0.6 m for 11 GHz and still smaller for K and Ka band. The downlink frequency band of 12.2 to 12.7 GHz spans a range of 500 MHz which accommodates 32 TV/FM channels, each of which is 24 MHz wide. Some overlapping may occur between the channels but these are alternately polarised left hand circular (LHC) and right hand circular (RGHC) on vertical/horizontal, to reduce interference to acceptable levels. This is known as polarization interleaving. The details of outdoor unit of DBS home receiver are shown in Fig. 4.9.

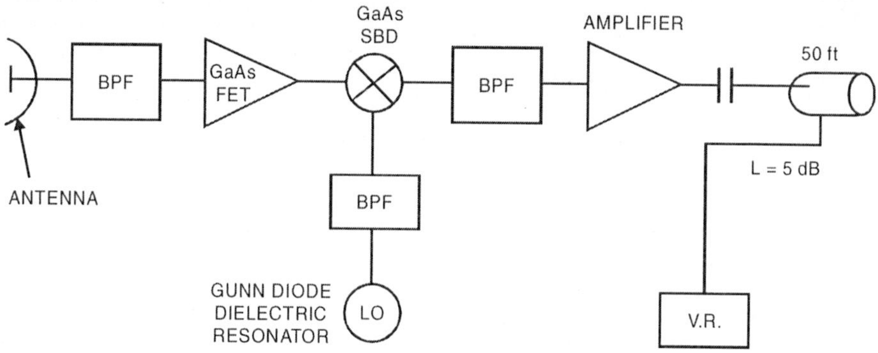

Fig. 4.9. Details of outdoor unit of a DBS home receiver.

The receiving horn feeds into a low noise converter (CNC) or a combination unit consisting of a low noise amplifier (LNA) followed by a converter. This combination is known as a low noise block (LNB). The low noise block provides gain for the broadband 12 GHz signal and then converts the signal to a lower frequency range so that a low cost coaxial cable can be used as feeder to the indoor unit. The converter output consists of a signal of VHF frequency ranging from 950–1450 MHz (500 MHz bandwidth). The advantage of using VHF frequency is that a low cost coaxial cable can be used as feed from the outdoor unit to indoor unit.

The low noise block cannot be kept inside unit because long cable between the horn and the first amplifier (LNA) will cause substantial degradation of overall noise figure of the set. Cable after the LNA cannot add any significant noise as the dominant noise is due to the first amplifier. The low noise block should be able to withstand with outside hostile environmental conditions.

Inside Unit : The signal fed to the indoor unit is normally a wideband signal converting the range 950 to 1450 MHz. This is amplified and fed to a channel selector circuit which selects the wanted channel. The selected channel is again down converted from the 950-1450 MHz range to a fixed intermediate frequency of 70 MHz by a local oscillator or mixer. The 70 MHz amplifier amplifies the signal up to the levels required for demodulation. The 70 MHz, frequency modulated IF carrier therefore noise be demodulated, and the baseband information used to generate a VSSB signal which is fed in to one of the VHF/UHF channels of a standard TV set.

There is a major difference between DBS TV and conventional TV is that with DBS, frequency modulation is used whereas with conventional TV, amplitude modulation in the form of vestigial single sideband (VSSB) is used.

4.8. DESIGN OF SATELLITE LINKS FOR SPECIFIED (C/N)

4.8.1. Complete Link Design

A single complete link consist of two earth stations, a satellite and a radio relay is done by satellite. So the complete link is made up of uplink and downlink. The information received on earth depends on the uplink, the satellite transponder and downlink. Let the carrier to noise power spectral density ratio for uplink or at the transponder input is $(C/N_0)_U$ and for down link is $(C/N_0)_D$. The carrier to noise power spectral density ratio $(C/N_0)_T$ at the receiving earth station can be derived in the following way.

The usefull carrier (signal) at the receiving earth station receiver input is given as

$$C = \frac{C_U G_S G_T G_R}{L} \qquad ...(4.37)$$

where, C_U = signal power at the satellite transponder input

G_S, G_T, G_R = satellite transponder gain, satellite transmitting antenna gain, satellite receiving antenna gain.

The noise power spectral density at the input to the receiving earth station receiver would be given by

$$N_0 = N_{OD} + N_{0u} \frac{(G_S G_T G_R)}{L} \qquad ...(4.38)$$

where, N_{0u} = noise power density at transponder input

N_{0D} = noise power density at the input to receiving earth station receiver.

Thus,
$$\left(\frac{C}{N_0}\right)_T = \frac{C}{N_0} = \frac{\dfrac{C_u G_S G_T G_R}{L}}{\dfrac{N_{OD} + N_{0u}(G_S G_T G_R)}{L}}$$

$$= \frac{C_u}{N_{0u} + \dfrac{N_{OD} \cdot L}{G_S G_T G_R}} \qquad ...(4.39)$$

If the bandwidth of transponder is B and it radiates a constant power P_T, its gain G_S will be expressed as

$$G_S = \frac{P_T}{C_u + N_{0u} \times B} \qquad ...(4.40)$$

for downlink only the signal power C_D is

$$C_D = \frac{P_T G_T G_R}{L} \qquad ...(4.41)$$

Substituting the value of G_S in Eq. (4.39)

$$\frac{C}{N_0} = \frac{C_u}{N_{0u} + \dfrac{N_{OD} . L\,(C_u + N_{0u} \times B)}{P_T G_T G_R}}$$

$$= \frac{C_u}{N_{0u} + \dfrac{N_{OD}\,(C_u + N_{0u} \times B)}{C_D}}$$

$$= \frac{\left(\dfrac{C}{N_0}\right)_u \cdot \left(\dfrac{C}{N_0}\right)_D}{\left(\dfrac{C}{N_0}\right)_D + \left(\dfrac{C}{N_0}\right)_u + B} \qquad ...(4.42)$$

In most of the cases B is much smaller than $\left(\frac{C}{N_0}\right)_u$ and $\left(\frac{C}{N_0}\right)_D$ so Eq. (4.42) gives,

$$\left(\frac{C}{N_0}\right)_T^{-1} = \left(\frac{C}{N_0}\right)_u^{-1} + \left(\frac{C}{N_0}\right)_D^{-1} \quad ...(4.43)$$

Normally, $\left(\frac{C}{N_0}\right)_u >> \left(\frac{C}{N_0}\right)_D$

so, $$\left(\frac{C}{N_0}\right)_T^{-1} \cong \left(\frac{C}{N_0}\right)_D^{-1} \quad ...(4.44)$$

So the complete link design depends on the **quality of downlink** and especially on $\left(\frac{C}{N_0}\right)_D$.

To include the effects of interfering signals Eq. (4.43) may be written as

$$\left(\frac{C}{N}\right)_{\text{Net uplink}} = \left[\left(\frac{C}{N}\right)_u^{-1} + \left(\frac{C}{I}\right)_u^{-1}\right]^{-1} \quad ...(4.45)$$

if I represents the noise power involved with the interfering signals.

Then $\left(\frac{C}{N}\right)_I$ = carrier to noise ratio of the interference signal for uplink.

Similarly for downlink,

$$\left(\frac{C}{N}\right)_{\text{Net downlink}} = \left[\left(\frac{C}{N}\right)_D^{-1} + \left(\frac{C}{I}\right)_D^{-1}\right]^{-1} \quad ...(4.46)$$

from Eq. (4.43) the **net carrier to noise ratio** is given by

$$\left(\frac{C}{N}\right)_{\text{Net}} = \left(\frac{C}{N}\right)_{\text{Net uplink}} + \left(\frac{C}{N}\right)_{\text{Net downlink}}$$

$$= \left[\left(\frac{C}{N}\right)_u^{-1} + \left(\frac{C}{I}\right)_u^{-1} + \left(\frac{C}{N}\right)_D^{-1} + \left(\frac{C}{I}\right)_D^{-1}\right]^{-1}$$

$$= \left[\left(\frac{C}{N}\right)^{-1} + \left(\frac{C}{I}\right)^{-1}\right]^{-1} \quad ...(4.47)$$

where, $$\left(\frac{C}{I}\right)^{-1} = \left(\frac{C}{I}\right)_u^{-1} + \left(\frac{C}{I}\right)_D^{-1} \quad ...(4.48)$$

Here, $\left(\frac{C}{N}\right)$ = carrier to noise ratio of overall link

$\left(\frac{C}{I}\right)$ = carrier to interference ratio of overall link if $$\left(\frac{C}{I}\right) > \left(\frac{C}{N}\right)$$

satellite link is called **noise dominant**

if $$\left(\frac{C}{I}\right) < \left(\frac{C}{N}\right)$$

satellite link is called **interference dominant.**

4.8.2. FM Improvement Factor

Threshold effect in the receiver demodulator is the limiting factor for (C/N) in almost all satellite system. A linear relationship exist between the signal to noise ratio (S/N) of the demodulator output and the (C/N) at the demodulator input. For a typical FM demodulator the output S/N after the demodulator is given by

$$\left(\frac{S}{N}\right)_{\text{out}} = \left(\frac{C}{N}\right)_{\text{in}} + \text{FM improvement} \quad ...(4.49)$$

From the above equation it is clear that adequate value of (S/N) can be obtain with low (C/N) at the receiving system and this is because of the FM improvement term. Rather than dB in terms of power ratios Eq. (4.49) can be rewritten as

$$\frac{S}{N_{\text{out}}} = \frac{C * \text{FM Improvement}}{N_{\text{in}}} \quad ...(4.50)$$

4.9. INTERFERENCE EFFECT ON COMPLETE LINK DESIGN

There are several types of interference which effect the link design equation because they contribute noise power.

(1) One kind of interference arises from terrestrial microwave links because 6/4 GHz frequency band is allocated to satellite communication and also to the terrestrial microwave links. Thus the transmitting earth station working at 6 GHz may generate interference into terrestrial earth station working at 6 GHz. Similarly receiving earth station may cause interference. There may be mutual interference between a earth station and a terrestrial microwave system.

(2) Second type of interference in the satellite link is the cross polarization interference and depolarization caused of rain. Cross polarisation interference arises in the satellite communication system that use frequency reuse techniques. Coupling/conversion of the some of the energy in a transmitted signal may exist from one polarisation to other. This arises from the finite cross polarisation discrimination of the earth station. **Cross polarisation discrimination** is defined as the ratio of power received in the principal polarisation to the power received in the orthogonal polarization from the same incident signal.

(3) Third type of interference is the **adjacent channel interference**. This is arises mainly in band limited channel. Mainly it occurs where the high power amplifier (HPA) and or the satellite (TWT) operate near or at saturation point.

Solved Questions

Example 4.1. A geostationary satellite is located at a distance of 3000 km with an operating frequency 14.25 GHz. The gain of transmitting and receiving antennas are 15 and 20 simultaneously. If the transmitter power is 200 kW. Calculate the power received by the receiving antenna.

Solution. Given that

$$G_T = 15, G_R = 20$$
$$P_T = 20 \text{ kW}, f = 14.25 \text{ GHz}$$
$$R = 3000 \text{ km}.$$

Received power is given by

$$P_R = P_T\, G_T\, G_R \left(\frac{\lambda}{4\pi R}\right)^2 \quad ...(4.51)$$

where, $\lambda = \frac{C}{f} = \frac{3\times10^8}{14.25\times10^9} = 0.021 \text{ m}$

Substituting the values in equation we get,

$$P_R = 200\times10^3\times15\times20\times\frac{0.021}{(4\times3.14\times3\times10^6)^2}$$

$$= 6\times10^7\times10^{-12}\times\frac{0.021}{1419.7824} = 0.887\times10^{-8}$$

$$\boxed{P_R = 8.87\times10^{-10} \text{ W}}$$ **Ans.**

Example 4.2. A satellite earth station antenna having a gain (max.) of 80 dB at the operating frequency of 14 GHz, from a power amplifier and it generate 10 kW at the output. If the feeder loss is 4 dB, determine earth station effective isotropic radiated power (EIRP).

Solution. Given that,

Power at the output of amplifier = 10 kW = 40 dB W

Antenna gain = 80 dB W

Feeder loss = 4 dB W

So, EIRP = [Output power] + [Antenna gain] – [Feeder loss] ...(4.52)

= 80 + 40 – 4 = 116 dB W.

$$\boxed{\text{EIRP} = 116 \text{ dB W}}$$ **Ans.**

Example 4.3. In a satellite communication link, the uplink carrier to noise ratio $(C/N)_U$ is 20 dB where as the downlink carrier to noise ratio $(C/N)_D$ is 25 dB. Find the link carrier to noise ratio.

Solution. The link carrier to noise ratio (C/N) is given by

$$\left(\frac{C}{N}\right) = \left[\left(\frac{C}{N}\right)_u^{-1} + \left(\frac{C}{N}\right)_D^{-1}\right]^{-1} \quad ...(4.53)$$

Given that, $\left(\frac{C}{N}\right)_u = 20 \text{ dB} = 100$

$$\left(\frac{C}{N}\right)_D = 25 \text{ dB} = 316.22$$

Substituting these values in Eq. (4.53) we get,

$$\left(\frac{C}{N}\right) = \left[\frac{1}{100}+\frac{1}{316.22}\right]^{-1} = \left(\frac{416.22}{316.22}\right)^{-1}$$

$$\boxed{\frac{C}{N} = 76 = 18.8 \text{ dB.}} \quad \textbf{Ans.}$$

Example 4.4. In the cascaded arrangement of three blocks with their gain and noise figure as $G_1 = 100$, $F_1 = 2$, $G_2 = 10$, $F_2 = 10$, $G_3 = 10$, $F_3 = 15$. Determine the noise figure of the cascaded arrangement.

Solution. The overall noise figure of a cascaded arrangement of three stage is given by

$$F = F_1 + \frac{F_2 - 1}{G_1} + \frac{F_3 - 1}{G_1 G_2}$$

$$= 2 + \frac{10-1}{100} + \frac{15-1}{100 \times 10} = 2 + 0.9 + 0.014$$

$$\boxed{F = 2.914} \quad \textbf{Ans.}$$

Example 4.5. Determine the power received by a satellite located at 40000 km from the surface of earth. Satellite is operating at a frequency of 11 GHz and has EIRP of 21 dB W. The gain of a receiving antenna is 50.5 dB.

Solution. Power received is given by

$$P_R = EIRP + G_R - \text{Path loss}$$

Given that $EIRP = 21$ dB

$G_R = 50.5$ dB

$h = 40000$ km

Path loss is given by

$$\text{Path loss} = \left(\frac{4\pi h}{\lambda}\right)^2 = \frac{4 \times 3.14 \times 4 \times 10^7}{(2.727 \times 10^{-2})}$$

$$= 20 \log_{10}\left(\frac{4 \times 3.14 \times 4 \times 10^7}{2.727 \times 10^{-2}}\right) \text{dB}$$

$$= 205.3 \text{ dB}$$

Therefore,

Power received, $P_R = 21.0 + 50.5 - 205.3$

$$\boxed{P_R = -133.8 \text{ dB W}} \quad \textbf{Ans.}$$

Example 4.6. In a satellite communication, the output power of a satellite transponder is 20 W with output backoff is –1.0 dB. The gain of satellite antenna and earth station antennas are 30.0 dB and 59.2 dB respectively. Given that the free loss at 4 GHz is 195.0 dB, beam loss is 2.0 dB, clear air atmosphere loss is 0.2 dB and other losses are 0.4 dB. Calculate the power received at earth station.

Solution. The satellite transponder output power is,

$$P_T = 20 \text{ W} = 10 \log 20 \text{ dBW} = 13.0 \text{ dB}$$

The power received at the earth station will be,

$$P_R = P_T + G_T + G_R + BO_i - \text{All losses}$$

Given that,

$$G_T = 30.0 \text{ dB}$$

$$G_R = 59.2 \text{ dB}$$

$$Bo_i = -1.0 \text{ dB}$$

$$\text{All losses} = \text{Free loss} + \text{Beam loss} + \text{Clear air atmosphere loss} + \text{Other losses}$$

$$= 195.0 + 2.0 + 0.2 + 0.4 = 197.6 \text{ dB}$$

Substituting all these values, we get

$$P_R = 13.0 + 30.0 + 59.2 - 1 - 197.6$$

$$\boxed{P_R = -96.4 \text{ dB W}} \quad \textbf{Ans.}$$

Example 4.7. Determine the value of overall carrier to noise ratio at the earth station in a satellite link. A signal is transmitted to a satellite transponder with a carrier to noise ratio of 15 dB and transponder transmits it with a ratio of 20 dB.

Solution. The overall C/N ratio is given by

$$\left(\frac{C}{N}\right)_T = \frac{1}{\dfrac{1}{\left(\frac{C}{N}\right)_{up}} + \dfrac{1}{\left(\frac{C}{N}\right)_{down}}}$$

Given that

$$\left(\frac{C}{N}\right)_{up} = 15 \text{ dB}$$

$$\left(\frac{C}{N}\right)_{down} = 20 \text{ dB}$$

Substituting these values, we get

$$\left(\frac{C}{N}\right)_{up} = \frac{1}{\dfrac{1}{15} + \dfrac{1}{20}}$$

$$= \frac{1 \times 60}{7} = \frac{60}{7} = 8.57 \text{ dB}$$

$$\boxed{\left(\frac{C}{N}\right)_T = 8.57 \text{ dB}} \quad \textbf{Ans.}$$

Example 4.8. An antenna design is matched into a receiver. The noise temperature of the antenna and receiver are 40 K and 100 K respectively. If the receiver bandwidth is 36 MHz, determine its noise power. ($K = 1.38 \times 10^{-23}$ J/K).

Solution. Noise power is given by

$$P_N = KT_N B_N$$

Given that,

$$T_a = 40 \text{ K}$$

and

$$T_R = 100 \text{ K}$$

Therefore total noise temperature,

$$T_N = 40 + 100 = 140 \text{ K}$$

Substituting the values, we get

$$P_N = 1.38 \times 10^{-23} \times 140 \times 36 \times 10^6$$

$$\boxed{P_N = 0.0696 \text{ PW}} \quad \textbf{Ans.}$$

Example 4.9. If a satellite is operating at a distance of 20000 km from the surface of earth and radiates a power of 4 W. If the gain of transmitting antenna is 20 dB, determine the flux density at the receiving point.

Solution. Flux density at the receiving point is given by

$$\phi = \frac{P_t G_t}{4\pi R^2}$$

Given that,

$$P_t = 4 \text{ W}$$

$$G_t = 20 \text{ dB} = 100$$

$$R = 20000 \text{ km} = 2 \times 10^7 \text{ m}$$

Substituting all these values, we get

$$\phi = \frac{4 \times 100}{4 \times 3.14 \times (4 \times 10^7)^2}$$

$$= 7.95 \times 10^{-14} \text{ W/m}^2$$

$$\boxed{\phi = 7.95 \times 10^{-14} \text{ W/m}^2} \quad \textbf{Ans.}$$

Example 4.10. A satellite is rotating at a distance of 20,000 km from a point on the earth surface. If the operating frequency of satellite is 4 GHz and receiving antenna gain is 61.4 dB, determine the power received.

Given that EIRP = 26.02 dB.

Solution. Received power is given by

$$P_R = EIRP + G_R - \text{Path loss}$$

Given that,

$$EIRP = 26.02 \text{ dB}$$

$$G_R = 61.4 \text{ dB}$$

$$\lambda = \frac{C}{f} = \frac{3 \times 10^8}{4 \times 10^7} = 0.075 \text{ m}$$

$$\text{Path loss} = 20 \log_{10} \left(\frac{4\pi h}{\lambda} \right)$$

$$= 20 \log_{10} \left(\frac{4 \times 3.14 \times 20 \times 10^6}{0.075} \right) = 190.50$$

Therefore,

$$\text{Power received} = 26.02 + 61.4 - 190.50$$

$$\boxed{P_R = (-103.08 \text{ dB})} \quad \textbf{Ans.}$$

Example 4.11. A satellite at a distance of 40,000 km from a point on the earth's surface radiates a power of 200 W from an antenna with again of 17 dB in the direction of the observer. Find the flux density at the receiving point and the power received by an antenna with an effective area of 10 m^2.

Solution. Given that,

$$P_t = 200$$
$$G_t = 17 \text{ dB} = 50$$
$$R = 40 \times 10^6 \text{ m}$$

Flux density at the receiving point will be,

$$F = \frac{P_t G_t}{4\pi R^2}$$

$$F = \frac{2 \times 50}{4 \times 3.14 \times (40 \times 10^6)^2}$$

$$\boxed{F = 4.97 \times 10^{-15} \text{ W/m}^2.} \quad \textbf{Ans.}$$

Therefore, the power received with an effective collecting area of 10 m^2 is,

$$\boxed{P_r = 4.97 \times 10^{-14} \text{ W.}} \quad \textbf{Ans.}$$

Example 4.12. For a 4 GHz receiver. The following parameters of gains and noise temperature are given as

T_{in} = **50 K**	G_{RF} = **23 dB**
T_{RF} = **50 K**	G_m = **0 dB**
T_m = **500 K**	G_{IF} = **30 dB**
T_{IF} = **1000 K**	

Calculate the system noise temperature.

Solution. The system noise temperature is given by,

$$T_s = T_{RF} + T_{in} + \frac{T_m}{G_{RF}} + \frac{I_{IF}}{G_m G_{RF}}$$

Given that, $G_{RF} = 23 \text{ dB} = 200$

$$G_m = 0 \text{ dB} = 1$$

Substitute all these values, we get

$$T_s = 50 + 50 + \frac{500}{200} + \frac{1000}{200}$$
$$= 107.5 \text{ K}$$

$$\boxed{T_s = 107.5 \text{ K.}} \quad \textbf{Ans.}$$

Example 4.13. An earth station antenna has a diameter of 30 m, has an overall efficiency of 68% and is used to receive a signal at 4150 MHz. At this frequency, when the antenna points at the satellite at an elevation angle of 28° the system noise temperature is 79 K. Calculate the G/T for this earth station. If heavy rain causes the sky temperature to increase so that the system noise temperature rises to 88 K. What is the new G/T value.

Solution. For a circular aperture the antenna gain will be

$$G_l = \eta \frac{4\pi A}{\lambda^2} = \eta \frac{\pi^2 D^2}{\lambda^2}$$

where, $D = 30$ m

$$\lambda = \frac{C}{f} = \frac{3 \times 10^8}{4150 \times 10^6} = 0.0723 \text{ m}$$

$$n = 68\% = 0.68$$

Therefore, $$G_r = 0.68 \times \left(\frac{3.14 \times 30}{0.0723}\right)^2$$

$$G_r = 1.16 \times 106$$

$$G_r = 60.6 \text{ dB}$$

Now, $$\frac{G}{T} \text{ at } T = 79 \text{ K}$$

$$= 10 \log 10\,(79) = 19.0 \text{ dBK}$$

$$\frac{G}{T} = 60.6 - 19.0$$

$$\boxed{\frac{G}{T} = 41.6 \text{ dB/K}}$$ **Ans.**

In heavy rain, $T = 88$ K

$$= 10 \log_{10}(88) = 19.4 \text{ dBK}$$

Thus, new $\frac{G}{T}$ will be

$$\left(\frac{G}{T}\right)_{\text{new}} = 60.6 - 19.4$$

$$\boxed{\left(\frac{G}{T}\right)_{\text{new}} = 41.2 \text{ dB/K.}}$$ **Ans.**

Example 4.14. Calculate the gain of 3 m paraboloidal antenna operating at a frequency of 12 GHz. If the aperture efficiency of antenna is 0.55.

Solution. Given that,

$$D = 3 \text{ m}$$

$$f = 12 \text{ GHz}$$

$$\eta = 0.55$$

The gain of a paraboloidal antenna is given as

$$G = \eta(10.472\, fD)^2$$

$$= 0.55\,(10.472 \times 12 \times 3)^2$$

$$G \cong (78/68)$$

$$G = 10 \log_{10}(78/68)$$

$$\boxed{G = 48.9 \text{ dB.}}$$ **Ans.**

Example 4.15. The range between a ground station and a satellite is 42,000 km. Calculate the free space lost at a frequency of 6 GHz.

Solution. Given that,

$$R = 42{,}000 \text{ km}$$
$$f = 6 \text{ GHz}$$

The free space loss is given by

$$L_{FS} = 32.4 + 20 \log R + 20 \log f$$
$$= 32.4 + 20 \log (42 \times 10^3) + 20 \log (6 \times 10^3)$$

$$\boxed{L_{FS} = 200.4 \text{ dB.}} \quad \textbf{Ans.}$$

Example 4.16. An antenna has a noise temperature of 35 K and is matched into a receiver which has a noise temperature of 100 K. Calculate

(*a*) Noise power density

(*b*) Noise power for a bandwidth of 36 MHz.

Solution. Given that,

$$B_N = 36 \text{ MHz}$$
$$K = 1.38 \times 10^{-23}$$
$$T = (35 + 100) = 135 \text{ K}$$

(*a*) Noise power density is given by

$$N_0 = KT$$
$$= 1.38 \times 10^{-23} \times 135$$

$$\boxed{N_0 = 1.86 \times 10^{-21} \text{ J.}} \quad \textbf{Ans.}$$

(*b*) Noise power is given by

$$P_N = KT\,B_N$$
$$= 1.38 \times 10^{-23} \times 135 \times 36 \times 10^6$$

$$\boxed{P_N = 0.067 \times 10^{-12} \text{ W.}} \quad \textbf{Ans.}$$

Example 4.17. An LNA is connected to a receiver which has a noise figure of 12 dB. The gain of LNA is 40 dB and its noise temperature is 120 K. Calculate the overall noise temperature referred to the LNA input.

Solution. Given that,

$$T_{LNA} = 120 \text{ K}$$
$$F = 12 \text{ dB} = 15.85$$

Therefore, $T_1 = (15.85 - 1) \times 290 = 4306$ K

Gain of LNA, $40 \text{ dB} = 10^4$

Thus,

$$T_{in} = T_{LNA} + \frac{T_1}{G_{LNA}}$$
$$= 120 + \frac{4306}{10^4}$$

$$\boxed{T_{in} = 120.43 \text{ K.}} \quad \textbf{Ans.}$$

Example 4.18.

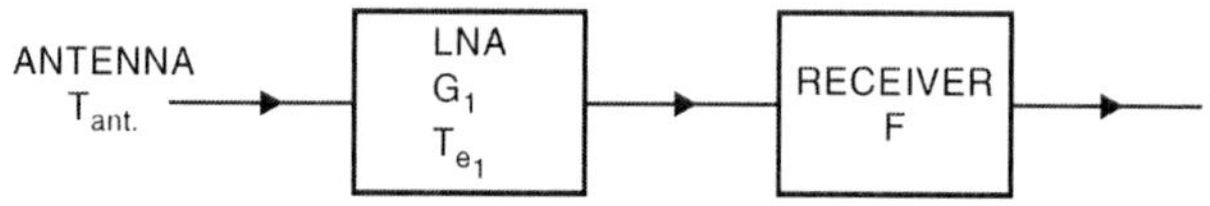

Fig. 4.10.

For the system shown in above Fig. 4.10, the receiver noise figure is 12 dB, the cable loss is 5 dB, the LNA gain is 50 dB and its noise temperature 150 K. Calculate the noise temperature referred to input if the antenna noise temperature is 35 K.

Solution. Given that

Gain of LNA, $G = 50 \text{ dB} = 10^5$

For the cable, $L = 5 \text{ dB} = 10^{0.5}$

$L = 3.16$

For the receiver, $F = 12 \text{ dB}$

$= 10^{1.2} = 15.85$

Noise temperature referred to the input will be

$$T_s = 35 + 150 + \frac{(3.16 - 1) \times 290}{10^5} + \frac{3.16 \times (15.85 - 1) \times 290}{10^5}$$

$$\boxed{T_s \cong 185 \text{ K.}} \text{ Ans.}$$

Example 4.19. For a satellite circuit the individual link carrier to noise spectral density ratios are : Uplink 100 dB Hz, downlink 87 dB Hz. Calculate the combined C/N.

Solution. Given that,

$$\left(\frac{C}{N}\right)_u = 100 \text{ dB} = 10^{10}$$

$$\left(\frac{C}{N}\right)_D = 87 \text{ dB} = 10^{8.7}$$

The combined $\left(\frac{C}{N}\right)$ ratio is given as

$$\left(\frac{C}{N}\right)^{-1} = \left(\frac{C}{N}\right)_u^{-1} + \left(\frac{C}{N}\right)_D^{-1}$$

$$= 10^{-10} + 10^{-8.7}$$

$$= 2.095 \times 10^{-9}$$

$$\frac{C}{N} = -10 \log_{10} (2.095 \times 10^{-9})$$

$$\boxed{\left(\frac{C}{N}\right) = 86.79 \text{ dB Hz.}} \text{ Ans.}$$

Example 4.20. For a satellite circuit the individual link carrier to noise spectral density ratios are : Uplink 23 dB Hz, downlink 20 dB Hz, intermodulation 24 dB Hz. Calculate the overall carrier to noise ratios in decibels.

Solution. Given that,

$$\left(\frac{C}{N}\right)_u = 23 \text{ dB} = 102.3$$

$$\left(\frac{C}{N}\right)_D = 20 \text{ dB} = 102$$

$$\left(\frac{C}{N}\right)_I = 24 \text{ dB} = 102.4$$

The overall carrier to noise ratio is given by

$$\left(\frac{C}{N}\right)^{-1} = \frac{1}{\left(\frac{C}{N}\right)_u + \left(\frac{C}{N}\right)_D + \left(\frac{C}{N}\right)_I}$$

$$= \frac{1}{10^{2.3} + 10^{2} + 10^{2.4}}$$

$$= 0.0019$$

$$\left(\frac{C}{N}\right) = -10 \log_{10} (0.0019)$$

$$\boxed{\left(\frac{C}{N}\right) = 17.2 \text{ dB.}} \quad \textbf{Ans.}$$

Example 4.21. Determine the rain absorption temperature if the rain absorption is 0.5 dB over 0.1% of the time over a year.

Solution. Given that,

$$\text{Rain attenuation ratio} = 10^{0.05}$$

$$= 1.122$$

Rain absorption temperature is given by

$$T_{\text{rain}} = T_0 \left[1 - \frac{1}{\alpha_r}\right]$$

$$= 290\left[1 - \frac{1}{1.122}\right]$$

$$= 290 \times 0.108$$

$$\boxed{T_{\text{rain}} = 31.53 \text{ K}} \quad \textbf{Ans.}$$

Example 4.22. A receiving system has antenna noise temperature 60 K and receiver noise figure 9 dB. Calculate the system noise temperature.

Solution. Given that

$$F = 9 \text{ dB} = 7.94$$

$$T_0 = 290$$

$$T_{\text{ant.}} = 60 \text{ K}$$

The system noise temperature is given by

$$T_S = T_{ant} + T_0(F - 1)$$
$$= 60 + 290\ (7.94 - 1)$$
$$= 60 + 290 \times 6.94$$
$$T_S = 2072.6\ °K$$
$$\boxed{T_S = 33.165 \text{ dB K}} \quad \textbf{Ans.}$$

Example 4.23. A satellite down link at 12 GHz operates with a transmit power of 6 W and an antenna gain of 48.2 dB. Calculate the EIRP in dBW.

Solution. Given that

$$P_T = 6 \text{ W}$$
$$G_T = 48.2 \text{ dB}$$
$$f = 12 \text{ GHz}$$

The EIRP is given by

$$\text{EIRP} = P_T G_T$$
$$\text{EIRP (in dB)} = (P_T) + (G_T) \text{ dBW}$$
$$= 10 \log (6) + 48.2$$
$$= 7.8 + 48.2$$
$$\boxed{\text{EIRP} = 56 \text{ dBW.}} \quad \textbf{Ans.}$$

Example 4.24. A satellite is positioned at a distance of 38000 km from the surface of earth and radiates a power of 3 W in the direction of earth station. If the satellite antenna gain is 20 dB. Determine :

(*a*) Flux density at earth station.

(*b*) Power received by antenna of effective area 8 m².

Solution. Given that,

$$R = 38000 \text{ km}$$
$$P_t = 3 \text{ W}$$
$$G_t = 20 \text{ dB}$$
$$A_e = 8 \text{ m}^2$$

(*a*) Flux density at the earth station is given by :

$$F = \frac{P_t G_t}{4\pi R^2}$$
$$= \frac{3 \times 20}{4 \times 3.14 \times (38 \times 10^6)^2}$$
$$= \frac{15}{3.14 \times (38 \times 10^6)^2}$$
$$\boxed{F = 1.65 \times 10^{-14} \text{ W/m}^2} \quad \textbf{Ans.}$$

(*b*) Power received by the antenna,

$$P_r = F \times A$$
$$= 1.65 \times 10^{-14} \times 8$$
$$= 1.32 \times 10^{-13} \text{ W}$$
$$\boxed{P_r = 1.32 \times 10^{-13} \text{ W}} \quad \textbf{Ans.}$$

Summary

1. Link budget calculation basically relate two quantities the transmit power and the receiver power.
2. The cost of launching a satellite is around $100,000 per kg per launch with final positioning and it limits the size and weight of satellite.
3. The quality of an earth station is specify by G_R/T_S ratio and it is known as figure of merit.
4. The upper layer of the atmosphere causes refraction and depolarization while lower layer causes absorption and diffusion.
5. Absorptive network is one which contains resistive elements. These introduces losses by absorbing energy from the signal and converting it to heat. Examples of absorptive networks are transmission line, wave guides and resistive attenuators etc.
6. The product of transmitted power with the gain of transmitting antenna is known as effective isotropic radiated power (EIRP).
7. Cross polarization, depolarization and adjacent channel interference effect the complete link.
8. Receiver noise includes contributions from thermal noise, shot noise and Flicker noise.
9. The 6/4 GHz bands have been the most popular because they offer the fewest propagation problems.
10. Gain of the receiver and FM improvement factor effect the signal to noise ratio at the output of the receiver's detector.
11. Design of downlink requires high *G/T* at the input of the receiver and use of FM to get improvement in signal to noise ratio at the output.
12. Design of uplink is not a problem as it is possible to use high power transmitters and big size antenna at the ground.
13. The carrier to noise ratio is related to the quantity of information that can be transmitted in unit bandwidth as shown by Shannon's law.
14. Carrier to noise ratio over the satellite link is of fundamental importance. This is a major determinant of the theoretical information carrying capacity of the link and quality of signal received at earth station.
15. The various gains and losses of a proposed system are calculated and added together the result commonly known as the link budget. This gives the information about the received signal strength under certain conditions and assumptions.
16. Noise density is defined as the power present in a 1 kHz bandwidth.
17. The most important factor in information channel is signal to noise ratio (S/N) and it depends on carrier to noise (C/N) ratio of the RF or IF signal in the receiver, type of modulation and the baseband channel bandwidths in the receiver.

18. The product term PTGT is known as effective isotropic radiated power and it gives a measure of power flux.

19. In decibels the received power is given by

$$P_R = (EIRP + G_R - LFS) \text{ dB}$$

20. The noise power in the receiver is given by

$$P_N = KT_n B$$

21. The First transmission equation is

$$P_R = P_T G_T G_R \left(\frac{\lambda}{4\pi R}\right)^2$$

22. The receiver system noise temperature is given by

$$T_s = T_{RF} + T_{in} + \frac{T_m}{G_{RF}} + \frac{T_{IF}}{G_m G_{RF}}$$

23. The relationship between a noise temperature and noise figure is given by

$$T_d = T_0 (NF - 1)$$

24. The satellite transponder is a quasilinear amplifier and the received carrier level determine the output level.

25. Rainfall rsults in attenuation of radio waves by scattering and by absorption of energy from the wave and it increases with increase in frequency.

26. In the C band, more especially the Ku band—rainfall is the most significant cause of signal fading.

27. For complete link design, the net carrier to noise ratio is given by

$$\left(\frac{C}{N}\right)_{Net} = \left[\left(\frac{C}{N}\right)^{-1} + \left(\frac{C}{I}\right)^{-1}\right]^{-1}$$

28. In the satellite communication system that use frequency reuse techniques, cross polarization interference arises.

29. Cross polarization descrimination is defined as the ratio of power received in the principal polarization to the power received in the orthogonal polarization from the same incident signal.

30. Adjacent channel interference arises mainly in band limited channel and occurs because of the high power amplifier used.

Review Questions

1. Derive general link equation. Find out the expressions for *C/N* and *G/T* ratio.
2. What is the system noise temperature ? Derive the expression for equivalent noise temperature.
3. Explain the various factors which affects the design of uplink and downlink.
4. What are the various interferences that may affect the satellite link performance ?
5. A receiving system consists of an antenna having a noise temperature of 60 K, feeding directly into a low noise amplifier (LNA). The amplifier has a noise temperature of 120 K and a gain of 45 dB. The coaxial feeder between the LNA and

the main receiver has a loss of 2 dB, and a noise figure of 9 dB. Calculate the system noise temperature referred to input.

6. Two amplifiers are connected in cascade. Each having a gain of 10 dB and a noise temperature of 200 K. Calculate

(*a*) the overall gain

(*b*) the effective noise temperature referred to input.

7. For an earth station having an antenna of diameter 30 m and overall efficiency of 68%, the working signal frequency is 4150 MHz. At this frequency the system noise temperature is 79 K when the antenna points at the satellite at an elevation angle of 28 K. Calculate G/T ratio of the earth station. In case the sky noise temperature rises to 88 K, what would be the new *G*/*T* ratio ?

8. Explain how intermodulation noise originates in a satellite link ? and describe how it may be reduced.

9. What is meant by *C*/*N* ratio ? The (*C*/*N*) values for a satellite circuit are uplink 25 dB, downlink 15 dB. Calculate the overall (*C*/*N*) value.

10. Derive the complete link design equation. Explain the FM improvement factor.

11. Explain what is meant by the *G*/*T* ratio of a satellite receiving system. *A* satellite receiving system employs a 5 m parabolic antenna operating at 12 GHz. The antenna noise temperature is 100 K, and the receiver front-end noise temperature is 120 K. Calculate *G*/*T* ratio.

12. Define the following terms :

(*a*) EIRP | (*b*) Saturation flux density

(*c*) Figure of meric | (*d*) FM improvement factor

(*e*) Antenna noise temperature.

13. Derive a relation for received power, P_r and determine path loss in dB in terms of frequency and other parameters.

14. What do you understand by *G*/*T* ratio ? Prove that *C*/*N* at the input of a detector in the receiver is proportional to *G*/*T*.

15. What is Friis transmission equation ?

16. How is uplink design different from the downlink design ?

17. Explain the importance of gain to equivalent noise temperature for an earth station.

18. Explain the parameters that control the design of each station. What is the optimum *G*/*T* ratio for a standard of earth station.

19. Write short notes on the following :

(*a*) Link performance

(*b*) Link budget.

20. Calculate the free space loss as a power ratio and in decibels for transmission at frequencies of

(*a*) 4 GHz | (*b*) 6 GHz

(*c*) 12 GHz | (*d*) 14 GHz

The range being 42,000 km.

21. The noise factor of an amplifier is 7 : 1. Calculate

(*a*) Noise figure

(*b*) Equivalent noise temperature

22. A receiver system consist of an antenna having a noise temperature of 60 K, feeding directly into a low noise amplifier (LNA). The amplifier has a noise temperature of 120 K and a gain of 45 dB. The coaxial feeder between the LNA and the main receiver has a loss of 2 dB and the main receiver has a noise figure of 9 dB. Calculate the system noise temperature referred to input.

23. In a satellite link the propagation loss is 200 dB. Margins and other losses account for another 3 dB. The receiver (G/T) is 11 dB and the effective isotropic radiated power is 45 dB. Calculate the received (C/N) for a system bandwidth of 36 MHz.

24. The following parameters apply to a satellite downlink :

Saturation EIRP	= 22.5 dBW
Free space Loss	= 195 dB
Other losses and margins=	1.5 dB
Earth station (G/T)	= 37.5 dB/K

Calculate the (C/N) at earth station. Assuming an output backoff of 6 dB is applied, what is the new value of (C/N).

25. An antenna having a noise temperature of 35 K is connected through a feeder having 0.5 dB loss to an LNA> The LNA has a noise temperature of 90 K. Calculate the system noise temperature referred to

(*a*) The feeder input

(*b*) The LNA input

26. A carrier to noise density ratio of 90 dB Hz is required at a receiver having a (G/T) ratio of 12 dB. Given that the total loss in the link amount to 196 dB. Calculate the required EIRP.

27. Compare the increase in interference levels expected. When satellite orbital spacing is reduced from 4° to 2° for earth station antenna sidelobe patterns by

(*a*) 32–25 log θ dB

(*b*) 29–25 log θ dB

28. An antenna has a gain of 46 dB at 12 GHz. Calculate its effective area.

Objective Type Questions

1. Earth station figure of merit is defined as

(*a*) $10 \log\left(\frac{G}{T}\right)$ (*b*) $10 \log_e\left(\frac{G}{T}\right)$

(*c*) $20 \log_e\left(\frac{G}{T}\right)$ (*d*) $20 \log\left(\frac{G}{T}\right)$

2. A satellite earth station antenna has a gain of 10^6 and a noise temperature of 100 K. The earth station (*G*/*T*) in dB/K is

(*a*) 40 dB/K (*b*) 80 dB/K

(*c*) Indeterminate (*d*) Zero

3. A 10 GHz signal was beamed up at a satellite at a distance of about 40,000 km from earth's surface. The signal undergoes a free space loss of

(*a*) 10 dB (*b*) 50 dB

(*c*) 200 dB (*d*) 385 dB

4. The percent satellite system for INTELSAT satellite requires a figure of merit of

(*a*) 40.7 (*b*) 38.5

(*c*) 60 (*d*) 0.1

5. In a satellite system, the upward frequency is greater than that of downward link. Is it true

(*a*) Yes (*b*) Not always

(*c*) No (*d*) None of these

6. Building of a geostationary satellite like INTELSAT-IV and its launching may involve costs of

(*a*) About 200 crores (*b*) About 50 crores

(*c*) About 100 crores (*d*) None of these

7. The overall noise temperature T_e, absolute temperature T and noise figure F are related as

(*a*) $T_e = T(F - 1)$ (*b*) $T_e = F(T - 1)$

(*c*) $T = T_e(F - 1)$ (*d*) $T_e = (T)(F)$

8. For a three stage cascaded amplifier, having gains of G_1, G_2, G_3 and noise figure of F_1, F_2, F_3 the overall noise figure F =

(*a*) $F = F_1 + \frac{F_2 - 1}{G_1} + \frac{F_3 - 1}{G_1 G_2}$ (*b*) $F = F_1 + \frac{F_1 - 1}{G_1} + \frac{F_2 - 1}{G_1 G_2}$

(*c*) $F = F_3 + \frac{F_1 - 1}{G_1} + \frac{F_2 - 1}{G_1 G_2}$

(*d*) None of the above

9. In a two stage cascaded amplifier the equivalent noise temperature are Te_1 and Te_2. If the gains of two stages are G_1 and G_2, the overall noise temperature T_e is

(*a*) $T_e = T_{e2} + \frac{T_{e1}}{G_1}$ (*b*) $T_e = \frac{T_{e1}}{G_1} + \frac{T_{e2}}{G_2}$

(*c*) $T_e = T_{e1} + \frac{T_{e2}}{G_2}$ (*d*) $T_e = T_{e1} + \frac{T_{e1}}{G_1}$

10. The noise figure of individual stages of a two stage amplifier is 2.03 and 1.54 respectively. If gain of first stage is 62, the overall noise figure is

(*a*) 2.03 (*b*) 2.055

(*c*) 3.05 (*d*) 6.05

11. A 10 GHz signal was beamed, up at a satellite, at a distance of about 40,000 km from the earth's surface. The signal undergoes a free space loss of about

(*a*) 10 dB (*b*) 50 dB

(*c*) 200 dB (*d*) 385 dB

ANSWERS

1. (*d*)	**2.** (*a*)	**3.** (*b*)	**4.** (*a*)	**5.** (*a*)
6. (*c*)	**7.** (*a*)	**8.** (*a*)	**9.** (*c*)	**10.** (*c*)
11. (*d*)				

❑❑❑

5

CHAPTER

Propagation Effects and Their Impact on Satellite Earth Links

INSIDE THIS CHAPTER

5.1. INTRODUCTION

By introducing the concept of link power budget, the power received is calculated by

$$P_r = EIRP + G_2 - L_P - L_a - L_{ra} \text{ dB}W \qquad ...(5.1)$$

This equation indicates that the received power in decible depends on the transmitter EIRP, the receiving antenna gain and the various losser in the system. All these factors are indpendent of time, except the atmospheric loss, L_a.

Generally L_a is written as the sum of **atmospheric absorption** (a constant term) and the **attenuation** (a variable term) A. At commercial interest of frequencies, the atmospheric absorption is relatively unimportant and is disappears into the general uncertainity. In clear wheather the attentuation is zero but it can increase to large values during unfavourable propagation condition. Rapid fluctuations in attenuation is known as **scintillations**, while longer term increases are known as **fades**. Above 10 GHz, rain fade is a seviour problem.

All types of satellite links can be affected by attenuation. Mainly links that employ orthogonal polarization to transmits two different channels on a common or overlaping frequency band are also degraded by **depolarization**. There is a conversion of some of the transmitted signal energy from one polarization to another. If the propagation conditions are not ideal it can cause **co-channel interference** and **crosstalk** in dual-polarized satellite links. Rain is a primary cause of depolarization.

Both attentuation and depolarization originate from interactions between the propagating electromagnetic wave and on a satellite to link and the atmospheric conditions at that time. Figure 5.1 shows the layers in the earth atmosphere.

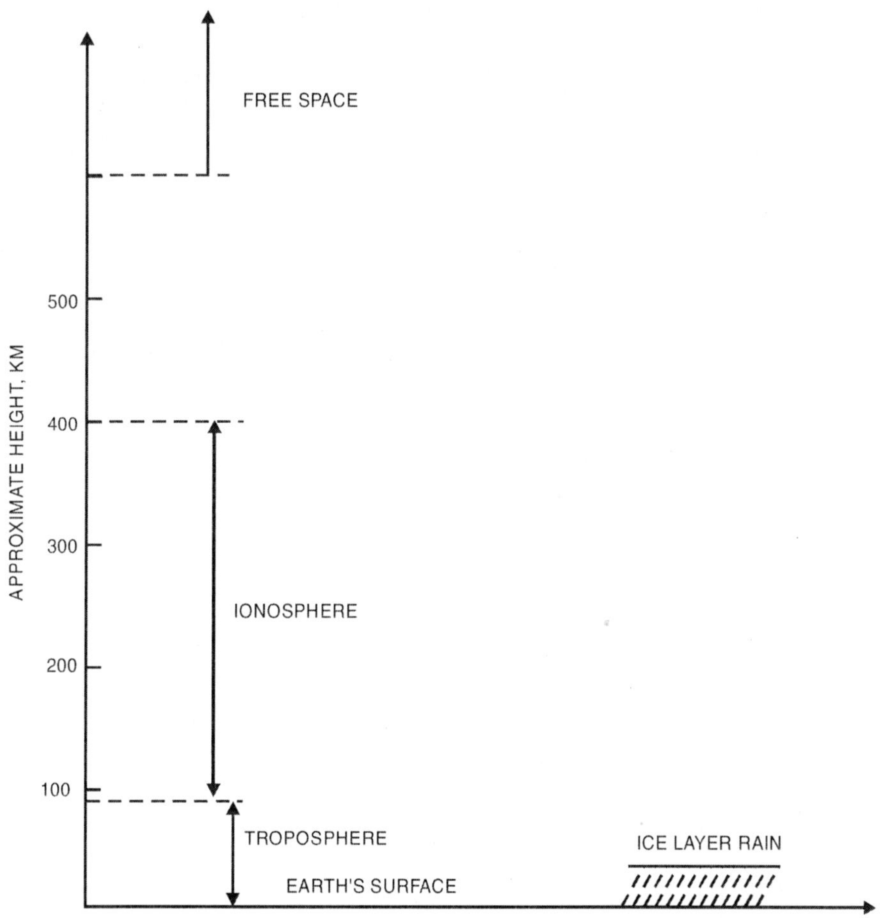

Fig. 5.1. Layers in the earth's atmosphere.

The atmospheric constituents may include free electrons, ions, neutral atoms, molecules and hydrometers. **Hydrometer** is a term that seems pedantic but that conveniently describes raindrops, snowflakes, sleet, hail, free-floating ice crystals. Their interactions with radio waves depend strongly on frequency and effect that dominates 30 GHz propagation.

As well as frequency increases, all propagation effects become more severe.

Table 5.1 summarized the certain propagation impairments and their causes.

Table 5.1. Propagation Concerns for Satellite Communications Systems.

S. No.	Propagation Impairment	Physical Cause	Prime Importance
1.	Attenuation and sky noise increases	Atmospheric gases, cloud, rain	Frequencies above about 10 GHz
2.	Signal depolarization	Rain, ice crystals	Dual-polarization systems at C and Ku bands (depends on system configuration)
3.	Refraction, atmospheric multipath	Atmospheric gases	Communication and tracking at low elevation angles
4.	Signal scintillations	Tropospheric and ionospheric refractivity fluctuations	Tropospheric at frequencies above 10 GHz and low elevation angles; iono-spheric at frequencies below 10 GHz
5.	Reflection multipath, blockage	Earth's surface, objects on surface	Mobile satellite services
6.	Propagation delays, variations	Troposphere, ionosphere	Precise timing and location systems; time division multiple access (TDMA) systems
7.	Intersystem interference	Ducting, scatter, diffraction	Mainly C band at present; rain scatter may be significant at higher frequencies

5.2. QUANTIFYING ATTENUATION AND DEPOLARIZATION

5.2.1. Attenuation

Attenuation is defined as the decible difference between the power received at a given time '*t*' and the power received under ideal propogation *i.e.* clear wheather condition.

$$A(t) = P_r(\text{ideal propogation})\ \text{dBW} - P_r(t)\ \text{dBW} \qquad ...(5.2)$$

To quantify depolarization is quite difficult and some background information are required for its description.

Orthogonal polarization works on **frequency sharing** because any transmitting antenna radiates waves with a polarization that is the characteristics of that antenna and when receiving the same antenna is blind waves whose polarization is orthogonal to the characteristic polarization. For example consider the pair of horn antennas as shown in Fig. 5.2.

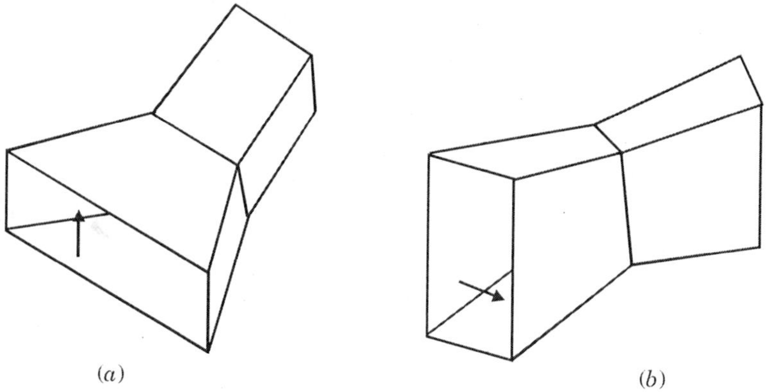

Fig. 5.2. Pair of horn antenna. (*a*) It transmits and receives vertically polarized waves. (*b*) It transmit and recives horizontally polarized waves.

The top horn transmits vertical polarization and is blind for horizontal and, the bottom horn transmits horizontal polarization and is blind for vertical. To illuminate a single reflector if the two horns were used, the result will be dual polarized antenna and is shown in Fig. 5.3.

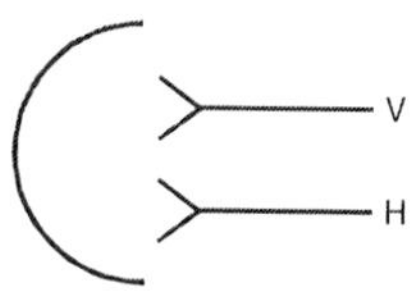

Fig. 5.3. Schematic representation of dual polarized antenna.

Two such antennas would allow two channels *i.e.* one for transmitting with vertical polarization and one for receiving with horizontal polarization, to share the same frequency. There would be no interference between the channels, if the antennas were ideal.

5.2.2. Depolarization

In this process **to calculate depolarization** let us consider a dual polarized antenna transmitting orthogonally polarized signals. The two polarizations *V* and *H* are for vertical and horizontal and these could be any orthogonal pair. Let '*a*' and '*b*' are the complex phasor amplitudes of transmitted electric field vectors with polarization V and H. It is shown in Fig. 5.4.

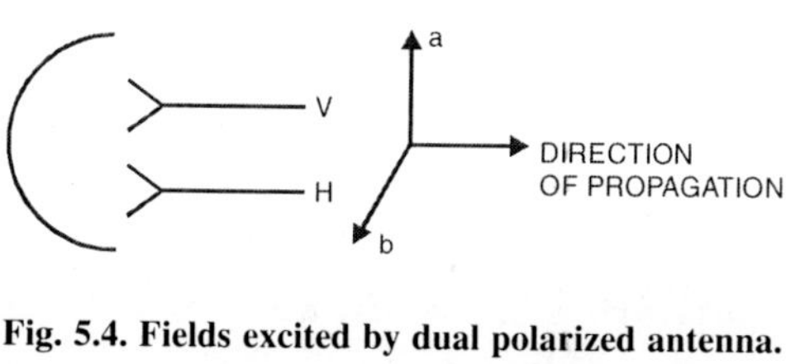

Fig. 5.4. Fields excited by dual polarized antenna.

The antenna is excited so that '*a*' and '*b*' are equal. If clear air is the transmission medium between antennas, the receiving antenna '*a*' would give rise to a V polarization wave of amplitude '*ac*' and '*b*' would be '*bc*' because of H-polarization. The subscript '*c*' denotes **copolarization.** These fields have the same polarization as in case of transmission and is shown in Fig. 5.5.

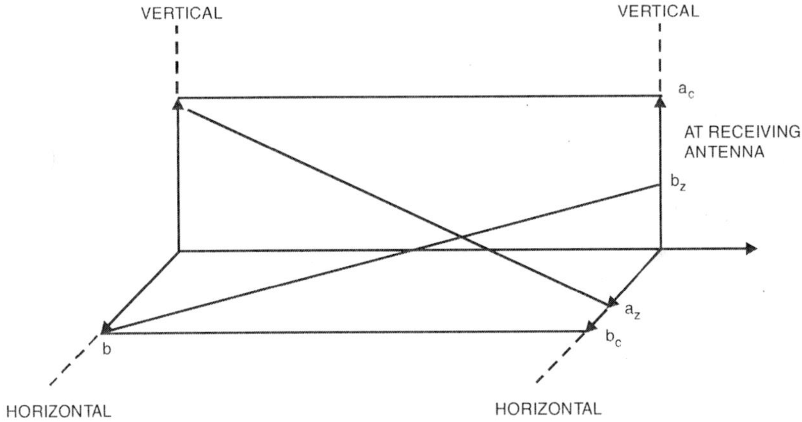

Fig. 5.5. The transmitted fields *a* and *b* produce co-polarized components *ac* and *bc* and cross-polarized components ax and bx at the receiving antenna. With perfect antennas and in the absence of depolarization ax and bx would be zero.

If there is some depolarizing medium example : rain or ice is the propagation medium, some of the energy will couple into a H-polarized field component and its amplitude at the receiving antenna is a_x. Similarly b write give rise to a small *V* polarized component b_x.

Thus the unwanted b_x will interference with wanted ac and unwanted a_x will interfere with wanted *bc*. This inteference will produce **crosstalk** on analog link and raise the BER on the digital link. This generation of unwanted cross polarized component is known as **depolarization.**

The **measurement of depolarization** is quite useful for analyzing communication system is the cross polarization isolation, *xpi*. In terms of complex phasor field amplitudes, for V and H polarized channel, it is given as :

$$XPI_V = \frac{a_c}{b_x} \quad \text{...(5.3)}$$

$$XPI_H = \frac{b_c}{a_x} \quad \text{...(5.4)}$$

These values are commonly expressed as in decibles. For example :

$$XPI_V = 20 \log_{10} \left| \frac{a_c}{b_x} \right| \text{dB} \quad \text{...(5.5)}$$

Physically **XPI** is the ratio of wanted power to unwanted power in the same channel. The larger its value, less interference and better a communication system will perform.

5.2.3. Crosspolarization Isolation and Descrimination

It is difficult to measure XPI, while alteast two groups can measure it with the COMSTAR family of satellites. Most satellite for propagation experimants transmit a single polarization (say V) and measure the copolarized and crosspolarized components of the of the received signal (say ac and ax). This gives to the **cross polarization**

discrimination, XPD. In case, when only V polarization is transmitted alone, as shown in Fig. 5.6.

$$XPD_V = \frac{a_c}{a_x} \text{ and } XPD_H = \frac{b_c}{b_x} \quad \text{...(5.6)}$$

or in decibles,

$$\boxed{XPD_V = 20 \log_{10} \left| \frac{a_c}{a_x} \right| \text{dB}} \quad \text{...(5.7)}$$

and

$$\boxed{XPD_H = 20 \log_{10} \left| \frac{b_c}{b_x} \right| \text{dB}}$$

Practically, *XPI* and *XPD* are equal and usually they are known as isolation.

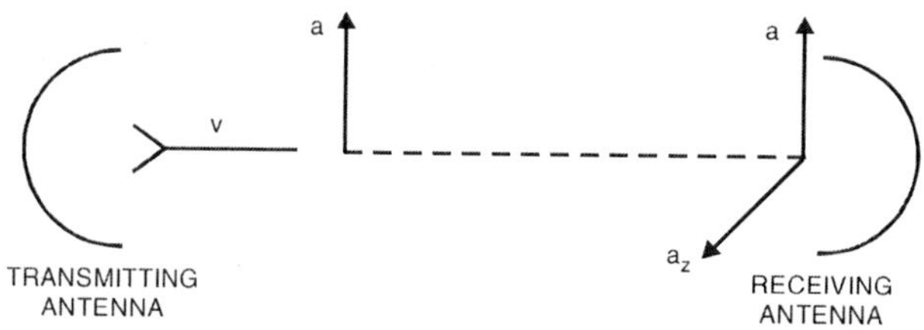

Fig. 5.6. With a vertically polarized transmitting antenna, the ratio of the wanted received field component *ac* to the unwanted component *ax* is called the cross-polarization discrimination, *XPD*.

Practically real antennas do not transmit polarization pair that are exactly orthogonal and isolation under the clear wheather condition is not infinite. Under operational conditions with precision antennas, clear wheather isolation of 30 to 35 dB may be realized.

5.3. PROPAGATION EFFECTS THAT ARE NOT ASSOCIATED WITH HYDROMETERS

Here, the propogation effects, that are not caused by raindrops or ice crystals will be discussed.

5.3.1. Atmospheric Absorption

Losses occur in the earth's atmosphere because of the energy absorption by the atmospheric gases. These losses are quite different from those which results from adverse wheather condition and are known as **atmospheric losses.** To distinguish between these, the wheather related losser are known as **atmospheric attenuation** and the absorption losses are known as **atmospheric absorption**.

Attenuation occur at sufficient high frequencies and the interaction of electromagnetic waves with molecules of atmospheric gases, occur at resonant frequencies, and are apparent in plots of zenith (90° elevation angle) attenuation versus frequency, as shown in Fig. 5.7.

Two **absorption peaks will be oversed :**

- the first one at a frequency of 22.3 GHz, resulting from resonance absorption in water vapour (H_2O)
- the second one at 60 GHz, resulting from resonance absorption in oxygen (O_2).

The frequency band allocation for satellite communication was carefully selected to avoid these frequencies and thus gaseous absorption is quite important in almost all commercial links. The resulting attenuation is usually less than 1 dB and is within the overall uncertainty in a power budget.

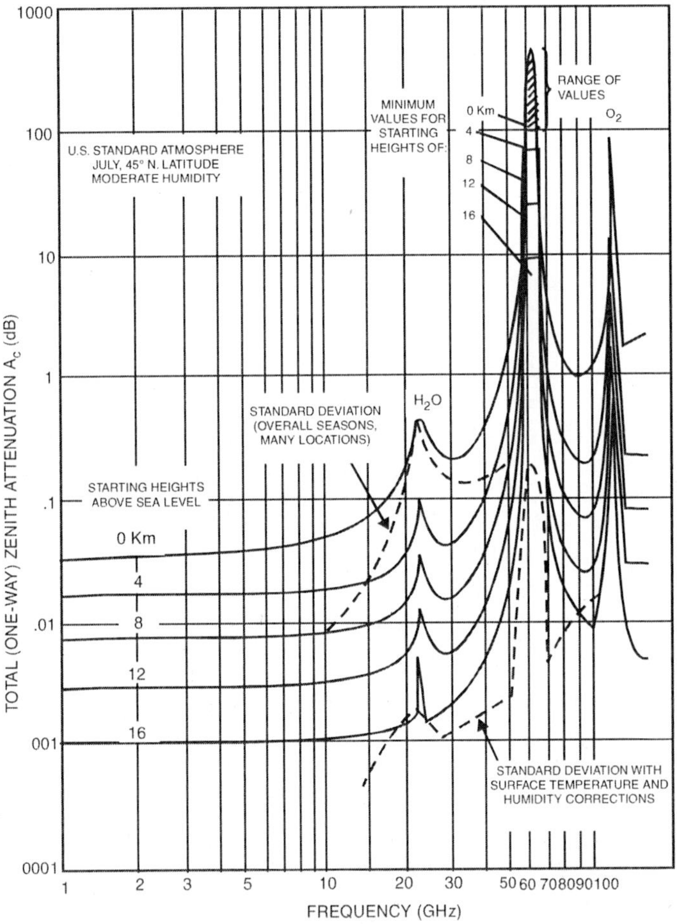

Fig. 5.7. Total zenith attenuation on a satellite path due to atmospheric absorption.

The graph shown is Fig. 5.7 is for an elevation angle of 90° at earth station antenna. Denoting this value of absorption loss at $[AA]_{90}$ dB, than for elevation angle down to 10°. The formula for the absorption loss in dB is given as

$$[AA] = [AA]_{90} \operatorname{cosec} \theta \qquad ...(5.8)$$

where, θ = Angle of elevation.

5.3.2. Tropospheric Multipath and Scintillation Effects

At elevation angle below than 5°, the tropospheric portion of the propogation path is so long that turbulence may cause rapid variations of amplitude and phase, and is known as **atmospheric scintillation.** This is a fading phenomenon and fading period is several tens as seconds. It is caused by the differences in the atmospheric refractive index, which in turn results in focusing and defocusing of the radio waves and follows different ray paths through the atmosphere Fig. 5.8 shows an example.

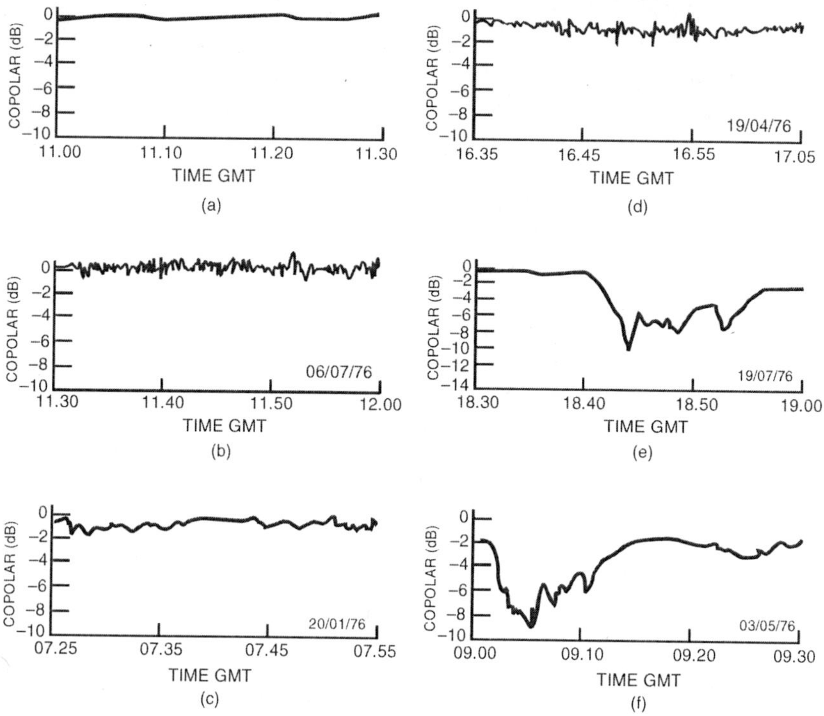

Fig. 5.8. Scintillations observed under a variety of weather conditions on a 30-GHz downlink from ATS-6.

Simultaneously waves may arrive at the receiving antenna via several propogation paths and by integering each other. It give rise to multipath fading and is known as **frequency selective fading.** For these reasons, commercial links try to avoid low elevation angles.

5.3.3. Land and Sea Multipath

At higher elevation angles, multipath propogation could result through reflection from surrounding terrain. These can be eliminated by using highly directional antennas at commercial earth stations. In addition multipath could become a major problem in land mobile satellite systems that will operate below 4 GHz frequency with low gain antennas and in low cost maritime systems that will also use small antennas.

5.3.4. Multipath Effects in System Design

For analog FM systems, **multipath fading** is similar to **rain fading.** Its effect can be eliminating by providing a clear wheather margin sufficient to overcome the fades that are anticipated for some percentage of time while the statistics are not well established for satellite link multipath fading. This procedure is routinely followed for terrestrial radio links and it should be possible to adapt it to satellite paths without difficulty.

The digital system response to multipath fading is different from analog and the terrestrial links designed according to the established procedures for analog systems have failed to provide the desired realibility.

5.3.5. Faraday Rotation in the Ionosphere

The ionosphere is the upper region of earth's atmosphere that contains large number of electrons and ions. It extends from about 100 km to over 400 km above the earth. The ionsphere has been ionized mainly by solar radiation. It completely dominates the radio propogation below about 40 MHz but effects on the frequencies used by most satellite systems.

Electrically the ionsphere is an inhomogeneous and unisotropic plasma, and it is quite difficult to analyse wave propogation through it. There exist two characteristic polarizations for a given frequency and direction of propogation with respect to the earth's magnetic field. With these polarization, waves are known as **characteristic waves** and it propogate with their unchanged polarization.

With the characteristic polarization, any wave entering the **ionosphere polarization can be resolved into two components :**

(1) Phase shift

(2) Attenuation.

- These two components can be calculated at any point in the propogation path. The total field can be calculated as the vector sum of the fields of the characteristic waves.
- When a linearily polarized (LP) satellite path signal reaches the ionosphere, it exites the waves either the two characteristic polarization. They travel at different velocities and they leave the ionosphere with a relative phase that is different from when they entered. Thus the wave that leaves ionosphere has a different polarization from the linearly polarized wave that was transmitted. This is known a **Faraday Rotation**.
- This effect is same as if the electric field vector of transmitted LP wave had been rotated by an angle ϕ. **Faraday rotation is inversely proportional to frequency squared** and is not considered to be a serious problem for frequnecies above about 10 GHz.
- For a path length *Zm* in the ionosphere, the **rotation angle ϕ** is given by

$$\phi = \int \left(\frac{23.6 \times 10^4}{f^2} \right) ZNB_0 \cos\theta . dz \text{ rad} \qquad ...(5.9)$$

where, θ = Angle between the geometric field and the direction of pogation.

N = Electron density in electrons/cubic meter.

B_0 = Geomagnetic flux density in teslas.

f = Operating frequency in Hz.

Suppose a linearly polarized wave produces an electric field E at the receiver antenna when no faraday rotation is present. The received power is proportional to E_2. A Faraday rotation of $\Delta\phi$ will result in the copolarized component of the received signal being reduced to $EC_0 = E \cos \Delta\phi$. In this case the received power is proportional to E_{CO}^2.

The **polarization loss** in decibles is

$$P_L = 20 \log_{10} \frac{E_{CO}}{E}$$

$$P_L = 20 \log_{10} (\cos \Delta\phi) \qquad ...(5.10)$$

To compensate the Faraday Rotation, the polarization of an earth station antenna can be adjusted. The XPD that results, when the polarization angle of LP wave changes by an amount $\Delta\phi$. A cross polarization component is given by, $EX = E \sin \Delta\phi$.

Hence the *XPD* is,

$$XPD = 20 \log_{10} \left[\frac{E_{CO}}{E_X}\right]$$

or

$$XPD = 20 \log_{10} (\cot \Delta\phi) \qquad ...(5.11)$$

Table shows the typical values for satellite paths

Table 5.2.

Country	Average	Peak Seasonal	Solar Flare
Northern Europe (at 11 GHz)	0.3°	1°	1.6°
United States (at 4 GHz)	2°	6°	8°

Maximum values quoted by Miya (1981) for faraday rotation are 9° at 4 GHz and 4° at 6 GHz. The 60° change from average conditions associated with a solar flare would reduce the XPD on a 4 GHz link to about 15.6 dB.

Thus to counter the depolarizing effects of Faraday rotation, cicular polarization may be used. Alternatively, if linear polarization is to be used, polarization tracking equipment may be installed at the antenna.

5.3.6. Ionospheric Scintillation

Ionospheric scintillations are variations in the amplitude, phase, polarization or angle of arrival of radio waves. They are caused by irregularities in the upper part of the ionosphere. With peak to peak fluctuations as large as 6 dB on 6 GHz downlinks to stations near the geomagnetic equator. The main effect of scintillation is fading of the signal. The fades can be quite severe and that may last up to several minutes. As with fading caused by atmospheric scintillations, it may be necessary to include a fade margin in the link power budget calculations to allow for ionospheric scintillation.

They tend to occur about one hour after local sunset and are most frequent about the spring and fall equinoxes. Generally they are not significant in the United States or in Europe.

5.4. RAIN AND ICE EFFECTS

Rain is the dominant factor in satellite propogation at frequencies above than 10 GHz. Intensively rain propogation has been studies in late 1960s. In this section we will discuss the characteristics of rain and ice.

5.4.1. Characterizing Rain

Rain attenuation is a function of rain rate. **Rain rate** is the rate at which rain water could accumulate in a rain guage situated at the ground in the region of interest. Metrologist want to know about rain accumulation because it is most important to farmers and hydrologist among others. But rain accumulation is of less use to satellite link designers because it is the rain fall rate rather than the accumulation that determines the propogation effects Fig. 5.9 shows a simple device for measuring rain rate.

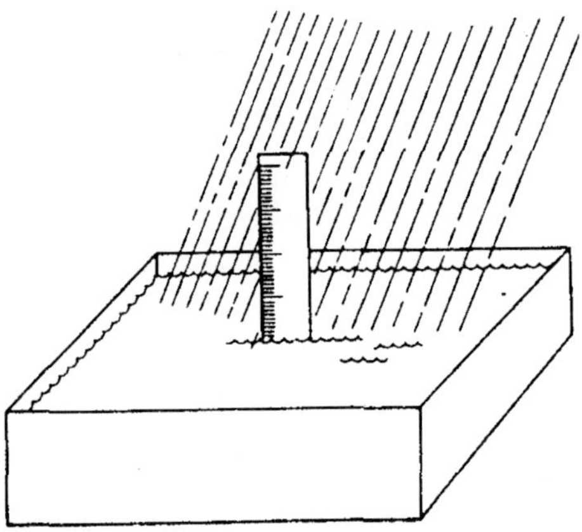

Fig. 5.9. A simple device for measuring rain rate.

Let us imagine a large flat bottomed container open to the rain, the rain rate is *R* and it is the rate at which the water level is rising in the container. Usually rain rate is expressed in units of millimeters per hour (mm/hr). Peak values of 100 to 150 mm/h may be expected for short periods during summer thunderstorms in the Mid-Atlantic region of United States.

The long term behaviour of rain rate is described by a **cumulative probability distribution** and popularly known as **exceedance curve**. This give the percentage of time that the rain rate exceeds a given value. Fig. 5.10 shows a typical plot of rain rate cumulative probability distributions for Blacksburg, virginia. Some measured values are as follows :

Table 5.3.

Percentage Time	Absolute Time out of 1 year	*R* (mm/hr) Exceeded for this % time
0.1	8.76 hr	8
0.01	52.56 min	43
0.001	5.26 min	106
0.0001	0.53 min	142

Mathematical models are variable that give the exceedance curve for a given location based on the total rain accumulation and other metrological data. Fig. 5.11 and 5.12 shows the rain climate regions for United States, Southern Canada and Europe.

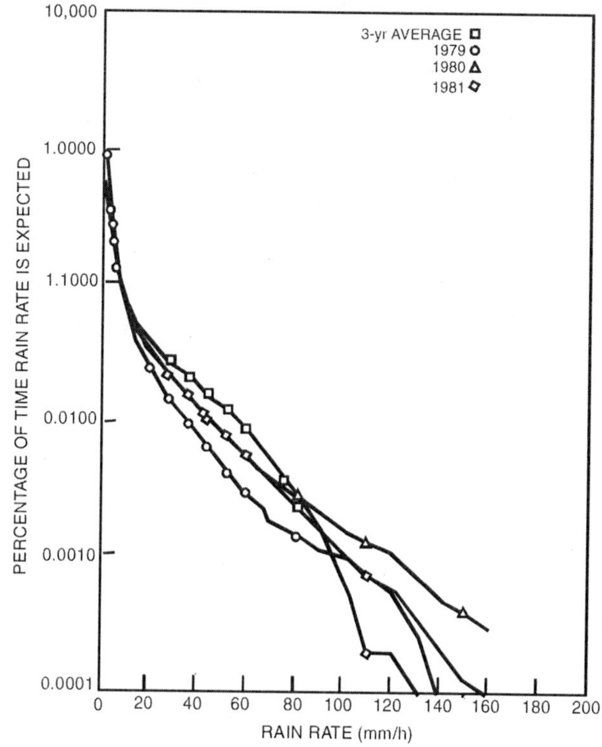

Fig. 5.10. Typical rain rate cumulative probability distributions or "exceedance" curves.

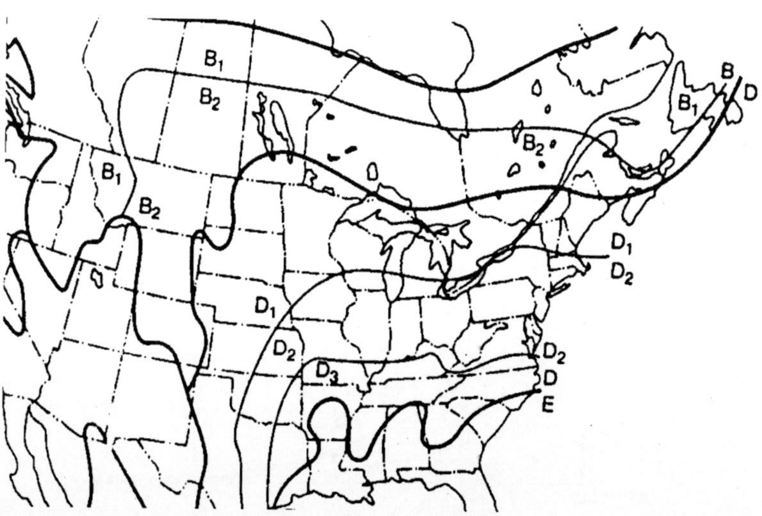

Fig. 5.11. NASA rain rate climate regions for Continental United States and Southern Canada.

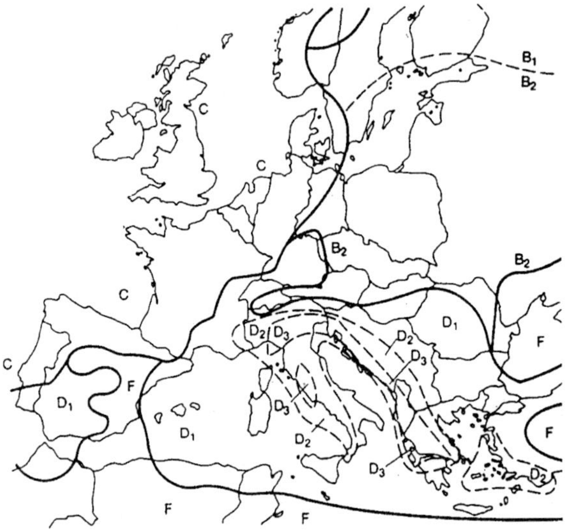

Fig. 5.12. NASA rain rate climate regions for Europe.

Figure 5.13 and Table 5.4 provide the rain rate statistics

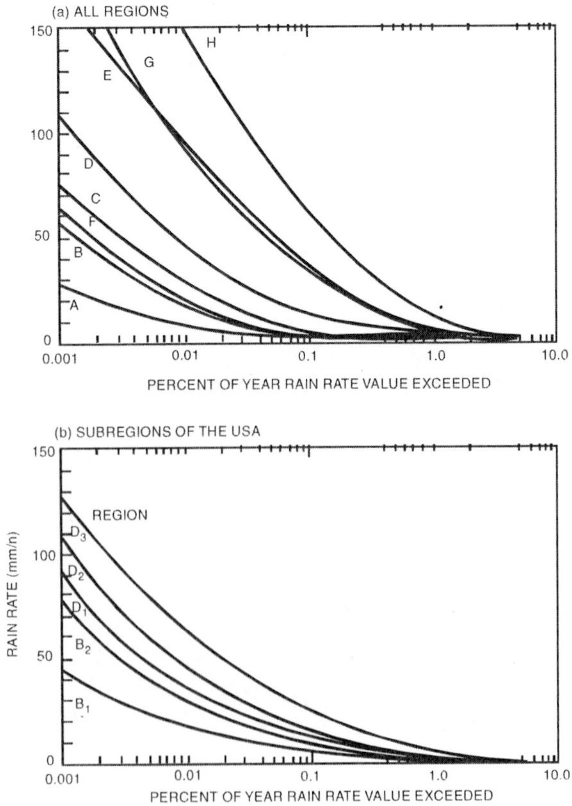

Fig. 5.13.

Table 5.4. Rain Rate Distribution Values Versus Percentage of Year that rain rate is Exceeded.

Percentage of Year	Rain Climate Region												Minutes per year	Hours per year
	A	B_1	B	B_2	C	D_1	$D = D_2$	D_3	E	F	G	H		
0.001	28.5	45	57.5	70	78	90	108	126	165	66	185	253	5.26	0.09
0.002	21	34	44	54	62	72	89	106	144	51	157	220.5	10.5	0.18
0.005	13.5	22	28.5	35	41	50	64.5	80.5	118	34	120.5	178	26.3	0.44
0.01	10.0	15.5	19.5	23.5	28	35.5	49	63	98	23	94	147	52.6	0.88
0.02	7.0	11.0	13.5	16	18	24	35	48	78	15	72	119	105	1.75
0.05	4.0	6.4	8.0	9.5	11	14.5	22	32	52	8.3	47	86.5	263	4.38
0.1	2.5	4.2	5.2	6.1	7.2	9.8	14.5	22	35	5.2	32	64	526	8.77
0.2	1.5	2.8	3.4	4.0	4.8	6.4	9.5	14.5	21	3.1	21.8	43.5	1052	17.5
0.5	0.7	1.5	1.9	2.3	2.7	3.6	5.2	7.8	10.6	1.4	12.2	22.5	2630	43.8
1.0	0.4	1.0	1.3	1.5	1.8	2.2	3.0	4.7	6.0	0.7	8.0	12.0	5260	87.7
2.0	0.1	0.5	0.7	0.8	1.1	1.2	1.5	1.9	2.9	0.2	5.0	5.2	10520	175
5.0	0.0	0.2	0.3	0.3	0.5	0.0	0.0	0.0	0.5	0.0	1.8	1.2	26298	438

5.4.2. Rain Drop Distributions

Rain attenuation and depolarization occurs because of two reasons :

(1) Individual rain drops absorb energy from radio waves (*i.e.* the drops are heated).

(2) Some energy in the wave is scattered out of propogation path.

These interactions depend on the number of raindrops encountered *i.e.* the rain rate and on their distribution of shapes and sizes. Generally, the **drop size distributions** are exponential and is given as,

$$N(D) = N_0 \exp\left(-\frac{D}{D_m}\right) \text{mm}^{-1}\ \text{m}^{-3} \qquad ...(5.12)$$

where, D_m = Median drop diameter.

$N(D).\ dD$ = Number of drops per cubic meter with diameters between D and $D + dD$ mm

The **rain rate *R*** is related to $N(D)$ and to terminal velocity $V(D)$ of falling drops in meters per second with diameter D, the given Eq. (5.13).

$$R = 0.6 \times 10^{-3}\pi \int D^3\ V(D)\ N(D)\ dD \text{ mm/h} \qquad ...(5.13)$$

5.4.3. Calculation of Attenuation

The attenuation calculation requires the details of scattering and absorption by a single raindrop and the summation over the drop population. In general, if in a percentage time, the measured rain rate is R.

The **specific attenuation α** is,

$$\alpha = aR^b \text{ dB/km} \qquad ...(5.14)$$

where a and b depend on frequency and polarization. Table shows the values of a and b. Subscripts h and v refers to horizontal and vertical polarization.

If the rain rate is constant over a path of length L km, the total **attentuation A called by the rain** is given by

$$A = \alpha L \text{ dB}$$

Therefore,

$$A = aR^b L \text{ dB} \qquad ...(5.15)$$

where, L is the **effective path length** of the signal through the rain. Since the rain density is unlikely to be uniform over the actual path length, an effective path legnth must be used rather than the actual length. This situation is shown in Fig. 5.14.

The rain rate changes on satellite path with position and time, and length L of the portion of the path that contains rain is also variable.

If this length is $L(t)$ and $R(y, t)$ is the rain rate in millimeters per hour (mm/h) at time t at a distance y km measured from the ground along the path.

Then the attentuation $A(t)$ is given by,

$$A(t) = \int_0^{L(t)} a\left[R(y,t)\right]^b dy \text{ dB} \qquad ...(5.16)$$

The Eq. (5.16) has no more practical use because there is no relation between attenuation $A(t)$ and point rain rate $R(t)$ measured on the ground near a receiving antenna.

This is because rain drops take a long time to fall from the upper part of a storm to a ground based rain guage. Typically the attenuation behaviour at time 't' is similar to the behaviour of the ground rain rate of time $t + T$. Where,

T = delay, it is variable and may be as long as 4 to 8 min.

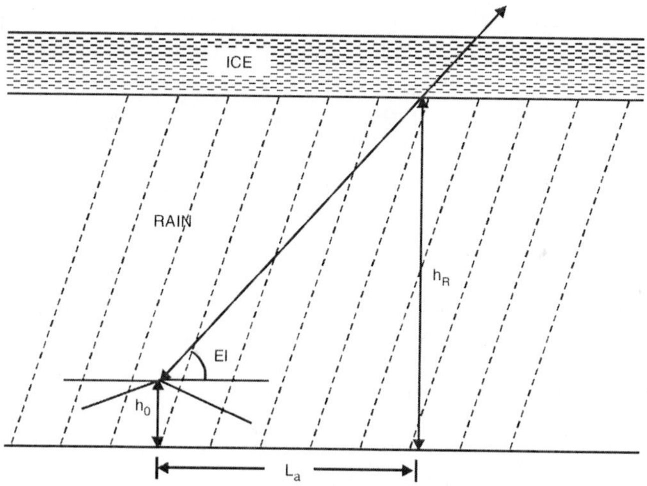

Fig. 5.14. Path length through rain.

Thus the predictive models for rain attenuation are statistical rather than instantaneous. Since attenuation is caused by rain, both A and R exhibits the same kind of statistical behaviour and is shown in Fig. 5.15

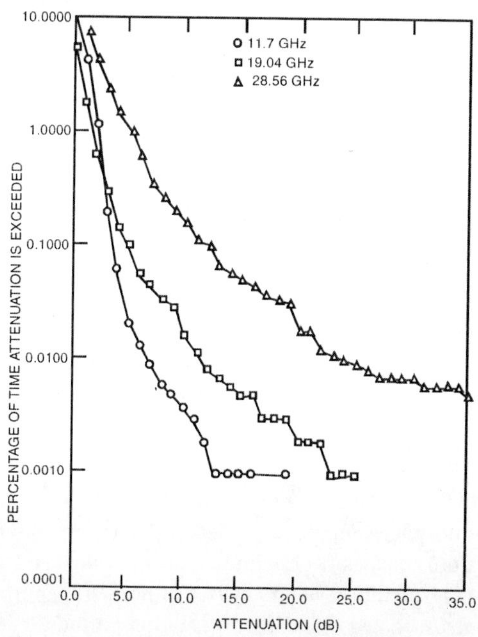

Fig. 5.15. Cumulative attenuation distribution for the calender year 1978.

If the path average rain rate is $R_P(t)$. Then,

$$A(t) = \int_0^{L(t)} a\left[R(y,t)\right]^b dy$$

$$= a\left[R_P(t)\right]^b L(t) \text{ dB} \qquad ...(5.17)$$

Since it is necessary that all the rain that passes through the propagation path should ultimately reach the ground, thus the exceedance curve of the path average rain rate should be the same as the exceedance curve of the ground rain rate. Therefore the attenuation value $A(P)$ equal or greater P percentage of the time should be proportional to the value of ground rain rate. $R(P)$ equal or exceeded for the same P percent of time

$$A(P) \propto a[R(P)]b \text{ dB}$$

$$\boxed{A(P) = a[R(P)]b\, L_{eff.} \text{ dB}} \qquad ...(5.18)$$

The proportionality factor is known as **effective path length**, $L_{eff.}$ and has a unit of kilometer. The effective path length may be measured by identifying and tabulating corresponding values of $A(P)$ and $R(P)$ with P as a parameter. The paired values of attenuation and rain rate is known as **equal probability values**. Fig. 5.16 shows the procedure to calculate them.

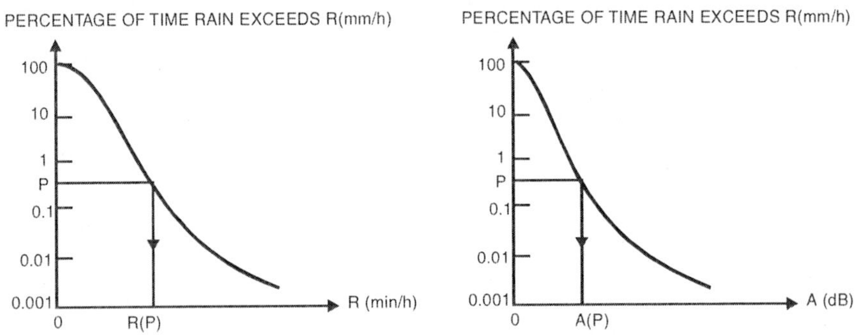

Fig. 5.16. The procedure for finding equal-probability values of attenuation *A* and rain rate *R*.

Equation (5.18) demonstrate the most predicctive models that give a relationship between $A(P)$ and $R(P)$. **There are two models :**

(1) The SAM model was developed with NASA support to provide simplified calculation technique. It assumes that rain extends from the earth stations elevation H_o km above sea level to an effective storm height H_e km, as shown in Fig. 5.17.

For an elevation angle El, path length L in rain is given by

$$\boxed{L = \frac{H_o - H_e}{\sin(El)} \text{ km}} \qquad ...(5.19)$$

In the lower atmosphere, temperature decreases with hight and the locus of points where the temperature reaches 0°C is known as **zero degree isotherm**. Its height is H_i^1, generally above H_i no liquid water exists. For prediction purposes, typical summer values of H_i may be estimated from an earth station latitude Λ_e degree by,

$$H_i = \begin{cases} 4.8, & |\Lambda_e| \le 30^\circ \\ 7.8 - 0.1|\Lambda_e|, & \Lambda_e \le 30^\circ \end{cases} \qquad ...(5.20)$$

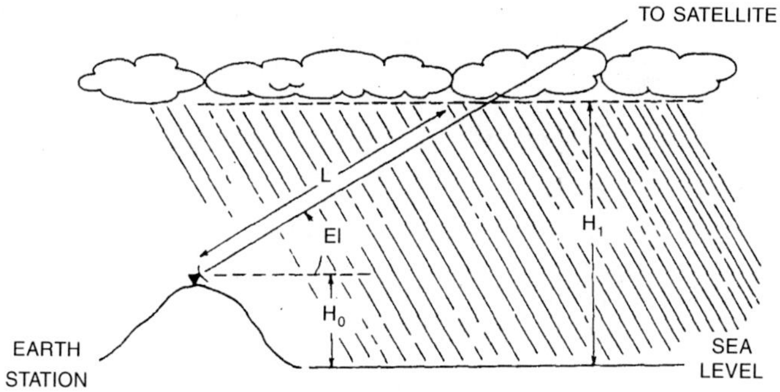

Fig. 5.17.

In the SAM model, the effective rain height H_e is related to H_i and the rain rate R by

$$H_i = \begin{cases} H_i^d, & R \le 10 \text{ mm/h} \\ H_i + \log_{10}(R/10), & R > 10 \text{ mm/h} \end{cases} \qquad ...(5.21)$$

When $R(P)$ is determined and L is calculated from Eq. (5.20), (5.22) and (5.22), the equal probability attenuation $A(P)$ is given by,

$$A(P) = \begin{cases} a[R]^b \; L, & R < 10 \text{ mm/h} \\ a[R]^b \left\{ \dfrac{1 - \exp\left[-\gamma b \log_e \left(\dfrac{R}{10}\right) L \cos(El)\right]}{\gamma b \log_e \left(\dfrac{R}{10}\right) \cos(El)} \right\}, & R > 10 \text{ mm/h} \end{cases} \qquad ...(5.22)$$

where the empirical quantity γ is 1.22.

(2) The **CCIR model** is based on predicting the attenuation expected for 0.01% of the time and then scaling it to other percentage. Here it is denonstrated for time percentage between 0.001 and 0.1%.

For earth station lattitude Λ_e, the rain height h_R is calculated by

$$h_R = 5.1 - 2.15 \log_{10} \left\{1 + 10^{[(|\Lambda_e| - 27)/25]}\right\} \text{ km} \qquad ...(5.22)$$

The slant path length L_S is calculated from h_R, the station height above sea level ho and the path elevation angle El by

$$L_S = \frac{2(h_R - h_0)}{\left[\sin^2(El) + \dfrac{2(h_R - h_0)}{8500}\right]^{1/2} + \sin(El)} \text{ km for } El < 10^\circ$$

$$L_S = \frac{h_R - h_0}{\sin(El)} \text{ km for } El \ge 10^\circ \qquad ...(5.23)$$

If r_p, is the path length reduction factor because of the non-uniformity of the rain rate. It is calculated from the horizontal projection of the slant path L_G is kilometers (km) by

$$r_P = \frac{90}{90 + 4L_G} \quad ...(5.24)$$

where, $L_G = L_S \cos(El)$ ...(5.25)

If R_p is the rain rate exceeded for 0.01% of an average year then the attenuation that will be exceeded for 0.01% of time, given by the CCIR model is

$$A_{0.01} = a(R_P)^b L_S r_P \text{ dB} \quad ...(5.26)$$

The attenuation value A_P predicted for any other percentage of time P may be calculated from $A_{0.01}$ by

$$A_P = A_{0.01}\left(\frac{P}{0.01}\right)^{-\alpha} \text{ dB} \quad ...(5.27)$$

where,

$$\alpha = \begin{cases} 0.33 & \text{for } 0.001 \le P \le 0.01 \\ 0.41 & \text{for } 0.01 < P < 0.1 \end{cases} \quad ...(5.28)$$

For percentage of time outside the 0.001 to 0.1 percentage range, a relative complicated procedure is there.

5.4.4. Scaling Attenuation with Elevation Angle and Frequency

In the previous sections two models were presented and both permit the calculation of attenuation for any elevation angle and frequency. It is quite useful, given an attenuation value for one path and frequency, to quickly estimate the effects of changing one or both of these factors. For doing this, **there are two rules of thumb.**

(1) For a uniform rain enviorment and a flat earth, attenution in dB is proportional to path length.

(2) If the rain height is constant, then path length is proportional to the cosecant of the elevation angle El.

Thus the attenuation in dB at the same frequency at elevation angles El_1 and El_2 are approximately related by

$$\frac{A(El_1)}{A(El_2)} = \frac{CSC(El_1)}{CSC(El_2)} \quad ...(5.29)$$

This formula breaks down at low elevation angles ($El < 10°$).

There are many formulas for scaling attenuation with frequency but a good opproximation attenuation in dB simply varies in proportional to the square of frequency in GHz. Thus if $A(f_1)$ and $A(f_2)$ are the attenuations that would be measured on the same path at frequencies f_1 and f_2 GHz, approximately they are related by

$$\frac{A(f_1)}{A(f_2)} = \frac{f_1^2}{f_2^2} \quad ...(5.30)$$

CCIR recommends using,

$$\boxed{\frac{A(f_1)}{A(f_2)} = \frac{g(f_1)}{g(f_2)}} \qquad ...(5.31)$$

where, $$g(f) = \frac{f^{1.72}}{\left[1 + 3 \times 10^{-7} \left(f^{1.72}\right)^2\right]} \qquad ...(5.32)$$

where the attenuations are in dB and the frequencies are in GHz.

5.4.5. Rain Depolarization

The ideal shape of a raindrop is spherical because this minimizes the energy required to hold the raindrop together. The shape of small raindrops is close to sphericals but larger but instead flatten out and are shaped more likely oblate spheroids because of the air resistance. Figure 5.18 shows these two types of raindrops.

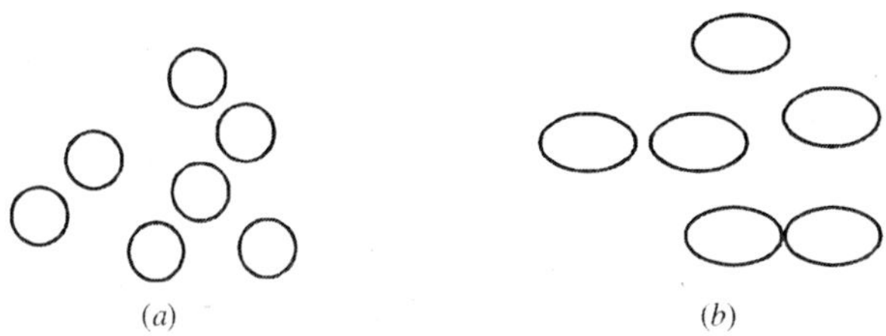

Fig. 5.18. Raindrops (*a*) small spherical, (*b*) flattening resulting from air resistance.

For vertically falling rain, the axis of the symmetry of raindrop is will be parallel to the local vertical as shown in Fig. 5.19 but more realistically, aerodynamic forces will cause some canting or tilting of the drops. Thus there will be a certain randomness in the angle of tilt as shown in Fig. 5.19.

Fig. 5.19. Raindrop : angle of tilt randomized through aerodynamic force.

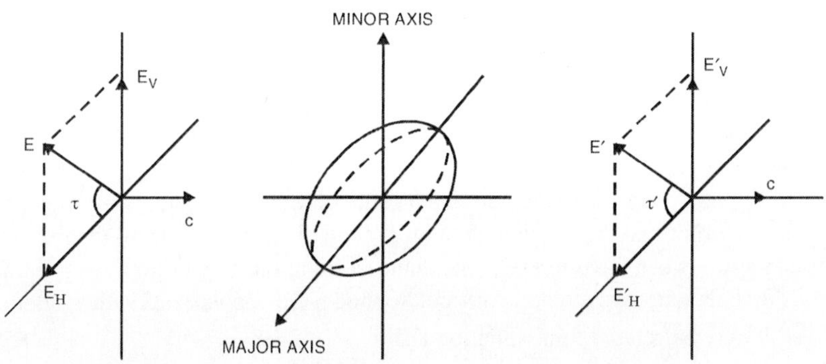

Fig. 5.20 Polarization vector relative to the major and minor axes of a raindrop.

A linearly polarized wave can be resolved into two component waves, one vertically polarized and other horizontal polarized. Consider a wave with its electric vector at some angle τ relative to a major axis of a raindrop, for clearity it is shown horizontal in Fig. 5.20.

The waves propagating with their electric field vector parallel to the drops minor axis would experience the minimum attenuation for that rain rate and waves propagating with their electric field vector parallel to the major axis would experience the maximum attenuation. In these two cases, no depolarization would occur. Similarly each electric field component also experienced difference in phase shift. These differences are known as **differential attenuation** and **differential phase shift.**

The angle of polarization of the wave emerging from the rain is altered relative to that of the wave entering the rain. The depolarization, because of differential phase shift is more significant than that of differential attenuation.

5.4.6. Calculating XPD

Let us consider a case of a wave whose linear polarization is intermediate between horizontal and vertical. It can be resolved into vertical polarized and horizontal polarized component as shown in Fig. 5.21.

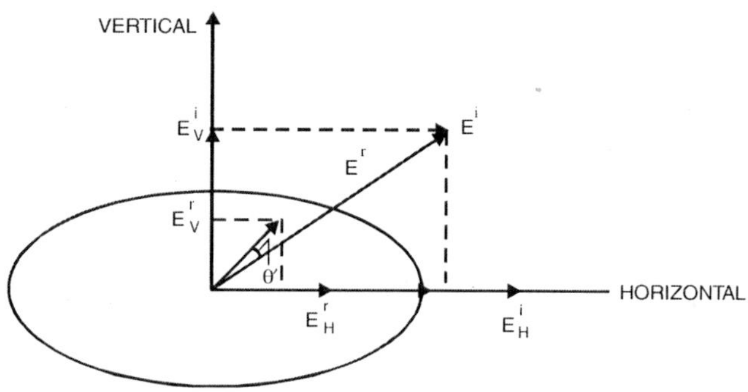

Fig. 5.21. Simplified explanation of rain depolarization.

E^i = Indicident electromagnetic wave with electric field vector E_i.

E^r = Received wave.

These components propogate through rain with their polarization inchanged but the horizontal component is more attenuated than vertical. To reconstruct the wave, if we recombine the vertical and horizontal compoments at any point, we find that its polarization has rotated towards the vertical and a cross polarized component is now present.

The explanation of depolarization is a simplification of a complicated problem in a electromagnetic wave scattering and was apparently developed by at least two research groups.

Mainly attenuation depends on the volume of water that is present in propgation path. Therefore it may be predictable with acceptable accurracy from the rainrate. But depolarisation is sensitive to size, shape and orientation distribution of the raindrops. These dependencies make it quite difficult to model the depolarization than attenuation.

XPD may be related to attenuation (A) by,

$$\boxed{XPD = U - V \log_{10} (A) \text{ dB}} \qquad \text{...(5.33)}$$

The nominal range of A is $3 \leq A \leq 30$ dB.

This equation does not predict the instantaneous values of *XPD*. It provides an estimates of the average values of *XPD* that will be associated with a given attenuation. U and V are the empirically determined coefficient. The following formula for U and V for the frequency range 8 to 35 GHz.

$$V = \begin{cases} 20 & \text{for } 8 \leq f \leq 15 \text{ GHz} \\ 23 & \text{for } 15 \leq f \leq 35 \text{ GHz} \end{cases} \quad \text{...(5.34)}$$

and
$$U = 30 \log_{10}(f) - 40 \log_{10} \cos(El) - 20 \log_{10}[\sin(2\tau)] \quad \text{...(5.35)}$$

where, El = Path elevation angle

τ = Angle between the received electric field and local horizontal. $\tau = 45$ for circular polarization

f = Frequency of in GHz.

5.4.7. Effect of Rain on Antenna Noise Temperature

By increasing the earth station antenna noise temperature, rain degrades the performance of a satellite communication system. In clear wheather antenna find the cold background of space but in rain it receives thermal radiations from the rain drops.

The increase in antenna noise temperature due to rain, T_b may be calculated as

$$\boxed{T_b = 280\left(1 - e^{-A/4.34}\right) K} \quad \text{...(5.36)}$$

where, A = Attenuation in dB. It should be added to clear wheather antenna noise temperature when calculating (C/N) in rain.

Number 280 = Effective temperature of the rain and it differs from true temperature because of scattering and antenna effect values ranging from 270 to 290 are used.

But under some conditions, the increase in antenna noise temperature can have more effect on a satellite link that does the associated attenuation. Exmaple : 1 dB rain fade will cause about 58 K increase in antenna noise temperature. In the clear wheather system noise temperature is 232 K, this increase will raise the noise power by 1 dB. The rain fade will be accompained by a 2 dB drop in (C/N), 1 dB for the fade and 1 dB for the increase in noise level.

5.4.8. Ice Depolarization

Intense depolarization in the absence of attenuation was first noted in 1976 during the European phase of NASA's ATS-6 expriment. An ice layer is present at the top of rain region and ice crystals can result in depolarization.

The experimental evidence demonstrates that the mechanism producing depolarization in ice is differential phase shift with little differential attenuation present. This is because ice is a good dielectric unlike water, which has considerable losses. Ice crystals have a tandency to be needle-shaped or platelike and, if randomly oriented have little effect but depolarization occurs when they become aligned. Sudden increase in XPD that coincides with lightning flashes are throught to be a result of the lightning producing alignment.

An international radio consultative committee (CCIR) recommendations for taking ice depolarization into account is to add a fixed decibel value to the XPD value calculated for rain. In North America and Europe satellite link performance is dominated by rain depolarization rather than ice. Values of 2 dB are suggested for North America and 4 to 5 dB for maritime regions. Further it is suggested that the effects of ice can be ignored for time percentage less than 0.1%.

5.4.9. Examples of Rain and Ice Effect

Figure 5.22 shows the typical rain effects on satellite links. It was measured on the CTS 11.7 GHz circularly polarized downlink and the COMSTAR linearly polarized 19.04 and 28.56 GHz downlinks.

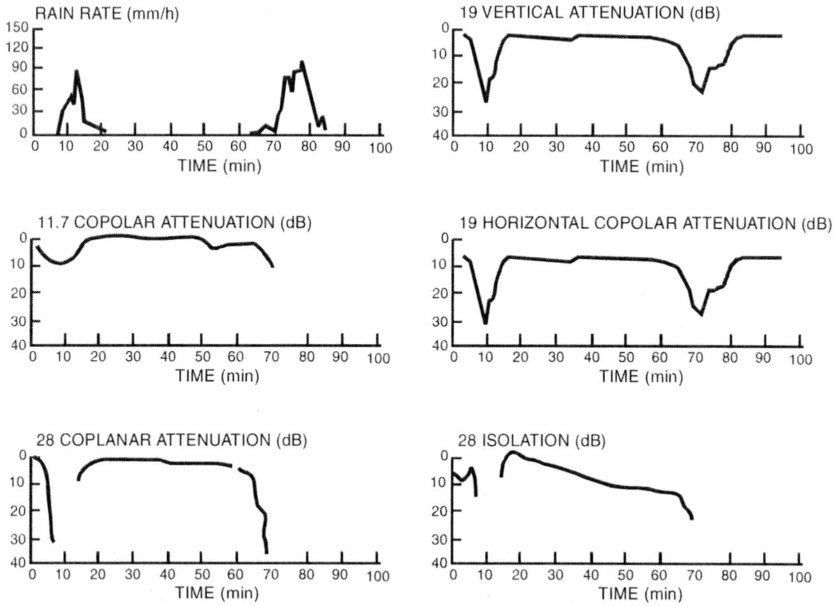

Fig. 5.22. Typical rain effects on 10–30 GHz satellite links. These plots show some of the effects of a thunderstom that occured on May 31, 1977, on signal from the CTS 11.7 GHz beacon and the COMSTAR D-2 19.04 and 28.56 GHz beacons.

5.5. ELIMINATING OR ALLEVIATING PROPAGATION EFFECTS

5.5.1. Attenuation

A simple technique to overcome the attenuation is to provide so much signal at the receiver so that fade does not matter. The **fade margin** is the difference in dB between the link (C/N) that is achieved in clear wheather and the minimum value of (C/N) that is required for satisfactory operation.

For a link reliability r, the required rain margin Mr in dB is equal to the attenuation $A(P)$ that will be equalted or exceeded for $P\%$ of the time, where,

$$\boxed{P = 100\% - r} \qquad ...(5.37)$$

Plus ΔN, the decibel increase in the link thermal noise due to the rain in the propagation path.

$$M_r = A(P) + \Delta N \text{ dB} \qquad \text{...(5.38)}$$

M_r can be achieved through a combination of increased receiving antenna gain and increased transmitter EIRP. But for a large rain attenuations, the necessary power and gain may be unavailable and it may be impossible to achieve the required realibility with a single downlink. Then the choice is between abandoning the realibility requirement or going to site-diversity system.

5.5.2. Site Diversity

In a site diversity operation, the two earth stations spaced a few kilometers apart are available to communication with the same transponder. Since heavy rain occur in small geographically 'cells' there is a probability that equally heavy rain will not affect both stations at the same time. Thus there is a probability that both sites will not experience the same rain attenuation at the same time. If the two site numbers are 1 and 2, and if the attenuations they experience at time t are $A_1(t)$ and $A_2(t)$ dB, then the joint attenuation $A_j(t)$ is defined by

$$A_j(t) = \text{Minimum } [A_1(t), A_2(t)] \text{ dB} \qquad \text{...(5.39)}$$

The **average single site attenuation $A_s(t)$** is the mean of $A_1(t)$ and $A_2(t)$.

$$A_s(t) = \left[\frac{A_1(t) + A_2(t)}{2}\right] \text{dB} \qquad \text{...(5.40)}$$

An **ideal diversity system** is one that monitors the received downlink signals at both sites and select the stronger one that would experience an attenuation of $A_j(t)$. **There are two statistical quantities of measurements :**

(1) Diversity gain

(2) Diversity improvements.

Diversity gain, $G_D(P)$ is the decibel difference between the average single site attenuation $A_s(P)$ equaled or exceeded P percent of the time and joint attenuation $A_j(P)$ equalled or exceeded P % of the time.

$$G_D(P) = A_s(P) - A_j(P) \text{ dB} \qquad \text{...(5.41)}$$

Diversity Improvement, $I_D(A)$ is the ratio between the % of time P_s that the average single site attenuation As exceeds A dB to the percentage of time P_j that the joint attenuation A_j exceeds A dB.

$$I_D(A) = \frac{P_s(A)}{P_j(A)} \qquad \text{...(5.42)}$$

Figure 5.23 shows the determination of these two quantities.

Diversity gain determines the system margin and it is the measure of diversity system performance. Diversity improvement is preferred by engineers designing terrestrial microwave links, they may employ diversity to overcome multipath fading.

The required margin M_r for rain fades in a site diversity system is calculated by

$$M_r = A(P) + \Delta N - G_D(P) \text{ dB} \qquad \text{...(5.43)}$$

where, $A(P)$ = Expected attenuation for a single site.

ΔN = Increased noise in decibles contributed by the increase in antenna temperature.

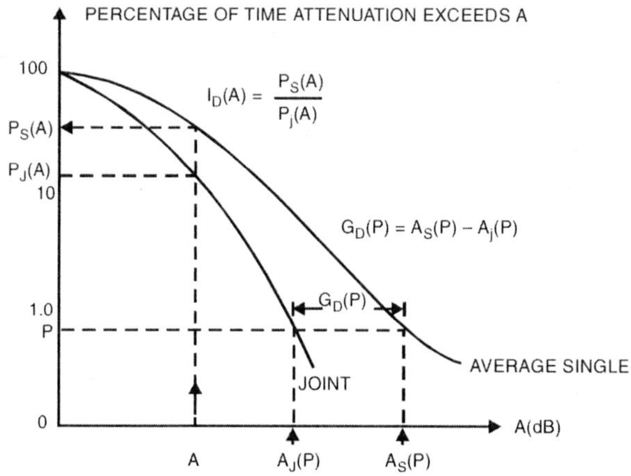

Fig. 5.23. Determining diversity gain and diversity improvement.

Diversity gain may be estimated by two empirical models developed by Hodge

(*a*) **In the earlier and simple model,** G_D is related to A by

$$G_D = a'\left(1 - e^{-b'd}\right) \text{dB} \quad ...(5.44)$$

where, the coefficients a' and b' are given by

$$a' = A - 3.6\left(1 - e^{-0.24A}\right) \quad ...(5.45)$$

$$b' = 0.46\left(1 - e^{-0.26A}\right) \quad ...(5.46)$$

and d is the site separation distance in kilometers.

(*b*) **The second and more detailed model** added the effects of frequency f in GHz, elevation angle El, and base line to path angle Δ. The base line is the straight line joining the two sites. The diversity gain is the product of terms containing the dependence an each factor.

$$G_D = G_d\, G_f\, G_E G_\Delta \text{ dB} \quad ...(5.47)$$

where,

$$G_d = a\left(1 - e^{-bd}\right) \text{dB} \quad ...(5.48)$$

$$a = 0.64\,A - 1.6\left(1 - e^{-0.11A}\right) \quad ...(5.49)$$

$$b = 0.585\left(1 - e^{-0.98A}\right) \quad ...(5.50)$$

$$G_f = 1.64\, e^{-0.025f} \quad ...(5.51)$$

$$G_E = 0.00492\, El + 0.834 \quad ...(5.52)$$

$$G_D = 0.00177\,\Delta + 0.887 \quad ...(5.53)$$

In above equations, Δ and El are measured in degrees and f in GHz.

Diversity gain depends strongly on the site separation distance *d*. Logically it should range from 0 dB at 0 km separation (where both sites experience exactly the same rain) to a value equal to the single site attenuation when d becomes large enough that the rain at the two sites was completely uncorrelated. There is also an average spacing of 10 to 30 km between thunderstorms.

Diversity system has not been widely used in commercial satellite systems. This is because only a few system are operating at 11/14 GHz and all the rest are 4/6 GHz where rain attenuation is not a problem. Now most of the systems at 11/14 GHz have chosen to accept a lower realibility or a higher single site margin.

5.5.3. Depolarization

Rain and ice crystals may cause depolarization. It is more difficult to be quantified than the attenuation. Dual polarization link must contend with both attenuation and depolarization. It is quite easy to calculate the amount of attenuation, from the min (C/N) and margin, that can be tolerated. But determining, minimum allowable polarization isolation value is a difficult problem.

With reference to one channel cross-polarized interference from the other channel is simply **co-channel interferences.** The extent to which this interference degrades link performance depends on the details of the modulation process employed and on the information carried by the two channels. Depolarization is perceived by uses as crosstalk in analog systems and by an increased bit error rate in digital systems. Unlike attenuation, depolarization can not be alleviated by adding margin to the link power budget. There are two potential remedies for this

(1) Site Diversity Reception : To overcome depolarization, the idea of diversity operation is the same as that for attenuation. Two propagation paths spaced few kilometers apart will not experience severe rain or ice crystal effect simultaneously. For the same link, at the cost of maintaining two earth stations, performance can be improved by monitoring depolarization at both sites and selecting the best one. This gives the concept of **polarization diversity gain**. It is the decibel difference between the single site and joint XPD values observed for the same percentage of time. The use of site diversity to improve the performance of dual polarized links.

(2) Adaptive Depolarization Compensation (Cancellation) : This remedy for depolarization is concellation. A suitable attenuated and phase shifted replica of wanted signal component is introduced at the receiver to cancel out the cross polarized signal that was generated in the propagation path. The amplitude and phase of depolarized signal vary with time, thus some sort of adaptive control is required to minimise the crosstalk.

Summary

1. A signal travelling between an earth station and a satellite must pass through the earth's atmosphere including the ionsphere.
2. The atmosphere introduces a time varying loss called attenuation into link power budget and it affects all satellite links.
3. Propagation effects of clear atmosphere are absorption, tropospheric and scintillation, multipath, ionspheric scintillation and Faraday roation.

4. Ionspheric scintillation can causes signal fluctuations as large as 6 dB for stations near the geomagnetic equator.
5. Faraday rotation varies inversary with frequency squared and its depolarization effects at 4 and 6 GHz are usually insignificant.
6. Wheather related losses are known as atmospheric attenuation and the absorption losses as atmospheric absorption.
7. The absorption loss (in decibels) in terms of elevation angle θ is given by

$$[AA] = [AA]_{90} \text{ cosec } \theta$$

8. An atmospheric scintillation is a fading phenomenon which is caused by differences in the atmospheric refractive index, which in turn results in focusing and defocusing of the radio waves.
9. Ionospheric scintillations are variations in the amplitude, phase, polarization or angle of arrival of radio waves. They are caused by irreqularities in the ionosphere which change with time.
10. Depolarization is a conversion of same of the energy in a transmitted signal from one polarization to another. Rain is a primary cause of depolarization.
11. Attenuation is the decible difference between the power received at a given time 't' and the power received under ideal propagational (clear wheather) condition.

    $A(t) = P_r(\text{clear wheather}) \text{ dBW} - P_r(t) \text{ dBW}$
12. Rain attenuation is a function of rain rate and rain rate is the rate at which rain water would accumulate in a rain guage situated at the ground in the region of interest.
13. The ideal shape of raindrop is spherical because this minimizes the energy required to hold the raindrop together. The shape of small raindrops is close to spherical while larger drops are as oblate spheriods.
14. There is a difference in attenuation and phase shift experienced by each of the electric field component. These differences are known as differential attenuation and differential phase shift.
15. The measure of depolarization that is useful for analyzing communication systems is the cross polarization isolation, xpi. It is the ratio of received copolar power to the received crosspolar power.
16. Cross polarization discrimination is the measure to quantify the effects of polarization interference.
17. XPD is related to attenuation A(dB) by

$$XPD = U - V \log_{10} (A) \text{ dB}$$

    for $3 \leq A \leq 30$ dB
18. For quick estimates with linear polarization, U and V in terms of elevation angle (El) and frequency (f) are given as

    $U = 30 \log_{10} (f) - 40 \log_{10} \cos (El)\ 20 \log_{10} [\text{Sin} (2\tau)]$

    $V = 20$
19. The rain rate, R in terms of $N(D)$ and $V(D)$ terminal velocity of falling drops in meters per second with diameter D is given by

$$R = 0.6 \times 10^{-3} \pi \int D^3 \ V(D) \ N(D) \, \text{dB mm/h}$$

20. The unit of specific attenuation α is decibles per kilometers and it is given by

$$\alpha = aR_b$$

21. If the rain rate is R and length of path is L km, the attenuation A caused by the rain is given by

$$A = aR_b L \text{ dB}$$

22. For an elevation angle (El), the path length L in rain is given by

$$L = \frac{H_e - H_o}{\sin(El)} \text{ km.}$$

23. In the SAM model, the effective rain height H_e is related to H_i and the rain rate R by

$$H_e = \begin{cases} H_i,\ R \leq 10 \text{ mm/h} \\ H_i + \log_{10}\left(\frac{R}{10}\right),\ R > 10 \text{ mm/h} \end{cases}$$

24. In CCIR model, the rain height, h_R in terms of earth station latitude Λ_e is given by

$$h_R = 5.1 - 2.15 \log_{10}\left\{1 + 10^{[\Lambda_e| - 27)/25]}\right\} \text{ Km.}$$

25. The increase in antenna noise temperature due to rain, T_b may be calculated as

$$T_b = 280\left(1 - e^{-A/4.94}\right) \text{K}$$

26. For a path length 2 m in the ionosphere, the rotation angle ϕ is given by

$$\phi = \int\left(\frac{2.36 \times 10^4}{f^2}\right) 2NB_0 \cos\phi\, dz \text{ rad.}$$

27. Hail, ice and snow have little effect on attenuation because of the low water content.

28. Wave may arrive simultaneously at a receiving antenna via several propagation paths and by interforming with each other, give rise to multi path fading. It is also known as frequency selective fading.

29. For analog FM systems, multi path fading is similar to rain fading and its effect can be eliminated by providing sufficient clear wheather margin to overcome to fades.

30. The cross polarization discrimination that results when the polarization angle of an LP wave changes by an amount $\Delta\phi$ is given by

$$XPD = 20 \log_{10} (\cot \Delta\phi)$$

31. The long term behaviours of rain rate is described by a cumulative probability distribution and populary known as exceedance curve.

32. Rain attenuation and depolarization occur because individual raindrops absorb energy from radio waves and because some energy in the waves is scattered out of the propogation path.

33. Temperature decreases with height in the lower atmosphere and the locus of points were the temperature reaches 0°C is known as zero degree isotherm.

34. At the same frequency, the attenuation indecibels at the elevation angles El_1 and El_2 are approximately related by

$$\frac{A(El_1)}{A(El_2)} = \frac{Csc(El_1)}{Csc(El_2)}$$

35. The attenuations $A(f_1)$ and $A(f_2)$ that is measured on the same path at frequencies f_1 and f_2 GHz, are related by

$$= \frac{A(f_1)}{A(f_2)}$$

36. The difference between the attenuations experienced by waves with horizontal and vertical polarization is small *i.e.* only a few decibels. This is known as differential attenuation.

37. In site diversity operation, two earth stations spaced a few kilometers apart are available to communicate with the same transponder.

38. The average site attenuation $A_s(t)$ is the mean of $A_1(t)$ and $A_2(t)$ *i.e.*,

$$A_s(t) = \frac{A_1(t) + A_2(t)}{2}$$

39. Diversity gain $G_D(P)$ is the decibel difference between the average single site attenuation $A_s(P)$ equaled or exceeded P % of the time and the joint attenuation $A_j(P)$ equalled or exceeded P % of the time.

$$G_D(P) = A_s(P) - A_j(P) \text{ dB}$$

40. Diversity improvement $I_D(A)$ is the ratio between the % of time P_s that the average single site attenuation As exceeds A dB to the percentage of time P_j that the joint attenuation A_j exceeds A dB

$$I_D(A) = \frac{P_s(A)}{P_j(A)}$$

41. In a diversity system, the required margin for rain fades is calculated from

$$M_r = A(P) + \Delta N - G_D(P) \text{ dB}$$

42. Two potential remedies are there to overcome the depolarization

- Site diversity reception.
- adaptive depolarization compensation (cancellation).

Review Questions

1. Draw and discess the layers in the earth atmosphere.
2. Discuss the various parameters that effect to received power.
3. Discuss the major effects of the ionosphere on the transmission of satellite signals at frequencies of 4 GHz and 12 GHz.
4. What do you understand by multiple propagation ? Discuss the land and sea multipath also.
5. Differentiate between cross polarization discrimination and cross polarization isolation.
6. Briefly discuss the various propagation effects that are not associated with hydrometers.
7. What do you meant by frequency selective fading.
8. Explain what do you understand by rain rate ? How this is related to specific attenuation ?
9. What do you understand by effective path lenght in connection with rain attenuation ?
10. What do you understand by scintillation.
11. What is faraday rotation ? Derive on expression of path lenght in terms of rotation angle.
12. Briefly describe the behaviour of rain rate with the help of exceedance curve.
13. Discuss the raindrop distribution in term of attenuation and depolarization.
14. Derive the relation of attenuation with the help of SAM model and CCIR model.
15. Briefly explain the relation between attenuation, elevation angle and frequency.
16. What is cross polarization discrimination ? Relate XPD with attenuation (A) and elevation angle (El).
17. Explain the effect of rain on antenna noise temperature.
18. Discuss the various examples of rain and ice effects.

19. Explain, what is orthogonal polarization and its importance in satellite communication ?
20. What do you understand by vertical and horizontal polarization of a satellite signal ?
21. What is site diversity ? Explain briefly.
22. Discuss the two major remedies of depolarization.
23. Explain how depolarization is caused by rain.
24. Why is Faraday rotation of no concern with circularly polarized wave ?

OBJECTIVE TYPE QUESTIONS

1. Rapid fluctuations in attenuation are known as
   (*a*) scintillation (*b*) depolarization
   (*c*) fades (*d*) cross talk.
2. To analyze a communication system, the most useful measurement of depolarization is know as
   (*a*) cross polarization discrimination
   (*b*) cross polarization isolation
   (*c*) attenuation
   (*d*) depoloarization.
3. The region of ionosphere is
   (*a*) 100 km to 400 km (*b*) 400 km to 600 km
   (*c*) 50 km to 100 km (*d*) none of these.
4. The long term behaviour of rain rate is described by
   (*a*) cumulative probability distribution
   (*b*) exceedance curve
   (*c*) both (*a*) and (*b*)
   (*d*) none of these.
5. The drop size distribution is given by
   (*a*) $N(D) = N_0 \exp\left(-\frac{D}{D_m}\right)$ (*b*) $N(D) = N_0 \exp\left(\frac{D}{D_m}\right)$
   (*c*) $N(D) = N_0 \exp\left(-\frac{D_m}{D}\right)$ (*d*) $N(D) = N_0 \exp\left(\frac{D_m}{D}\right)$.
6. Effective path length is
   (*a*) the proportionality factor between $A(P)$ and a $[R(P)]b$
   (*b*) the ratio of $A(P)$ and a $[R(P)]b$
   (*c*) the product of $A(P)$ and a $[R(P)]b$
   (*d*) none of these.
7. With elevation angle (El), the path length, L in the ratio is given by
   (*a*) $L = \frac{H_o - H_e}{\sin(El)}$ (*b*) $L = (H_o - H_e)\sin(El)$
   (*c*) $L = \frac{H_e - H_o}{\sin(El)}$ (*d*) $L = (H_e - H_o)\sin(El)$.
8. The locus of points where the temperature reaches 0°C is known as
   (*a*) zero degree attenuation (*b*) zero degree isotherm
   (*c*) zero degree polarization (*d*) none of these.

9. The increase in attenna noise temperature due to rain T_b is calculated by

(*a*) $T_b = 280\left(1 - e^{-A/4.34}\right) K$ (*b*) $T_b = 300\left(1 - e^{-A/4.34}\right) K$

(*c*) $T_b = 280\left(1 - e^{-A/5.34}\right) K$ (*d*) none of these.

11. Diversity improvement factor *ID*(*A*) is

(*a*) $\frac{P_j(A)}{P_s(A)}$ (*b*) $\frac{P_s(A)}{P_j(A)}$

(*c*) $P_j(A) . P_s(A)$ (*d*) none of these.

12. In the SAM model, the empirical quantity is

(*a*) $\frac{1}{12}$ (*b*) $\frac{1}{20}$

(*c*) $\frac{1}{22}$ (*d*) $\frac{1}{32}$

13. The ionosphere completely dominates radio propagation below :

(*a*) 5 MHz (*b*) 10 MHz

(*c*) 20 MHz (*d*) 40 MHz.

ANSWERS

1. (*a*)	**2.** (*b*)	**3.** (*a*)	**4.** (*c*)	**5.** (*a*)
6. (*a*)	**7.** (*c*)	**8.** (*b*)	**9.** (*a*)	**10.** (*a*)
11. (*b*)	**12.** (*c*)	**13.** (*d*)		

❑❑❑

6

CHAPTER

EARTH STATION TECHNOLOGY

INSIDE THIS CHAPTER

6.1. INTRODUCTION

Earth station is very essential part of satelite communication which transmits to and receives from satellite. Earth station may be located on a ship at sea or on an aircraft and it maintains communication link between earth and space. Figure 6.2 indicates the major sub-systems of an earth station. The reliability requirements of an earth station is quite high (about 99.99%). To achieve this reliability, it is essential that adequate redundancy is provided in the system. Normally complete chains are kept as hot standbys.

If an earth station uses more than one transmit or receive chains, then the one to one redendancy may be relaxed to (2 + 1) or (3 + 1) configurations. Proper switching arrangements between the standby and on line chains should be effected. The basic elements for a redundant earth station are shown in Fig. 6.1.

The block diagram is shown in more detail in Fig. 6.2 where, for clarity, redundant units are not shown. Starting at the bottom of the diagram, the first block shows the interconnection equipment required between satellite station and the terrestrial network.

Earth station can be classified according to the **variety of equipment** needed depending upon :

(*a*) Types of service

(*b*) Frequency band

(*c*) Function of the earth station

(*d*) Transmitter

(*e*) Receiver

(*f*) Antenna characteristics

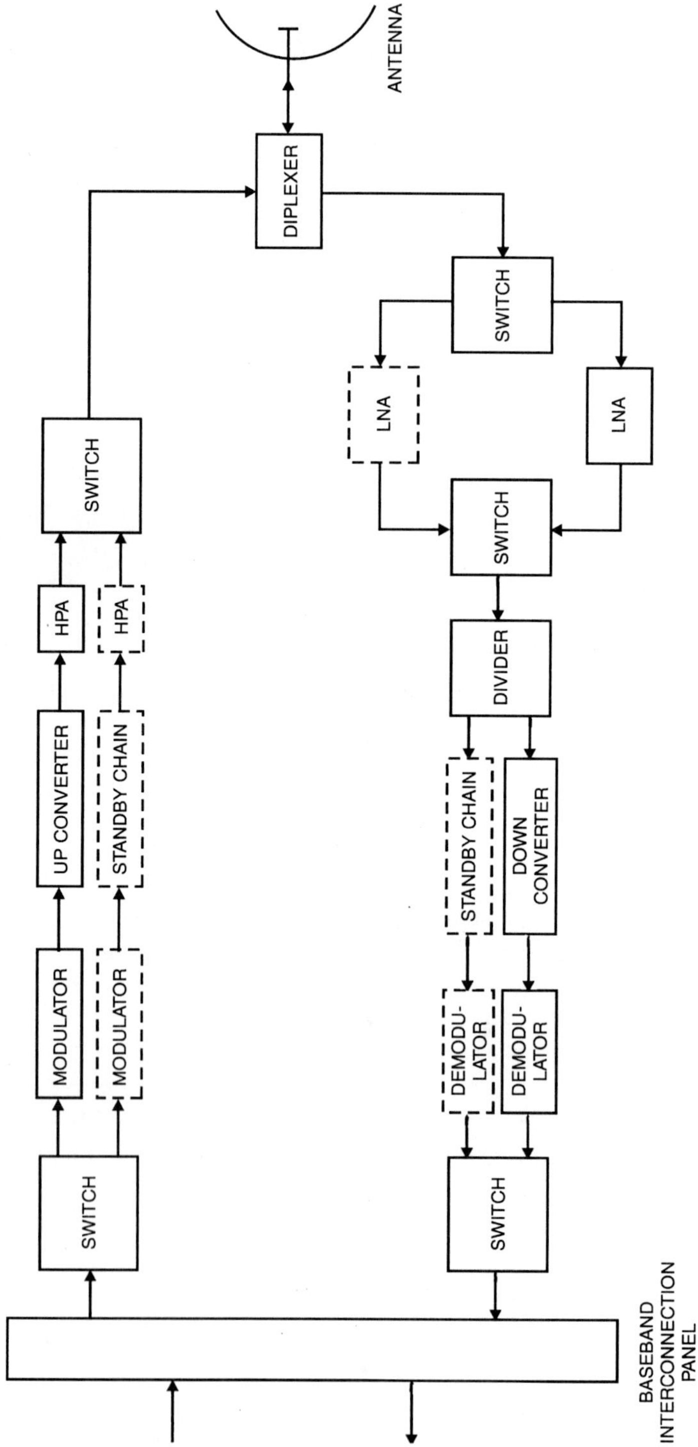

Fig. 6.1. Basic elements of a redundant earth station.

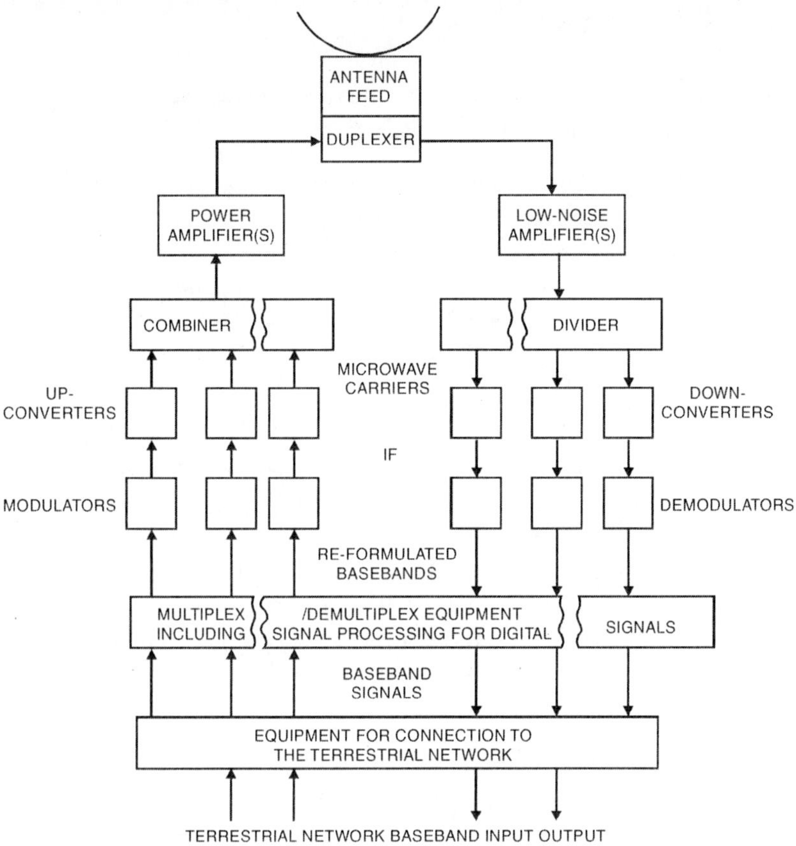

Fig. 6.2. More detailed block diagram of a transmit-receive earth station.

Mainly there are three **categories of an earth station :**

(*a*) Transmit and receive : used in two way communication.

(*b*) Transmit only : used in data collection system.

(*c*) Receive only : used in CATV and direct TV broadcast satellite system.

6.2. EARTH STATION DESIGN REQUIREMENTS

Eearth station receivers are extremely sensitive and hence are highly susceptible to interference from other system. Also the transmitted power is considerably high and earth stations are a potential source of interference to other system operating in the same frequency band. The interference can be minimized by the careful selection of earth station site.

Some of the critical **earth station parameters** are :

(*a*) High gain in the direction of wanted signals and low gain in the direction of unwanted signals.

(*b*) Effective noise temperature of the receiving system should be low.

(*c*) High antenna efficiency.

(*d*) Linearity of transmit and receive chain.

(*e*) Minimum feeder line losses.

(*f*) High discrimination between orthogonally polarized signal.

(*g*) Overall frequency stability of the transmit and receive chains.

(*h*) Ease of operation and maintenance.

The frequency stability of the system depends on :

- Modulator frequency stability.
- Up and down converter frequency stability.

This combined effect should be within limits which in the case of INTELSAT is ± 250 kHz for TV signals and ± 50 kHz for telephone carriers.

Frequency stability can be classified as :

- Short term stability.
- Long term stability.

While short term stability leads to increased noise contribution to the base band signal long term instability. Long term stability will determine how often one should turn up the system.

In case of INTELSAT earth stations, group delay equalization can be done for the individuals chains. Since the satellite transponder is already equalized for best performance. In case of domestic satellite systems the transponder may not be equalized to save spacecraft weight and hence full loop equalization at the ground terminal will be required. **Group delay equalization** is normally done in the IF chains. In case the satellite transponder is not equalized, it may be better to do it after the high power amplifier. But in this case design and realization of the equilizers will be more difficult since they have to handle large power, and also the insertion loss should be very low.

There are important considerations in frequency co-ordination.

- Out of band emission.
- Frequency co-ordination.

It is normally specified that the spurious emission should be below 60 dB from the main carrier. The main contribution to the spurious emission are :

- Local oscillator multiplier chains.
- Intermodulation in the signal path.
- Harmonic generations in HPA.

A harmonic rejection filter is invariable used after the high power amplifier to give sufficient harmonic suppression. There are quite a few components between the radiating element and the low noise amplifiers. These are the duplexer, rotary joint or flexible waveguide, bends, waveguide sections, etc. Similarly, between the radiating element and the high power amplifier (HPA) output, many elements such as the combiner network, harmonic rejection filter, diplexer, isolator, rotary joints, bends, good length of waveguide runs etc. The losses in these components and waveguide runs should be minimized.

6.3. EARTH STATION ANTENNA DESIGN

Earth station operation particularly in case of commercial stations can become a routine affair, but expert maintenance is a pre-requisite to meet the reability requirements unattended earth stations are coming into vogue. In these stations all monitoring is done remotely and the system is engineered for very high realibility. The station is attended only in case of failure or for routine maintenance. Some of the operational considerations of large earth station antenna and small earth station antennas are discussed here.

6.3.1. Antenna Fundamentals

An antenna is a **'transducer'** between electromagnetic waves in space and voltages or currents in a transmission line. When transmitting, the antenna converts the electrical signals into radio waves; a receiving antenna reverses the process and transforms radio waves back into the electrical signals. **Most antennas are passive**, simply metal structure that lauch or collect radio waves.

- All passive antennas can transmit as well as receive and a principle called **reciprocity** allows a passive antenna's transmitting properties to be derived from its receiving characteristics and vice versa.
- All antennas are directional, transmitting more power in some direction than in others. The direction of maximum transmission or reception is called the **boresight direction.** The directional properties of an antennas are described by its radiation pattern and a pictorial representation of relative radiated power versus direction, radiation patterns are there dimensionals.
- In practice, radiation patterns are usually drawn to show relative radiated power in two orthogonal planes, horizontal plane containing the antenna (the azimuth plane) and a vertical plane passing through the antenna (the elevation plane).
- Radiation pattern exhibits a lobe structure in which the largest lobe in three dimensional space constituties the beam through which the antenna radiates and/or receive most of its power. In this context, the locus of half power points surrounding its peak is taken to be perimeter of the main beam. The intersection of the main beam with the ground constitutes the **foot print** of an antenna on a satellite.
- The tendency of the antenna to cencentrate its radiated power is called **gain**. The anguler width of the main beam measured between the half power points is called the **half power beam width** (HPBW) or beamwidth of the antenna. Beamwidth and gain vary inversaly with each other, a high gain antenna has a narrow beamwidth and vice versa.
- **Gain** is proportional to the physical area of an antenna expressed in square wavelengths. The size of an antenna is often stated in terms of its 'aperture', which can be either the physical area of the radiating part of the antenna or the diameter of a circle having this area.
- Radiation pattern lobes other than the main beam are called **sidelobes.** The higher the sidelobe level, the more likely an antenna is to interfere with or by a receiver in the direction of the largest sidelobe. For this reason the sidelobe envelops of ground antennas used in the U.S. must meet specifications set by federal communications commission (FCC) to insure that an antenna pointed at one satellite does not have an unacceptably high sidelobe pointed at another satellite separated from the first satellite by 2° or more in the geostationary arc.

- Electromagnetic waves are vector quantities, that is they are polarized. The desired polarisation radiated by an antenna is termed as **co-polarization**; the orthogonal polarisation is termed as **cross polarization**. If polarization can be made pure enougth, orthogonally polarized wave can travel together without interference and be separated by properly designed receiving antennas. The ability of an antenna to discriminate in this manner is measured by the **cross polarization ratio** or **polarization isolation.**
- An antenna has a **center frequency** or **operating frequency** the frequency at which the antenna exhibits optimum performance and a bandwidth, the frequency large over which the antenna impedance and radiation pattern remain within some required tolerance.
- **A good antenna design** is one that achieves the required gain, sidelobe level, polarization isolation and bandwidth specifications in the smallest aperture size consistent with less limitations in case of manufacture, all at an acceptable price.

6.3.2. Large Earth Station Antennas

Large antennas are used at earth stations that require a high *G/T* ratio and are capable of carrying large number of telephone, television or data channels simultaneously. To cover the cost of a large earth station antenna more traffic is required, which implies operation with wideband carriers. For a example : consider the case of an earth station designed to operate within a zone coverage region of an INTELSAT V satellite. The satellite is designed for operation with standard A earth stations having a minimum *G/T* of 40.7 dB K^{-1} using FDM/FM/FDMA. The satellite cannot be used with earth stations having a *G/T* less than 40 dB K^{-1}.

Suppose a FM carrier is being sent by a 4 GHz transponder of an INTELSAT V satellite so that it occupies the full 72 MHz of a wideband transponder with an EIRP of 26 dBW in the direction of the receiving earth station. The received $\frac{C}{N}$ is given by :

$$\boxed{\frac{C}{N} = \frac{P_t G_T}{KB}\left[\frac{\lambda}{4\pi R}\right]^2 \cdot \frac{G_r}{T_S}} \qquad ...(6.1)$$

Ignoring the interference and uplink noise contributions, we get From Eq. (6.1).

$$\frac{C}{N} = (26 + 228.6 - 78.6) - 196 + \frac{G}{T}\text{ dB}$$

$$\boxed{\frac{C}{N} = \frac{G}{T} - 20\text{ dB}} \qquad ...(6.2)$$

INTELSAT V links operate with a 7 dB margin over a $\left(\frac{C}{N}\right)_{\min}$ of 11 dB. So for 18 dB, the value of $\left(\frac{G}{T}\right)$ should be of 38 dB. This is $\left(\frac{G}{T}\right)_{\min}$ that can be used for any earth station to communicate via an INTELSAT V zone beam transponder. When the transponder is fully loaded, if we use several narrower bandwidth FM carriers, the transponder transmitted power must be shared among the carriers in proportion to their bandwidths and the incident flux density at the earth's surface remains the same. So only standard A earth station can join the INTELSAT global network.

INTELSAT has made provision for smaller earth stations in its network by using some transponders in its satellite in a TDMA node. Standard *B* earth stations have 19-m diameter antennas but achieves the same noise temperature as standard *A* antennas. Transponders in standard *B* earth stations can be operated in or close to saturation, increasing the transmitted power and allowing a reduction in antenna diameter.

6.3.3. Design of Large Antennas

Large antennas are expensive to construct and install. Fully steerable antennas achieve the highest possible aperture efficiency and the lowest possible noise temperature so that *G*/*T* is maximum. The cost of fully steerable antenna is given as,

$$\boxed{\text{Cost} = Y(D)^{2.7}} \qquad \text{...(6.3)}$$

where, D = diameter of the antenna aperture infect

Y = depends on the currency used and inflation.

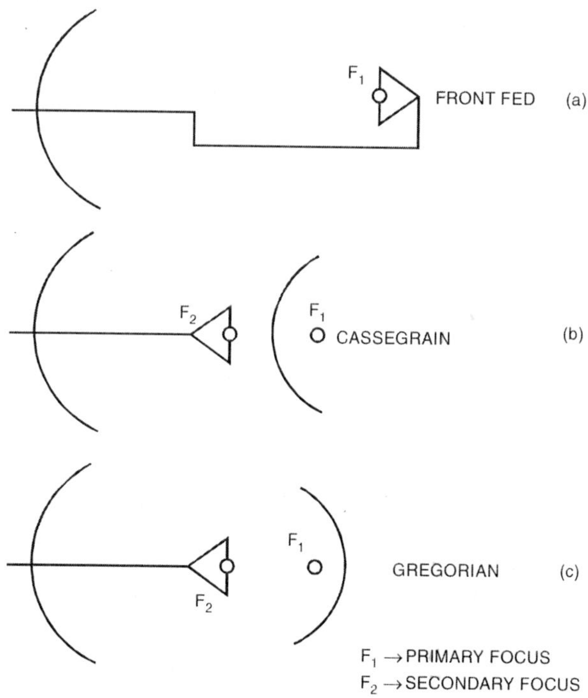

Fig. 6.3. Front fed, cassegrain and gregorian configurations of reflector antennas.

The front fed paraboloid is widely used as a configuration for smaller earth stations, but has an excessively long waveguide. The loss in this waveguide gives a major contribution to the system noise temperature, reducing the *G*/*T* ratio. The **cassegrain antenna is preferred** for large earth stations because of several reasons :

(1) The gain can be increased by approximately 1 dB, relative to a front fed reflector, by shaping the dual reflector system.

(2) Low antenna noise temperature can be achieved by controlling spillover and using short waveguide feeds or beam waveguide feeds.

(3) Beam waveguide feeds can be placed in the low noise amplifier of the receiver and other complex feed components in a convenient stationary position. The geometry of front fed, cassegrain and gregorian configurations are shown in Fig. 6.3.

Large Cassegrain Antennas

The geometry of the cassegrain antenna is shown in Fig. 6.3(*b*). The main reflector is paraboloid, which reflects incoming rediation towards the prime focus. The hyperboloid secondary reflector, often called a **subreflector** or **subdish** has one focus coincident with the paraboloid reflector's focus, and a feed system is placed at the second focus, such that the phase center of the feed coincides with the hyperboloid's second focus. Incoming plane wave is converted to a spherical wave converging toward the prime focus by a paraboloid convertor, which is then reflected by the subreflector to form a spherical wave converging on the feed.

In the design of cassegrain antenna it is assumed that the feed is transmitting. According to the reciprocity same antenna pattern and aperture efficiency will be achieved when the antenna is receiving. So there is a need of feed system with a radiation pattern such that the subreflector intercepts most of the energy transmitted by the feed. The subreflector edge is illuminated with a power level 10 to 15 dB below that at the center. This keeps the spillover from the feed. spillover reduces the effeciency of the antenna and increases its noise temperature so it should be minimum. Spill over can be reduced around 3% with a corrugated horn as the feed.

6.3.4. Small Earth Station Antennas

Basically small earth stations are those which uses antenna less than 60 wavelength in diameter. The broader beamwidth of a small antenna allows fixed pointing of the antenna when the satellite is kept within ± 0.1° of its nomial position. It eliminates the need of autotrack equipment, dual axis drivers and servosystem. Thus fixed pointing small aperture antenna can be used in design where the lower G/T ratio can be compensated by an increase in satellite transmitted power or reducation in RF bandwidth. Small earth stations were developed basically for TVRO (Television receive only) system but these can also be used in the communication networking system, mobile services or maritime services.

For such type of design front feed paraboloid reflector is being used. Cassegrain or dual reflector is not used because the blockage losses are very high. Offset feed antennas employing single and dual reflector configurations may also be used.

6.3.5. Design of Small Earth Station Antennas

Small earth station antenna corresponds to 5 m dishes at 4 GHz and 1.6 m dishes at 11 GHz with gains below about 44 dB. Brokage losses become expensive in symmetrical dual reflector antennas when the main reflector diameter is below 60λ, because the subreflector diameter must be 8λ or 10λ to obtain good control of the operture illumination.

Front Feed Paraboloidal Reflector Antennas

The simplest and lowest cost antenna for a small earth station is the conventional circular paraboloidal dish with a scalar feed. The scalar feed produces a symmetrical radiation pattern that is broad, flat topped and close to gaussion in shape. It is frequency independent over a wide bandwidth. Its design is simple. The half angle of the reflector is chosen to obtain a particular edge illumination with a given feed and a scalar feed can be selected to provide a required edge illumination with a given reflector. The geometry of the paraboloid reflector is shown in Fig. 6.4. The distance ***OF*** is called the **focal length** of the reflector, denoted by F. The curvature of the reflection can be specified by :

- Angle θ_0.
- $\frac{F}{D}$ ratio.

F/D and θ_0 are related by

$$\tan\theta_0 = \frac{1}{2}\left[\frac{\frac{D}{F}}{1-\left(\frac{\Delta^2}{16F^2}\right)}\right] \quad ...(6.4)$$

The surface of the reflector is given by

$$x^2 + y^2 = 4FZ \quad ...(6.5)$$

or

$$r^2 = 4FZ \quad ...(6.6)$$

where, r = radial distance from the axis OF to the point (x, y) on the reflector.

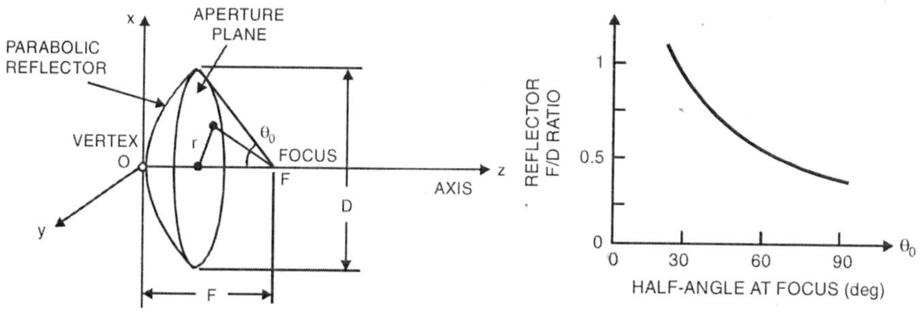

(a) Geometry of a parabolic reflector. (b) Variation of half angle at focus to the reflector ratio.

Fig. 6.4.

Most front fed paraboloid antennas have reflectors with F/D in the range 0.25 to 0.4 corresponding to half angles of 90° to 64°. The F/D = 0.25 reflector is known as **focal plane reflector** because the focus of the reflector lies in the aperture plane.

Scalar feed tends to have narrower beams and are best suited to reflectors with F/D ratios in the 0.3 to 0.4 range. Smaller antennas have relatively broad antenna patterns. Satellites spaced close to the wanted satellite will radiate interfering signals into the sidelobes of the earth station antenna when it is receiving and will be effected by interference from the antenna sidelobes when the earth station is transmitting. The smaller the antenna, the major this proble becomes.

Offset-fed Antennas

Blockage of the aperture by the feed and its supports can be avoided in an offset-fed antenna Fig. 6.5 shows the blockage shodow pattern for a cassegrain antenna with a tripod log arrangement to support the subreflectors.

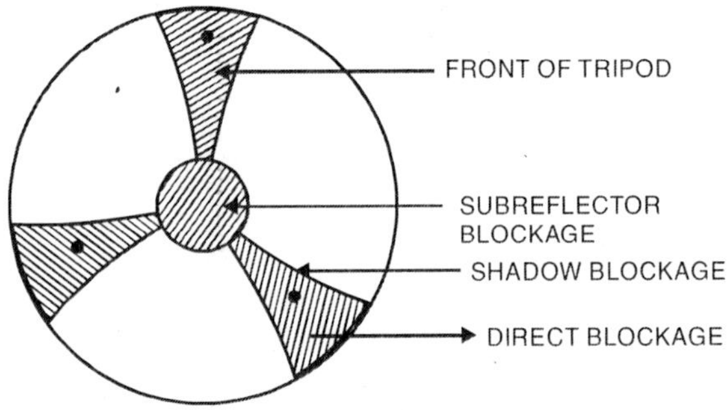

Fig. 6.5. Aperture blockage in a cassegrain antenna.

There is shadowing of both the main reflector aperture and also the subreflectors radiation pattern, leading to significant loss of gain and high near in sidelobes. Front feeds also need supports but because of smaller size and weight of their feeds the support can be much thinner than needed for a subreflector.

The Fig. 6.6 shows offset fed antenna configuration. In the front fed offset antenna the feed is located below the axis and in the open cassegrain arrangement the feed is located within the main reflector aperture, leading to slight blockage losses.

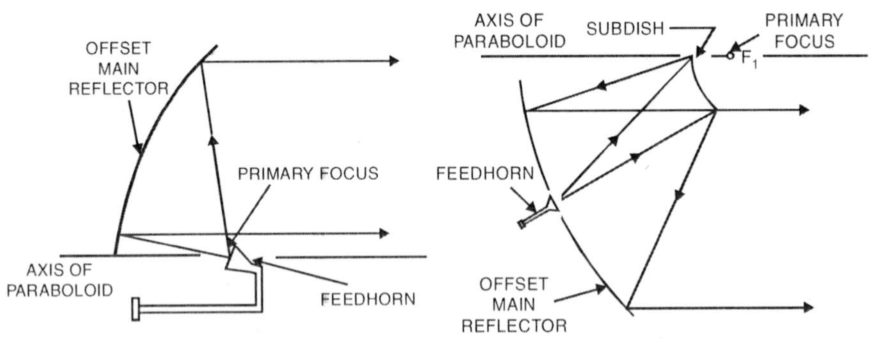

(a) Front fed offset antenna. *(b) Open cassegrain offset antenna.*

Fig. 6.6. Geometry of two offset reflector antenna configurations.

The performance of an offset antenna is better than the symmetrical antennas but they are more expensive. Symmetrical reflecters can be made up of a large number of identical 'petals' which have the same curvature. An offset reflector possesses symmetry only around the axis of the generating pataboloid and requires a number of differently shaped sections with the corresponding increase in tooling costs. Dual reflector offset cassegrains can use shaped reflectors to obtain increased gain by improved illumination efficiency.

6.4. EARTH STATION SUB-SYSTEM

6.4.1. Transmitter

In this section of an earth station the signal which is to be transmitted is converted to the uplink frequency. After the appropriate encoding and modulation it is then amplified and directed to the polarisation port of antenna feed. Simple block diagram of an earth station transmitter is shown in Fig. 6.1. In a large earth station many transmitters and receivers are required, and they are multiplexed onto one antenna. They provide channelized communication through satellite transponders. Modified version of an earth station transmitter is shown in Fig. 6.7.

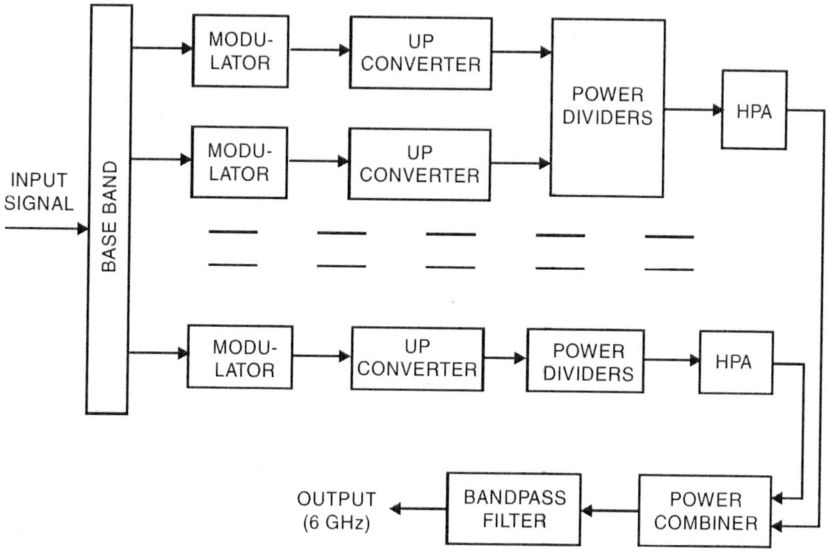

Fig. 6.7. Block diagram of transmiiter.

The important factors of transmitters to avoid interference with other channels and satellites are :

- Frequency stability.
- Power control.
- Out of band emission.

Because of these factors the transmitters are very expensive and cost increases with the increase in transmitted power. Its range may vary from 10 to 1000 of watts. High power amplifiers are used in earth station transmitters because they require transmission of microwave power. Examples of HPAs are travelling wave tubes and multicavity keystrons. Travelling wave tubes prefered over keystron tubes because they allow high power over a wide bandwidth.

Travelling wave tube gives the output power of 100-2500 W with a bandwidth of 500 MHz while klystron tube gives output power of 500-5000 W with 40 MHz band width. Here the modulation is done at 70 MHz IF and it is then upconverted. Outputs of several high power amplifiers may be combined by power combiners and bandpass filters. In present configurations one HPA for each transponder is used. In case of failure of event to improve realibility some redundancy configurations with automatic switch is used.

6.4.2. Receiver

An earth station receiver mainly consist of low noise amplifier (LNA), down converter, demodulators, decoder and some baseband treatment equipments. Basic block diagram of an earth station receiver is shown in Fig. 6.3, and modified version is shown in Fig. 6.8.

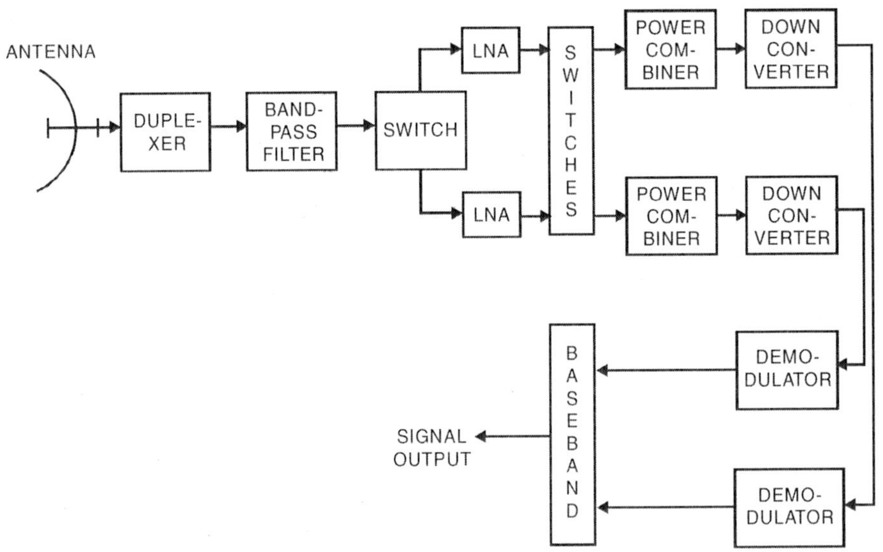

Fig. 6.8. Block diagram of receiver.

Duplexers are used to enhance the separation in the frequency domain as they are different for the uplink and downlink band. Because of the noise considerations of the downlink, first stage of receiver should have low noise amplifier with high gain. For large earth station cryogenically cooled parametric amplifiers are widely used. With liquid helium cooling at 4 K to achieve noise temperature of 20 to 40 K at 4 GHz.

Medium and small earth stations use GaAs FET amplifiers with no cooling and they achieve noise temperature of 50 K to 120 K at 4 GHz and 120 to 300 K at 11 GHz. Lownoise amplifiers used in earth stations usually cover the 500 MHz fixed service band at 4 GHz and 750 MHz at 11 GHz.

6.4.3. Low Noise Amplifiers

Receiver of an earth station requires low noise amplifiers and large earth stations need very low noise amplifiers. Cryogenically cooled parametric amplifiers are widely used. The LNA used in earth stations usually cover the 500 MHz fixed service band at 4 GHz and 750 MHz at 11 GHz. Some redundancy arrangements are used in case of the failure of one LNA and unused LNA is kept on test with input signal so that its present state can be monitored continuously. Dual polarisations earth stations need two RF receive channels and may use one for two redundancy in their LNAs. A comparative chart among all types of low noise amplifiers used in the earth station receivers are shown in given table.

Table 6.1. Comparison Chart of LNAs used in Earth Station Receiver.

S. No.	Type of amplifiers	Frequency	Noise temperature	Cost	Use
1.	Cryogenic parametric	4 GHz	15 K	High	Large earth station
2.	Cooled parametric	4-12 GHz	35-100 K	Medium	Standard A, B & C
3.	Uncooled	4 GHz	50 K	Medium	Large domestic
4.	Parametric	12 GHz	120 K	Medium	Eutelsat station
5.	Cooled FET	4 GHz	50 K	Low	Domestic
6.	GaAs	12 GHz	130 K	Low	Domestic
7.	FET GaAs	12 GHz	250 K	Low	Small domestic

6.4.4. High Power Amplifiers

Mostly large earth stations use a number of high power amplifiers with output power upto 8.5 kW. The configuration used depends on the number of carriers to be transmitted and whether these are FDM or TDM. One HPA for each transponder is used. In large earth stations at 6 GHz, HPAs having bandwidth of 40 to 80 MHz are used. To avoid intermodulation, FDM transmission of several carriers to one or more transponder requires a linear high power amplifier. A combining network is required to sum the outputs of several HPAs into a single transmit waveguide. Frequency selective networks and hybrid waveguide junctions are used to couple the HPAs with a loss of 4 dB per HPA. A comparative table among different HPAs are shown in given Table 6.2.

Comparison Chart of HPAs used in Earth Station Transmitter

In mobile earth stations solid state HPAs are attractive. Solid state HPAs have the advantage that they do not require very high voltages because high voltage tends to suffer more failure than lower voltage equipment and it is bulky and heavy also.

Table 6.2.

S. No.	Amplifier type	Output power	Bandwidth	Cost
1.	Klystron	500-5000 W	40 MHz	Medium
2.	TWT	100-2500 W	500 MHz	High
3.	FET (6 GHz)	5-50 W	Large	High
4.	BJT	upto 50 W	Wide	High

Non-Linear Behaviour in HPAs

Since each transponder will be processing a very large number of messages simaltaneously, any non-linearity in the transponder amplifier will lead to **intermodulation** which causes interference between the message signals by transferring modulations from one frequency range to another. The Fig. 6.9 shows a non-linear amplifier **voltage transfer characteristics** and the way it leads to signal distortion. The distortion is normally represented in terms of additional harmonics of the input signal, which are introduced by the amplifier. The non-linearity may also be represented in terms of the amplifier **power transfer** characteristics, which also shows the **saturation** and **saturation power** of the amplifier.

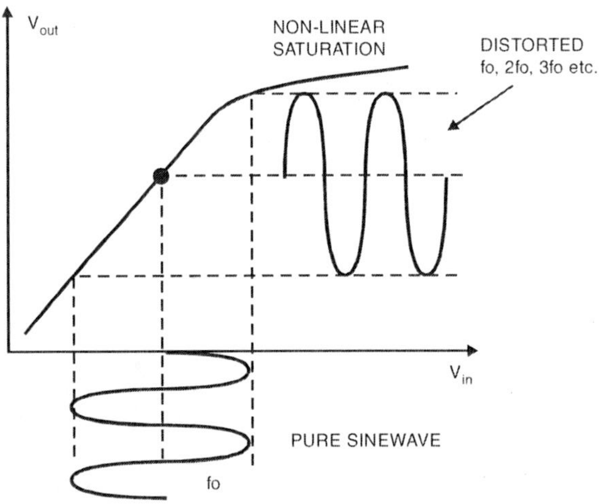

Fig. 6.9. The diagram shows the non-linear (in the saturation region) V_{out} vs V_{in} curve for an amplifier. If a sine-wave is applied to the input the non-linearity will distort the amplified output sinewave as shown.

Intermodulation can be reduced using back off as shown in Fig. 6.10. The input signal power is reduced to more below the non-linear segment of the characteristic. The amount of back off can be expressed in terms of either the **input signal back off** or the **output signal back off.**

A disadvantage of using backoff is that it reduces the efficiency of the amplifier because the RF output from the amplifier is reduced while it is still consuming the same DC power.

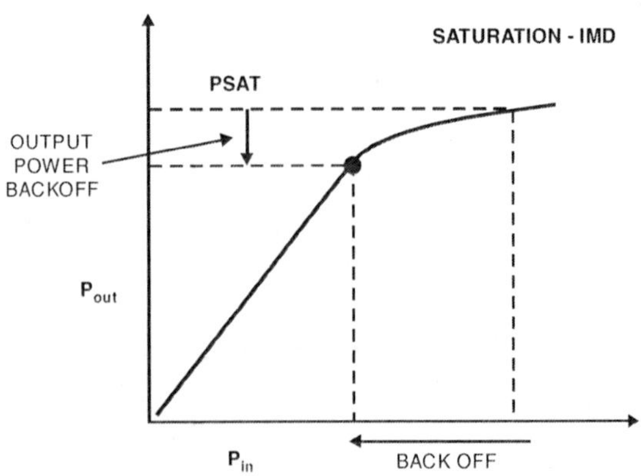

Fig. 6.10. The diagram shows how distortion can be reduced by backing off the input signal from the saturation region to the linear region.

The amount of back off needed to avoid intermodulation increases with the number of messages in the signal which is applied to the transponder. One possible solution is to increase the number of transponders on board the satellite so that each need only to handle a restricted bandwidth and number of carriers. This of course increases the satellite mass, so a suitable compromise must be reached between the number of transponders and the intermodulation.

Solid state amplifiers are superior to TWT amplifiers in their linearity. Considerable attention has been devoted to techniques for linearizing HPAs to improve their effeciency. This involves extending the linear part of the power amplifier characteristics.

6.4.5. FDM Systems

In FDM system many voice and data signals can be transmitted and it is being done by allocating separate frequencies to each signal, either at baseband or RF or both.

Normally a satellite link relay many signals from a single earth station and to avoid interfering with each other these must be separated. This seperation is called multiplexing.

The earth station operating in an FDM mode, a terrestrial link is used to send and receive FDM or TDM groups of channels. The link may be microwave, coaxial cable or optical fibre. The terrestrial interface may separate incoming channels back to baseband for reassembly as new FDM groups for transmission via the satellite link or it may assemble incoming FDM groups. Large earth stations using FDM/FDMA operation may carry several thousands 4 kHz voice channels. These channels are sent via a gateway and are collected from a wide geographical area. In an international system, this gate way is usually located in a major city and all traffic must be routed through this one point.

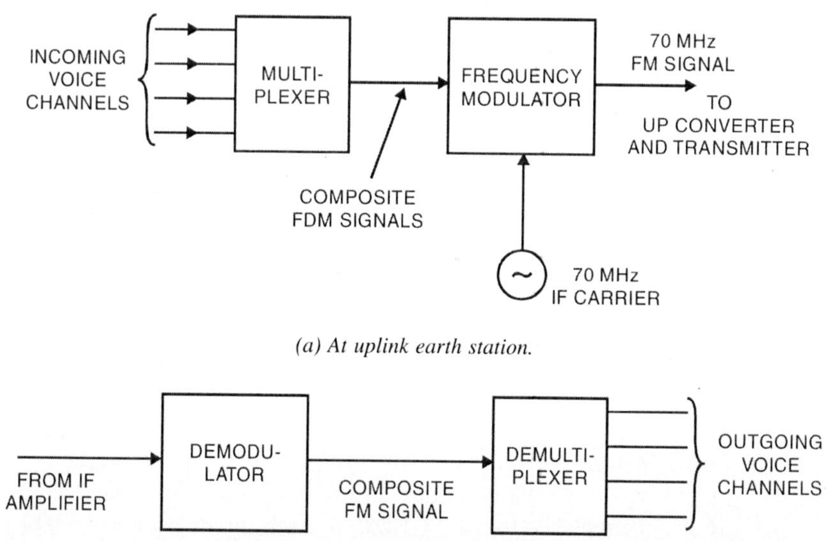

(a) At uplink earth station.

(b) At downlink earth station.

Fig. 6.11. Transmitting and receiving ends of a typical FDM system.

Since this communication is two way process but it is not essential that both the path should be via satellite. For one path under sea or terrestrial cable or terrestrial microwave links can be used and for other a satellite link. Figure 6.11 sketches a typical FDM system for uplink and downlink earth station.

The typical arrangement for a FDM/FDMA earth stations transmitter section and receiver section are shown in Fig. 6.11.

The transmitting section shown in Fig. 6.12 accepts baseband signals from the terrestrial interface and assembles these into FDM groups for different destinations. In this system each route has an allocated frequency and destinations are associated with RF frequencies at a transmitting station, so baseband channels must be translated to specific RF frequencies. This transmitter section uses a double frequency conversion with two IF frequencies 70 MHz and 770 MHz. The FDM signal consisting of 12 to 1872 telephone channels, is frequency modulated onto a 70 MHz IF carrier. The bandwidth of 70 MHz IF filter lies between 1.25-36 MHz and to compensate group delay in uplink, a variable group delay equalized may be included.

The 70 MHz IF carrier is than upconverted to a 770 MHz, IF band, where other carrier are added to form a composite FM/FDMA signals. Finally the 770 MHz broadband signal is finally upconverted to 6 GHz for amplification by the HPA.

The downlink arrangement for receiving portion of an earth station is shown in Fig. 6.13.

The downlink earth station contains the some equipment except that the downlink frequency converter. Carrier are amplified at 4 GHz in low noise amplifier and then it is down converted to 770 MHz IF by using appropriate local oscillator. Received 4 GHz carriers are separated by oscillator into individual 770 MHz receivers with 40 MHz bandwidth so that each receiver corresponds to one transponder. Another down conversion at 70 MHz allows accurate filtering and groups delay equalization and it is to be performed before demodulation. The bandwidth of IF stage depends upon the number of channels carried by the transponder and it is set between 1.25-36 MHz. After demodulation the voice channel signals are translated to baseband and band limited to 3.1 or 3.4 kHz and then the signal is passed to the terrestrial interface.

6.4.6. TDM Systems

TDM systems are time division and it is used to interlace digital signals into frames, which are transmitted sequentially through separate transponders on a satellite. For high speed data transmission terrestrial interconnections should be TDM. Figures 6.14 and 6.15 shows the equipments of an earth station using TDM/PSK/TDMA technique.

In case of FDM 70 MHz IF is used, while in TDM it must be replaced by a 140 MHz IF, when 120 Mbps dater is sent by 2 PSK using an 80 MHz bandwidth. The 770 MHz IF may be replaced by a 1.2 GHz IF in earth station using the 14/11 GHz band, where the RF bandwidth is 750 MHz. Signals are transferred by the terrestrial interface to four phase PSK modulators driven by 140 MHz IF carriers. The voice

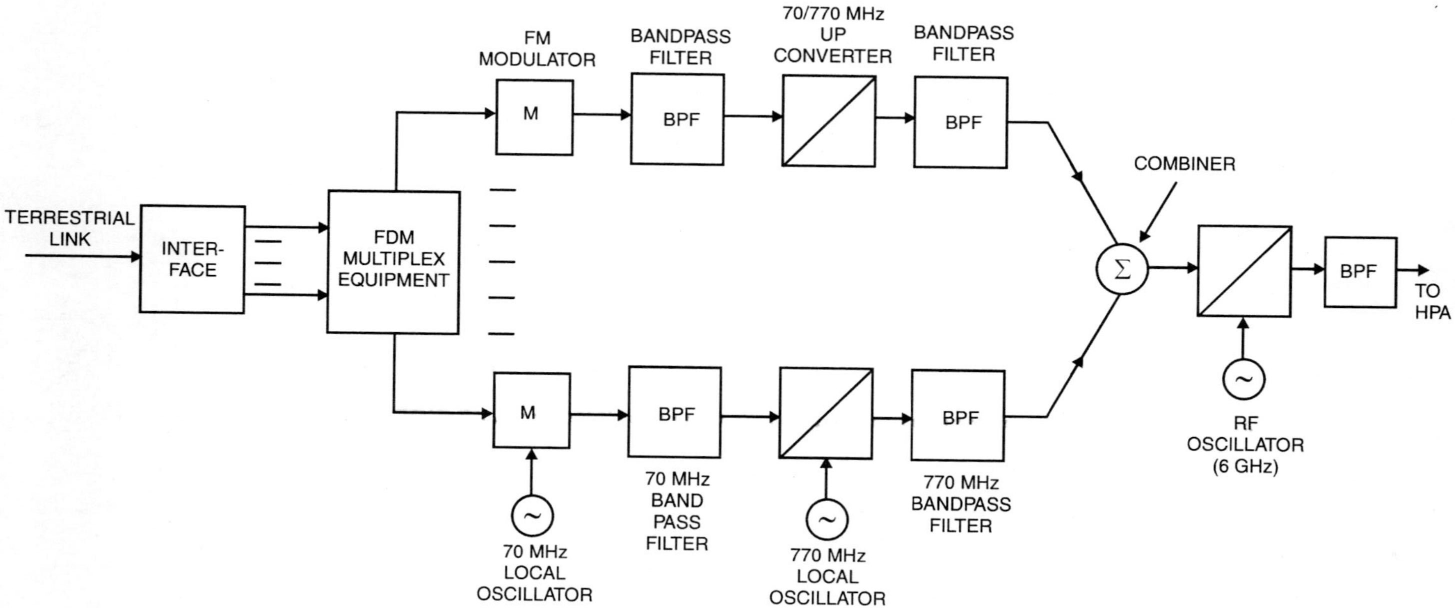

Fig. 6.12. Block diagram of an uplink equipment for transmitting portion of an FDM/FM/FDMA earth station.

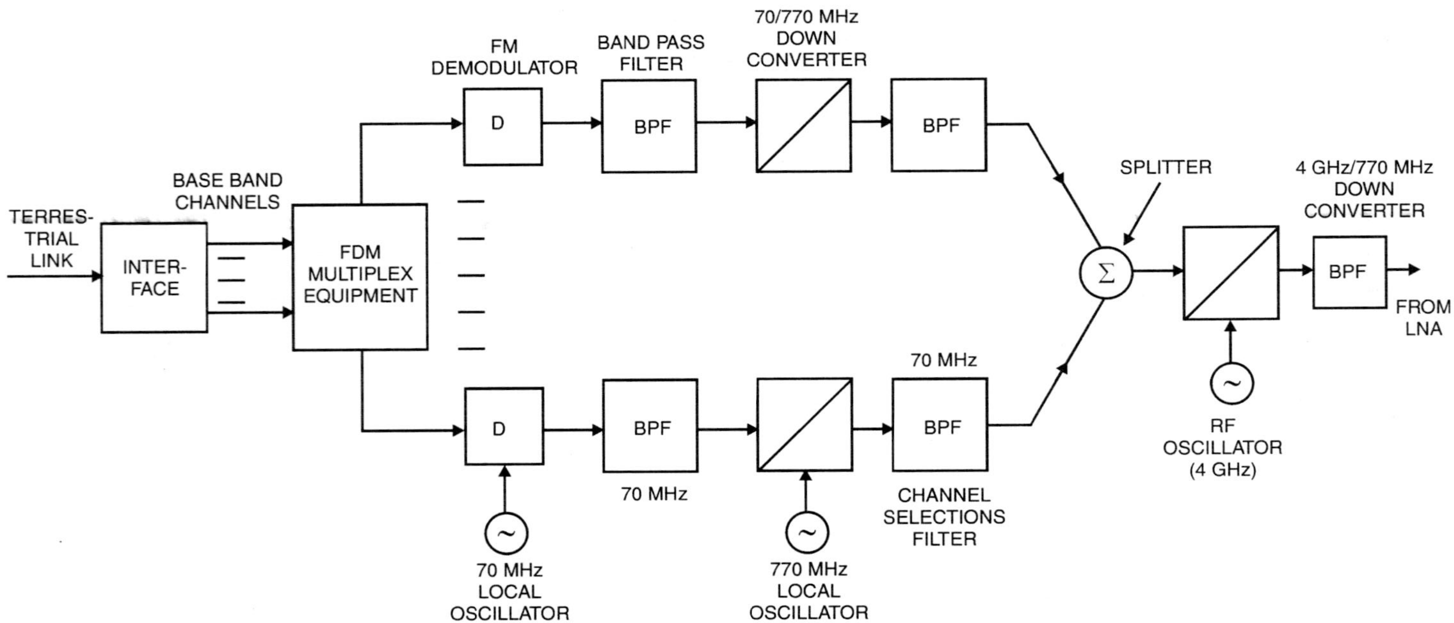

Fig. 6.13. Block diagram of an downlink equipment for the receiving portion of an FDM/FM/FDMA earth stations.

and data signals supplied by the interface are simply binary bit streams in TDM format. These data must be formatted into frames according to the system requirements, with one frame per transmission.

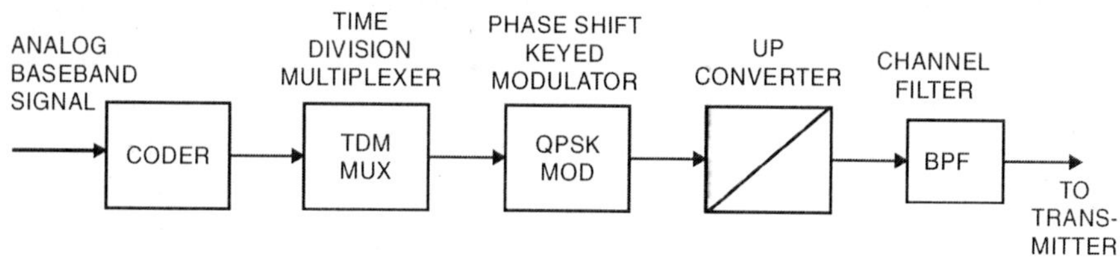

Fig. 6.14. Earth station transmitter equipments.

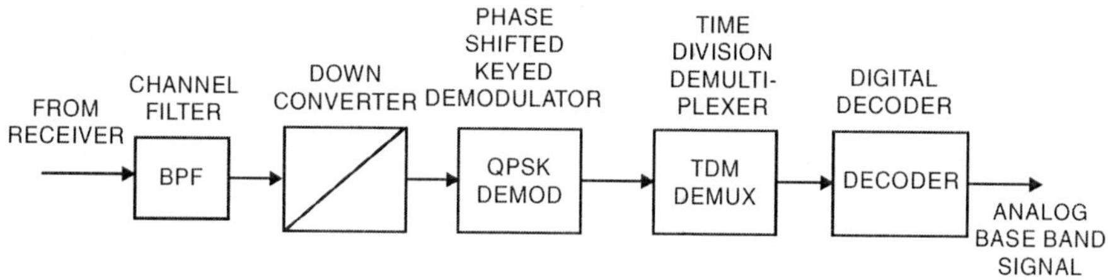

Fig. 6.15. Earth station receiver equipments.

Usually earth station transmits a burst of QPSK signal at the appropriate time because single station does not file a complete frame in a TDMA system. At the earth station accurate timing is required to ensure that when the burst of RF energy arrives at the satellite it is in exactly the correct position (in time) to interface between similar bursts arriving from other stations. So a synchronization and timing equipment is needed in the transmit portion of the GCE of a TDMA earth station.

Digital speech transmission efficiency can be increased by using digital speech interpolation (DSI). With DSI, any speech channels that is unused can be seized by another active channels for transmission of its data. On transmit and receive sides of GCE, DSI equipment is required to insert and separate the individual speech channels. A common mapping channels is transmitted along side DSI channel to inform the receive end of the position of each channel within the time frame sent by the transmitter.

6.5. TRACKING

Tracking consists of determination of the current orbit of the satellite. This is done by determining the range, which in turn determines the orbital elements. The beamwidth of a typical INTELSAT station in the transmit frequency band is only of the order of 0.15 degree. To realise the advantage of full gain of the antenna it is essential that the antenna beam is correctly pointed towards the satellite. The required pointing accuracy is of the order of ± 0.01 degrees. Even in the case of antennas with smaller diameters, the pointing accuracy required is of the order of ± 0.03 degree. A part from initial pointing the antenna has to keep a track of the satellite, since even a geostationary satellite makes certain excursions relative to the earth, because of the orbit inclination.

The absolute pointing accuracy will depend on many factors such as the alignment, wind velocity, structural rigidity etc. For tracking the satellite there are several methods available.

6.5.1. Manual Tracking

In manual tracking mode the antenna position is manually adjusted as per the orbit predictions or by looking at the signal quality. This kind of tracking is quite adequate for stations with fairly large beamwidths of the order of 0.4 degree and above, with satellites of low inclination.

6.5.2. Auto Tracking

Auto tracking involves generation of an error signal for each axis, in a monopulse or conical scan tracking feed and receiver system, and then feeding this error signal to the servo control system to null the error.

6.5.2.1. Monopulse

In amplitude monopulse system four beams are generated by using four feeds that are physically offset from the main beam. The beams intersect at the antenna satellite axis. The monopulse comparater produces the sum and difference signals. A null is obtained only for the signals arriving along the axis line. All signals arriving off axis produces an error signal which is used to drive the servo system.

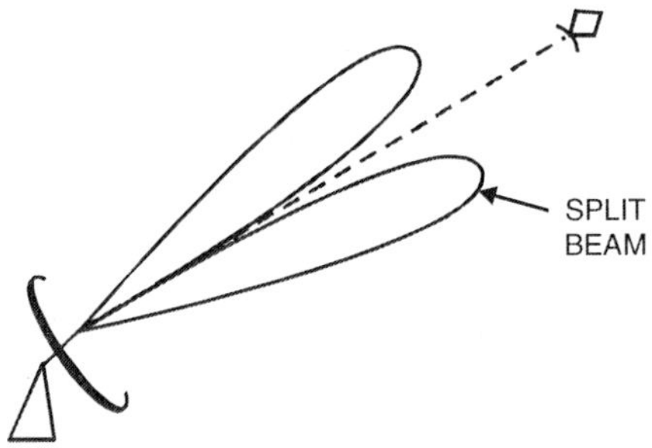

Fig. 6.16. Monopulse.

6.5.2.2. Conical Scan

The conical scan system uses a single feed, but the beam is rotated at a speed considerably faster than the expected satellite movement. This rotation produces an amplitude modulation on the incoming signal. The depth of modulation depends on how far the source is from the axis. The error signal thus derived is used to drive the antenna. The sense of the error is obtained by synchronously detecting the modulating envelope. Automatic tracking gives the minimum pointing error.

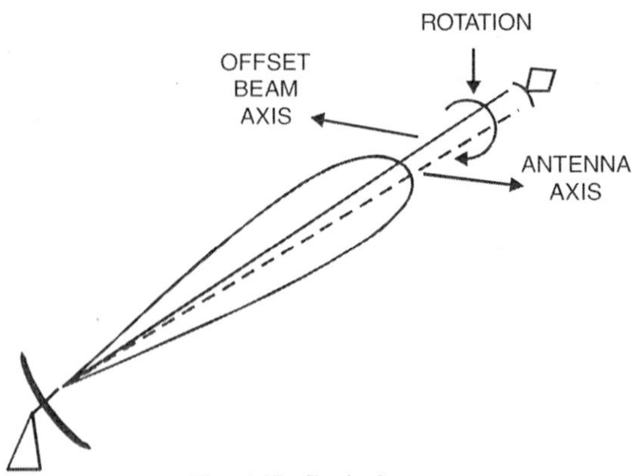

Fig. 6.17. Conical scan.

6.5.3. Programme Tracking

Programme tracking depends upon the accuracy of orbit predictions. In this system the antenna is controlled by a computer to which are fed the orbital parameters. The computer computes the antenna pointing angles and periodically moves the antenna. Jerky motions of the antenna is avoided by integrating the computer output and driving the servo system.

6.5.4. Step Tracking

In step tracking system the antenna is stepped intermittently in azimuth and elevation, the steps continue in the same direction if the received signal increases, but are automatically reverse if it falls. As a refinement the antenna can be made to describe a small close path. Over this path the signal may be integrated to produce an error signal to drive the servo. This is also known as **Hill climbing.**

While deciding upon a particular tracking system, the main point to keep in mind is the required tracking accuracy. INTELSAT requires that the EIRP (Effective isotropic radiated power) in the direction of the satellite should be maintained to within ±0.5 dB, except under adverse whether conditions. This limit includes power fluctuations resulting from antenna pointing errors. Inherently monopulse tracking system can give a better tracking accuracy than the step track system. However in many cases where the wind velocity is not excessive for most of the time, step track will be adequate, particularty as tracking errors can only result in signal reductions, not increases, and therefore can not adversely affect the transmission of other satellite users.

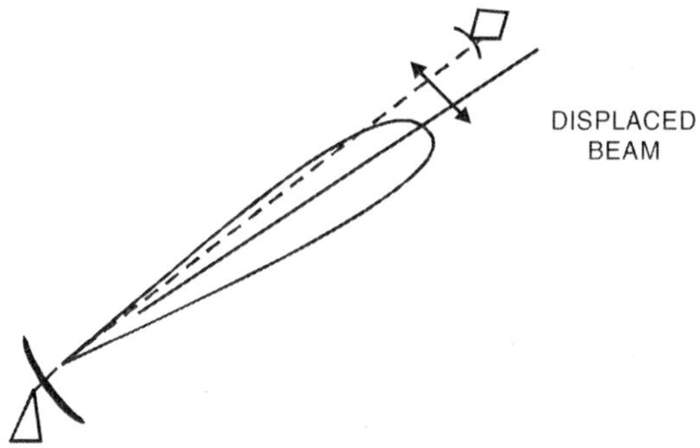

Fig. 6.18. Step track.

6.6. MONITORING AND CONTROL

In the previous sections the importance of low noise amplifier and high power amplifier has been mentioned. Redundancy means during the failure of on line equipment, the standby equipment has to switch over automatically and become the online equipment. This complete process is known as **monitoring and control**. In satellite communication, where several earth stations are involved, monitoring and control is quite important. This involves many levels such as RF terminal monitoring and control, remote monitoring and control of the earth station from the network control center via the satellite link. This system should be capable of collecting status data for classification, to convey status data to network operators, to switch over the redundant equipment, to initiate fault isolations for traffic assignment, antenna pointing and basically to maintain the equipment.

6.7. FREQUENCY COORDINATION

Satellite communication system and terrestrial microwave links share common frequency allocations in a number of bands. The major concern during the establishment of new earth station or terrestrial link, it should not interfere with the existing systems. This process is called **frequency coordination** and it has been specified by CCIR. It is particularly important for frequency bands of 4 and 6 GHz. The area around an earth station site within which interference with another link must be considered is called the **coordination area**. The curve binding this area is called **coordination contour** and the distance along any given azimuth from the earth station to the coordination contour is called the *coordination distance*.

All transmitting and receiving stations that shade the earth stations frequencies and lie within the coordination contour is identified and the probable interfering signal levels are calculated. If these are less than allowed maxima, the actual interference environment at the site may be measured using a transportable terminal.

While computer programs are available commercially to make interference calculations.

The process of calculation of coordinate distance is discussed in various CCIR reports, particularly 724-1 [33] and 382 [34]. The first step in the process is to determine the basic transmission loss L_b in decibels by the formula :

$$\boxed{L_b = P_T + G_T + G_R - A_S - P_R \text{ dB}} \quad \text{...(6.7)}$$

where, P_T = transmitted power in dB W

G_T and G_R = transmitted and receiver antenna gain in applopiate direction

P_R = maximum allowed received power from the interferer

A_S = site shielding factor.

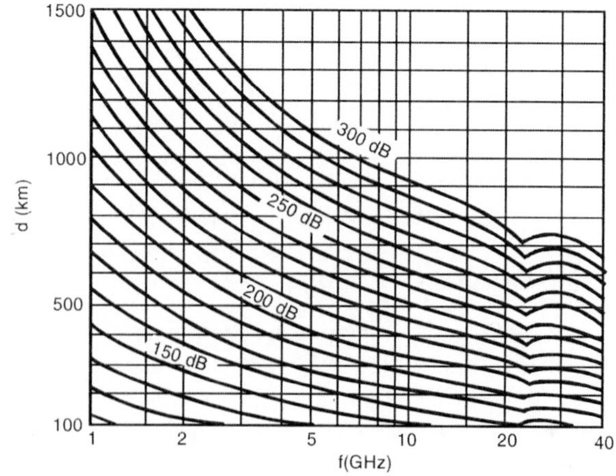

Fig. 6.19. CCIR for coordination distance (Mode A propagation).

The second step in the process involves the determination of distance that corresponds to L_b. The curves and equations to calculate this distance have been provided by *CCIR*. Basically there are two modes of propagation, mode *A* and mode *B*. Mode *A* corresponds to great circle path mechanisms and mode *B* corresponds to scattering by rain. In the Fig. 6.19 the *CCIR* curves for mode *A* display coordination distance versus frequency and $L_b - A_s$. A_s is calculated from the frequency *f* in GHz and the devation angle θ.

$$A_s = 20 \log_{10}\left[1 + 4.5\theta f^{0.5}\right] + \theta f^{0.33}$$

for $\theta \geq 0°$

for $\theta < 0°$ ...(6.8)

where θ is in degree.

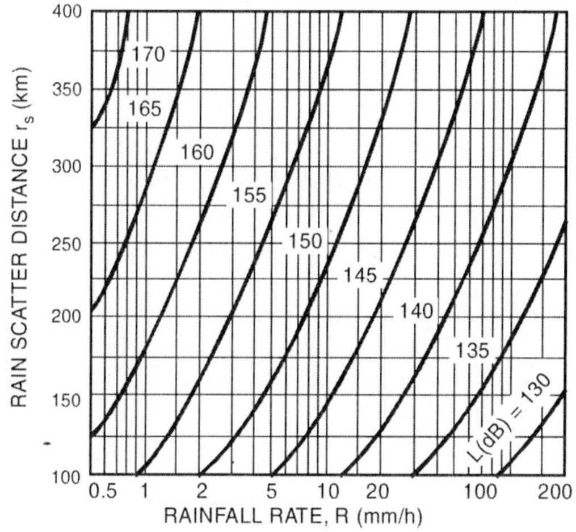

Fig. 6.20. Rain scatter distance as a function of basic transmission on loss and rainfall rate at a frequency of 4 GHz.

Site shielding is not a factor in rain scatter interference and the CCIR curves for rain scatter coordination distance involve rainfall rate and frequency directly.

Summary

1. The major subsystems of an earth stations are transmitter, receiver, antenna and tracking equipment alongwith terrestrial interface equipments and power supply.
2. There are three categories of earth station : transmit only, receive only and transmit and receive.
3. In satellite communication system, earth stations are characterized by their *G*/*T* ratio, which combines antenna gain with system noise temperature.
4. The cost of earth station is high because it carrying many telephone signals, high bit rate data links or several television channels simultaneously.
5. Small antennas often use a front fed configuration with scalar feed. In large antennas the gain is maximized by using the shaped cassegrain design.
6. Because of the narrow beams of large antennas, they are equipped with facilities to automatically track movement of a satellite.
7. It is necessary that earth stations do not suffer interference from other communication systems operating in the same frequency band.

8. Tracking may be achieved by either of three methods, namely monopulse or multihorn tracking, conical scanning and step by step technique.
9. In mobile and transport services small earth stations are used which are particularly portable or transportable kind.
10. The efficiency of earth station depends as to how efficiently it points the antenna beam to the satellite both in the transmit and receive mode.
11. Large antennas are expensive to construct and install. Fully steerable antennas achieves the highest possible aperture efficiency and lowest possible noise temperature.
12. In FDM system many voice and data signals can be transmitted and it is being done by allocating separate frequencies to each signal, either at baseband or RF or both.
13. In case of linearity, solid state amplifiers are superior to travelling wave tube amplifiers.
14. The efficiency of digital speech transmission can be increased by using digital speech interpolation (DSI).
15. Satellite communication system and terrestrial microwave links share common frequency allocation in a number of bands. This process is known as frequency coordination.
16. Earth station is very essential part of satellite communication which transmits to and receives from satellite. Each station may be located on a ship at sea or on an aircraft.
17. The realiability requirement of an earth station is quite high i.e. about 99.99% and this can be achieved if adequate redundancy is provided in the system.
18. Earth station receivers are extremely sensitive and hence are higher susceptible to interference from other system.
19. The frequency stability of the system depends on both the modulator frequency stability and, the up and down converter frequency stability.
20. Frequency stability can be classified as
    - Short term stability
    - Long term stability.
21. An antenna is a transducer between electromagnetic waves in space and voltage or currents in a transmission lines.
22. All passive antennas can transmit and receive called reciprocity allows a passive antenna's transmitting properties to be derived from its receiving characteristics and vice versa.
23. The direction of maximum transmission or reception is known as the boresight direction.
24. Practically radiation patterns are drawn to show relative radiated power in two orthogonal planes, horizontal plane containing the antenna and a vertical plane passing through the antenna.
25. The tendency of the antenna to concentrate its radiated power is known as gain. Gain is proportional to the physical area of an antenna expressed in square wavelengths.
26. The angular width of the main beam measured between the half power points is known as the half power beam width (HPBW).
27. The desired polarization radiated by an antenna is termed as co-polarization. The orthogonal polarization is known as cross polarization.
28. The received C/N is given as

$$\frac{C}{N} = \frac{G}{T} - 20 \text{ dB}$$

29. The cost of fully steerable antenna is given as

$$\text{cost} = Y(D)^{2.7}$$

30. The $\frac{F}{D}$ and θ_0 of an earth station antenna are related by

$$\tan \theta_0 = \frac{1}{2}\left[\frac{\left(\frac{D}{F}\right)}{1-\left(\frac{D^2}{16F^2}\right)}\right]$$

31. Blockage of the aperture by the feed and its supports can be avoided in an offset fed antenna.
32. High power amplifiers are used in earth station transmitters because they require transmission of microwave power.
33. Duplexers are used to enhance the separation in the frequency domain as they are different to the uplink and downlink band.
34. Large earth stations using FDM/FDMA operation may carry several thousands 4 kHz voice channels.
35. The non-linearity may be represented in terms of the amplifier power transfer characteristics. It also shows the saturation and the saturation power of the amplifier.
36. Tracking consists of determination of the current orbit of the satellite. This is done by determining the range.
37. The absolute pointing accuracy will depend on many factors such as the alignment, wind velocity, structural rigidity etc.
38. Basically there are four types of tracking techniques :
    - Manual tracking
    - Auto tracking
    - Programme tracking
    - Step tracking
39. In manual tracking mode the antenna position is manually adjusted as per the orbit predictions or by looking at the signal.
40. Automatic tracking gives the minimum point error.
41. The area around an earth station sit within which interference with another link must be considered is known as the coordination area.
42. The curve binding the coordination area is known as coordination contour.

Review Questions

1. What are the equipments that an earth station requires ? Discuss their design requirements also.
2. Explain, why the low noise amplifier in a satellite receiving system is placed at the antenna end of the feeder cable ?
3. Explain, what is meant by the term redundant earth station ?
4. Explain with the block diagram the working of transmitter part of earth station. Compare the main characteristics of HPA's being used here.
5. Discuss the antenna requirements for large and small earth stations.

6. Describe briefly the offset feed used with paraboloidal reflector antennas, stating its main advantages and disadvantages.
7. Explain, why double reflector antennas are often used with large earth stations ?
8. Describe briefly the main advantages to be gained in using an antenna array.
9. What is meant by tracking and pointing ? Explain its significance also.
10. Explain the monitoring and control requirements for earth station equipments. Discuss its importance also.
11. What is the need of frequency coordination among earth stations ? Discuss the technique to achieve then.
12. What do you understand by LNA ? Why is GaAs FET suitable as LNA in microwave systems ?
13. Cassegrain antenna is preferred for large earth stations. Why ?
14. Discuss the non-linear behaviour of the high power amplifier. How is it being removed.
15. Discuss the classification of an earth station on the basis of the variety of equipment required.
16. What do you understand by redundancy ? Why is it required at an earth station ?
17. What is tracking ? Briefly discuss its types also.
18. What do you meant by frequency stability ? Explain the factors on which it depends.
19. Derive an expression of carrier to noise (C/N) ratio for a large earth station antenna.
20. Why cassegrain antenna is preferred for large earth station antenna ? Explain it.
21. Write an expression which defines the cost of a fully steerable antenna in terms of its diameter.
22. Discuss the geometry of front feed paraboloidal reflector antenna and offset fed antenna.
23. Draw and discuss the block diagram of an earth station.
24. Illustrate the non-linear behaviour in high power amplifiers.
25. What do you understand by FDM system ? Discuss the uplink and downlink arrangements of an earth station.
26. Compare the advantages and disadvantages of FDM and TDM systems.
27. What do you meant by Digital speech interpolation (DSI) ? Why is it required ?
28. What do you understand by monitoring and control ?
29. What is hill climbing ?
30. Define the following :
    (*a*) Redundancy
    (*b*) Boresight direction
    (*c*) Half power beamwidth
    (*d*) Blockage loss
    (*e*) Antenna gain
    (*f*) Auto tracking
    (*g*) Step tracking
    (*h*) Coordination distance
    (*i*) Coordination contour.

Objective Type Questions

1. Satellites allow earth stations to share the satellite equipment through
    (*a*) Time division multiplexing (TDM)
    (*b*) Frequency division multiplexing (FDM)
    (*c*) Multiple frequency UP and DOWN links

(*d*) Focussed spot beams and zone beams

(*e*) All of above.

2. Higher the frequency of satellite or earth station transmitter

(*a*) Higher is the potential bandwidth that can be used for communications

(*b*) Closer the satellites can be spaced without interfering with each other.

(*c*) Narrow is the radiation beamwidth for a given earth station and satellite antenna diameter

(*d*) All of above

(*e*) None of these.

3. For the earth station antennas to be 6 feet in diameter, the satellite frequency bands must be in

(*a*) $\frac{4}{6}$ GHz range (*b*) $\frac{12}{14}$ GHz range

(*c*) $\frac{20}{30}$ GHz range (*d*) Both (*a*) and (*b*).

4. Earth station figure of merit is defined as

(*a*) $10 \log\left(\frac{G}{T}\right)$ (*b*) $10 \log (GT)$

(*c*) $20 \log\left(\frac{G}{T}\right)$ (*d*) None of these.

5. Which of the following are useful in comparing the noise performance of receivers

(*i*) input noise voltage (*ii*) noise figure

(*iii*) noise resistance (*iv*) noise temperature

Select answer as per the following codes

(*a*) (*i*), (*ii*) & (*iii*) (*b*) (*i*), (*iii*) & (*iv*)

(*c*) (*ii*), (*iii*) & (*iv*) (*d*) (*i*), (*ii*) & (*iv*).

6. The maximum range of transmitter depends on

(*a*) Its frequency

(*b*) Its power

(*c*) Both frequency and power

(*d*) None of the above.

7. A helical antenna is used for satellite tracking due to its

(*a*) Good front to back ratio (*b*) Broad bandwidth

(*c*) Circular polarization (*d*) Both (*a*) and (*b*)

8. The amplifier contributing maximum noise in a receiver is

(*a*) LNA (*b*) IF

(*c*) Video (*d*) Audio

9. Cost of a fully steerable antenna is given by

(*a*) Cost = $Y(D)^{2.7}$ (*b*) Cost = $Y(D)^{1/2.7}$

(*c*) Cost = $Y(D)^{3.7}$ (*d*) Cost = $Y(D)^{1/3.7}$

10. Step tracking is also known as

(*a*) Hill climbing (*b*) Programme control

(*c*) Manual recording (*d*) None of these

ANSWERS

1. (*e*)	**2.** (*d*)	**3.** (*b*)	**4.** (*a*)	**5.** (*d*)
6. (*b*)	**7.** (*d*)	**8.** (*a*)	**9.** (*a*)	**10.** (*a*)

□□□

CHAPTER 7

INTRODUCTION TO VARIOUS SATELLITE SYSTEMS

INSIDE THIS CHAPTER

7.1. INTRODUCTION

In the last chapters we have studied different aspects of satellite and communication through satellite. Generally the satellite can provide variety of services but broadly they are categorised as :

Geostationary Satellites

- Direct Broadcast Satellites (DBS)
- Very Small Aperture Terminals (VSATs)
- Mobile Satellite Services (MSAT)

In the field of satellite communication applications, satellite television, telephone and data communication are the major application areas. Satellite can be used as navigational aids by vehicles on land, in the air and on the sea. Some of the metrological applications like wheather forecasting, flood forecast etc. are also possible by the use of satellite.

In this chapter, we shall discuss the different applications of satellites.

7.2. VERY SMALL APERTURE TERMINAL (VSAT)

VSAT stands for very small aperture terminal. In satellite communication, the development and technology has allowed, data and voice communication cost effective for a broader range of users. In VSAT the earth station antenna size is typically less than 2.4 m in diameter. The trend is toward even smaller dishes not more than 1.5 m in diameter.

7.2.1. VSAT Structure

The basic structure of a VSAT network is shown in Fig. 7.1. A VSAT consists of two parts, a **transceiver** that is placed outdoors in direct line of sight to the satellite and a **device** that is placed indoors to interface the transceiver with the end communications device, such as a PC. The transceiver receives or sends a signal to a satellite transponder in the sky. The satellite sends and receives signals from a ground station computer that acts as a hub for the system.

Each end user is interconnected with the hub station via the satellite, forming a star topology. The hub controls the entire operation of the network. For one end user to communicate with another, each transmission has to first go to the hub station that then retransmits it via the satellite to the other end user's VSAT.

It consists of a hub station which provides a broadcast facility to all the VSATs in the network and the VSATs themselves which access the satellite in some form of multiple access mode.

The hub station is operated by the service provider and it may be shared among a number of users. Time division multiplex is the down link mode of transmission from hub to VSAT, and the transmission can be broadcast for reception by all the VSATs in a network, or address coding can be used to direct messages to selected VSATs.

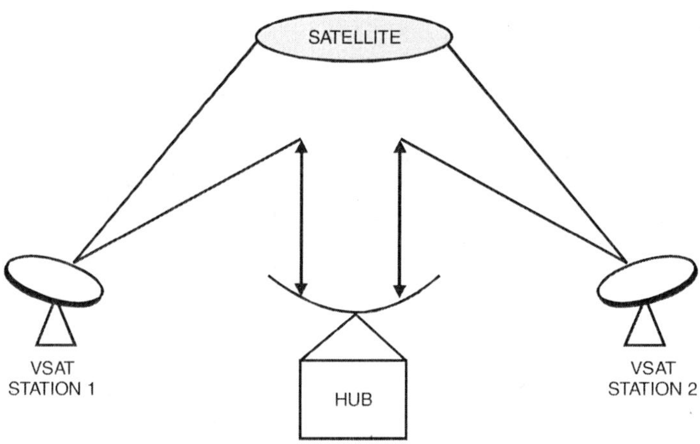

Fig. 7.1. VSAT Network.

7.2.2. VSAT Network Architecture

VSAT systems can be implemented in three basic architectures.

- One-way implementation.
- Split-two-way implementation.
- Two way implementation.

7.2.2.1. One-Way Implementation

This is the mode of a satellite used in the broadcast satellite service (BSS). The introduction of digital telephony allows the provider and user much greater flexibility in the operation of a broadcast network. Figure 7.2 shows the schematic of a BSS.

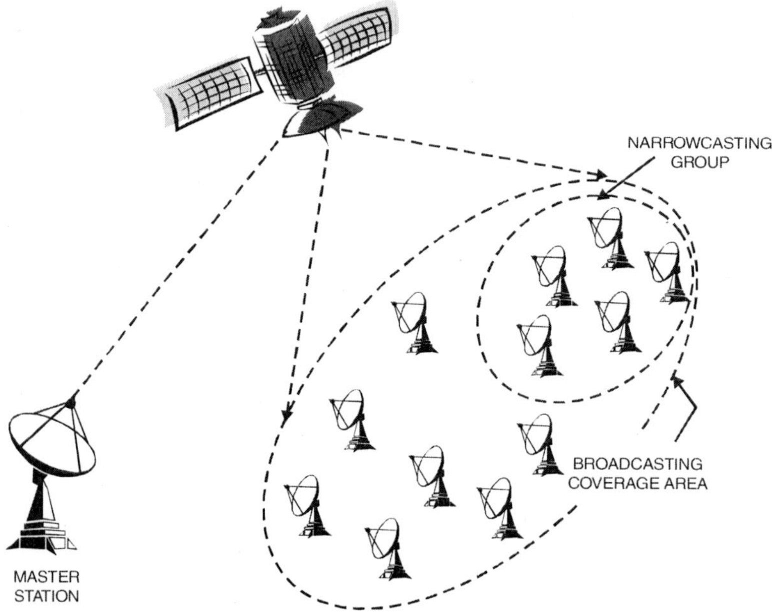

Fig. 7.2.

The Fig. 7.2 above shows a schematic of one-way implementation.

7.2.2.2. Split-Two-Way Implementation

This implementation is used when there is no normal return channel (*E.g.*, with Ku band broadcast satellite service systems that carry internet traffic). The relatively

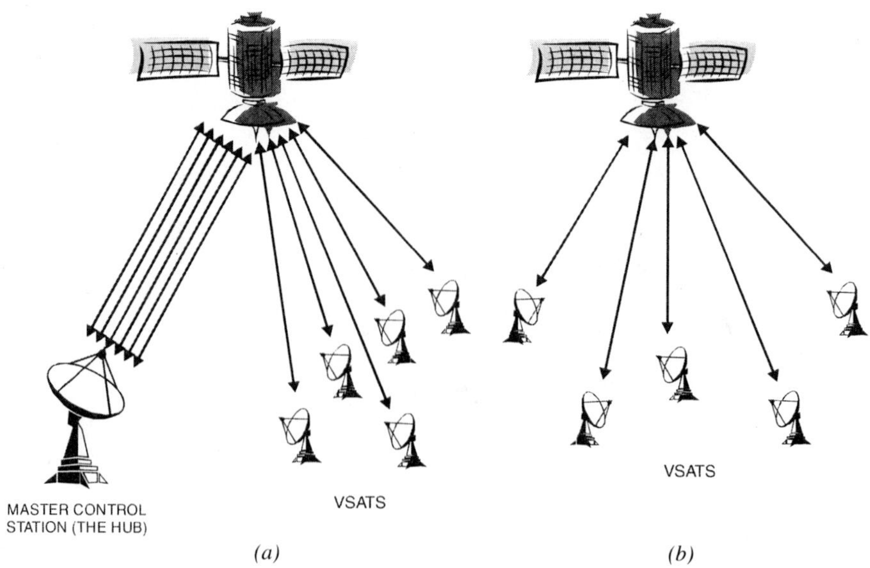

Fig. 7.3

high capacity downlink stream is not complemented by an uplink capability from the user terminal. If the BSS downlink is used as the download channel from an internet service provider, the only option the user has for a return link is via another telecommunications channel, such as a standard telephone line. The Internet Protocol (IP) is therefore split between a satellite downlink (outbound) channel and a terrestrial telephone (inbound) channel: hence the term split IP for this implementation. The advantage of this approach is that the VSAT terminal does not require a transmit capability, which significantly reduces its cost and complexity. The disadvantage is that the telephone line connection must usually be through a modem, with a bit rate generally restricted to 56 kbps.

7.2.2.3. Two Way Implementation

The service includes a return link so as to establish two-way communication over the same satellite, from the hub to the user and from the user back to the hub. Two topologies are implemented: Star topology and Mesh topology.

Star Topology

Here, all of the traffic is routed via the master control station, or hub. If a VSAT wishes to communicate with another VSAT, they have to go via the hub, thus necessitating a "double hop" link via the satellite. Since all of the traffic radiates at one time or another from the hub, this architecture is referred to as Star network.

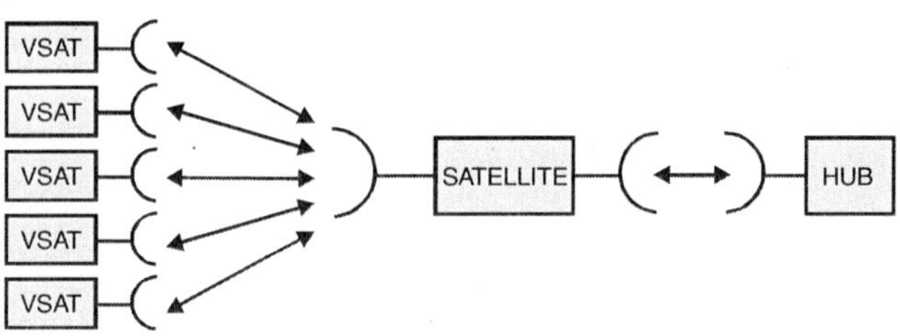

Fig. 7.4. Star topology

Disadvantage: (1) Communication between two VSATs requires a double hop thus increasing the time delay. (2) Since the low receive G/T of the VSATs, coupled with their limited transmit EIRP is compensated for by using a large hub with high G/T and EIRP, the cost of the hub used is very high.

Mesh Topology

Here, each of the VSATs has the capability to directly communicate with any of

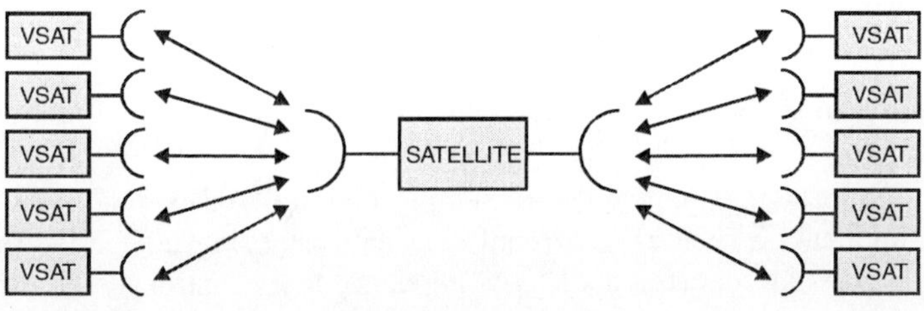

Fig. 7.5. Mesh topology

the other VSATs. The network control and the duties of the hub is either handled by one of the VSATs or the master control station functions are shared among the VSATs. Since the traffic can go to or from any of the VSATs, this architecture is referred to as Mesh network.

Figures 7.4 and 7.5 illustrate these two topologies.

7.2.3. Specifications

Some typical **specifications of VSAT** are :

- Antenna size : 1.2 m to 3.5 m
- Antenna gain : 40 dB
- Power uplink : 0.5–5 W
- G/T : 20 dB/°K

Typical **specifications of Hub stations** are :

- Antenna size : 3.5–10 m
- Antenna gain : 58 dB
- Power uplink : 50–200 W
- G/T : 30 dB/°K

Most VSAT systems operate in the Ku band (12-14 GHz) because it is less congested although there are some C-band systems in existance. For fixed area coverage by the satellite beam, the system performance is essentially independent of the carrier frequency. For fixed area coverage, the beamwidth and hence the ratio λ/D is a constant. The satellite antenna gain is therefore constant and for a given high power amplifier output, the satellite EIRP remains constant.

7.2.4. Applications of VSAT

Typical Applications of VSAT are :

- Computer communication
- Reservation system
- Database enquiries
- Billing system
- File transfer
- E-mail
- Video conferencing
- Stock control and management
- Credit card verification
- Point of sale transaction

Figure 7.6 shows some of the application areas of VSAT and data broadcast satellite.

- Figure 7.6(*a*) indicates an example of intelligent service which has an application such as in a retail data network using interactive transaction processing for credit cord verification, inventory control and broadcast video for branch store training, sales and product promotions.

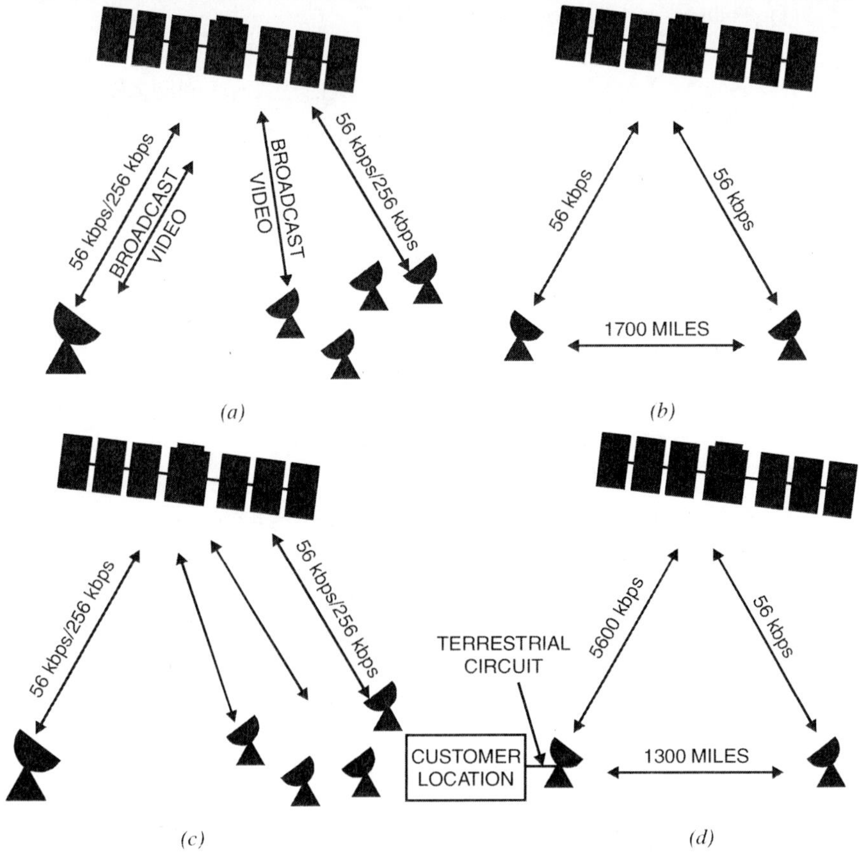

Fig. 7.6. Some application services of VSAT and DBS.

- Figure 7.6(*b*) indicates an example of 56 Kbps transmission between two customer locations. An application area is such as that of transfer and updates of data files between a company's corporate headquaters and one of its district offices.
- Figure 7.6(*c*) also indicates an example of intellegent service. Typical application area is that of a corporate data network using interactive transaction processing for inventory, payroll and other adminstrative requirements.
- Figure 7.6(*d*) indicates the service of 56 Kbps between customer promise and the company's national network is shown. Typical application is data transfers between a company's MIS center and corporate headquarters.

The basic requirements of such VSAT network require the use of low power transmitter with small antenna but relatively high EIRP form the satellite. The VSAT network is generally operated in a frequency division multiple access (FDMA) mode with several narrow bandwidth carriers. Because of the bursty and diverse nature of VSAT data traffic multiple access schemes such as FDMA and TDMA do not utilize the available satellite channel effeciently. Dynamic accessing may be useful but pocket oriented multiple access schemes are most suitable. These **schemes may be classified into two categories.**

(1) Contention or random access schemes.

(2) Reservation schemes.

The contention schemes suitable for use in VSAT based on the ALOHA concept of which there are three variations :

(1) pure ALOHA,

(2) slotted ALOHA and

(3) reservation ALOHA.

The major shortcomings of present day VSAT systems are :

- the high initial costs
- the tendency towards optimizing systems for large networks (typically more than 500 VSATs),
- the lack of direct VSAT to VSAT links.

Most of the shortcomings of VSAT system can be overcome by the use of microwave technology and digital signal processing.

7.2.5. Advantages and Disadvantages of VSAT

Advantages

1. Cost insensitive to distance.
2. Easy for installation and maintenance.
3. Flexibility.
4. Upgradeable.
5. Large number of VSAT networks can be integrated into the single network to provide cost effective expansion.
6. Mean time between failure

   (MTBF) > 25,000 hours (3 years)
7. Reliable service : 99.9% for data and 99.5% + for voice.

Disadvantages

1. High initial cost (hubs and basic elements must be place before the services can be provided).
2. Lack of adequate coverage in some parts of the world.
3. Some service quantity limitations such the high signal delays (latency).
4. Unavailability of direct links between various VSATs.
5. Rain attenuation might affect the performance of VSAT communications under rainy conditions.

7.3. RADARSET

Radarset is an earth resrouces remote sensing satellite which is a part of the Canadian space program. According to the condition space agency, the objective of the radarset program is to provide applications benefits for resource management, make SAR data available for research, establish a control facility etc.

Raderset is planned to fly in a low earth near circular orbit. Radarset flies at an altitude of 798 km, indication 98.6° and with a time period of 100.7 min. Radarset carries only C-band radar. C-band radar is not the optimal sensing mechanism for all the applications, the rationale for selecting it is that it does penetrate cloud cover, smoke, haze and it does not operate in darkness. Most of the sensing is required at high lattitudes, where solar illumination of the earth can be poor and where there can be persistent colour cover.

Radarset is a comparatively large spacecraft, the total mass in the orbit is about 3100 kg. The radar works at a carrier frequency of 5.3 GHz, which can be modulated with three different pulse width, depending on resolution requirements. The satellite completes 14 + 7/24 revolutions per day.

The **applications of radarset** are :

- Crop monitoring.
- Forest management.
- Shipping and fisheries.
- Ocean feature mapping.
- Oil pullution monitoring.
- Geological mapping.
- Topogrophic mapping.
- Land use mapping.
- Sea ice mapping.
- Iceberg detection.

7.4. MOBILE SATELLITE COMMUNICATION SYSTEM (MSAT)

In recent years of satellite communication, which have been gaining major thrust are ultra small aperture terminal mobile satellite (MSAT) communication systems. MSAT terminals can be called as VSAT-on-wheels, have several features common with VSATs. MSAT terminals give the communication services to the moving vehicles.

In such mobile satellite communication system, the direct service from the satelite to mobile phone requires one or more satellite in geosynohronous orbit. Satellites should be operated in the 800 MHz range with high effective radiated power. Efficient narrowband power modulation techniques are also required. The mobile units require front ends with low noise receivers and antennas with modest gain. Direct communication links can be established between a satellite and a velocity of stations : mobile units, portable units, gate way earth stations, base earth stations and a control earth station.

Figure 7.7 shows the network structure of MSAT.

Basically **this network structure is capable to accommodate four kinds of communications :**

- Mobile to mobile.
- Mobile to dispatcher.
- Mobile to public switches telephone network.
- Satellite and network control.

The network is capable of serving many mobile stations, a number of base earth station, gateway earth station and control earth station. For MSAT, satellite should provide sufficient radiated power in the direction of coverage area. Radiated powre depends the available power for the satellite and the gain of the satellite antenna. **Mobile stations use two types of antennas.**

- Steered antennas.
- Non-steered antennas.

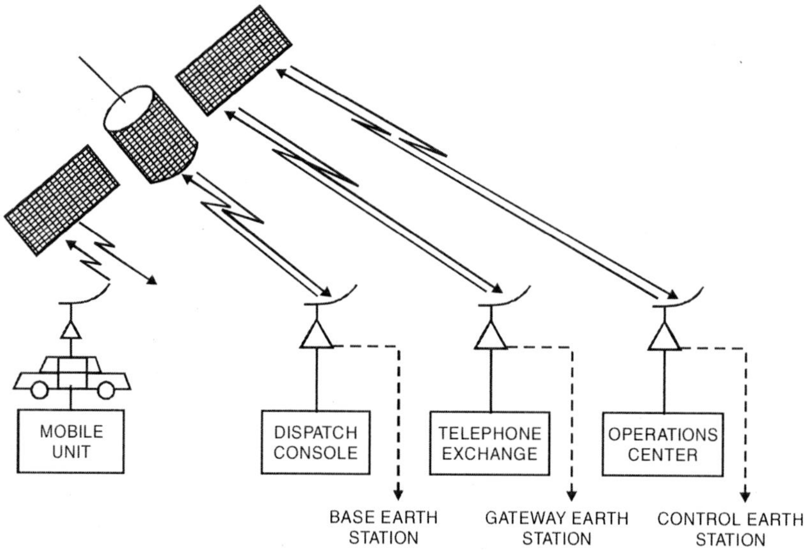

Fig. 7.7. Network structure of MSAT.

Steered antennas are directional w.r.t. azimeth with a gain more than 8 dB. Figure 7.8 shows the possible structure of stered antennas.

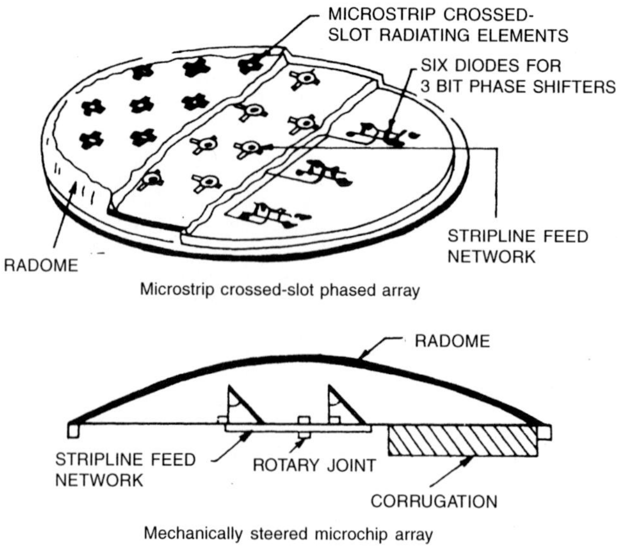

Fig. 7.8. Steered antennas.

The non-steered antennas are omnidirectional w.r.t. azimuth, with a gain of 4 to 8 dB. Some possible structure of non-steered antennas are shown in Fig. 7.9.

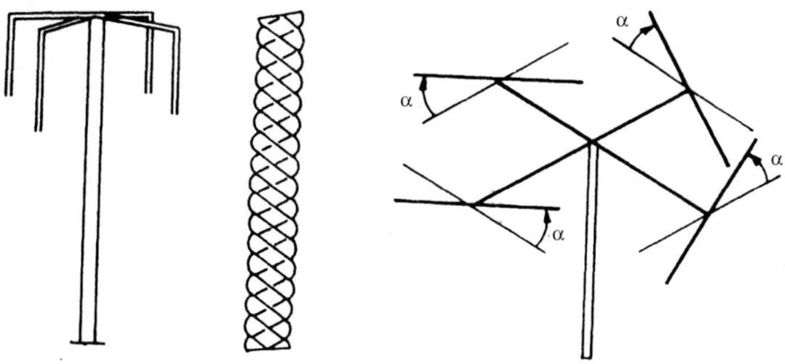

Fig. 7.9. Non-steered antennas.

7.5. INTELSAT

The International Telecommunications Satellite Organization (ITSO) is the world's largest commercial satellite communications services provider. It is an international consortium that owns and manages a constellation of communications satellites (Intelsats) to provide international broadcast services. Ownership and investment in ITSO (measured in shares) is distributed among ITSO members according to their respective use of services. Investment shares determine each member's percentage of the total contribution needed to finance capital expenditures. The organization's primary source of revenue comes from satellite usage fees which, after deduction of operating costs, are redistributed to ITSO members in proportion to their shares as repayment of capital and compensation for use of capital. Satellite services are available to any organization (both ITSO members and non-members), and all users pay the same rates.

The consortium began on August 20, 1964 as the International Telecommunications Satellite Consortium with 11 participating countries. In 1973, the name was changed and there were 80 signatories. ITSO currently has over 100 members and provides service to over 600 Earth stations in more than 149 countries, territories and dependencies. Intelsat maintains it headquarters in Washington, DC.

Spacecraft operations are controlled through ground stations in Fucino (Italy), Clarksburg (Maryland, USA), Beijing (PRC), Raisting (Germany), Perth (Australia), and Paumalu (Hawaii, USA).

Since its inception, ITSO has used several versions (blocks) of its dedicated Intelsat satellites. ITSO competes each block of spacecraft independently, leading to a variety of contractors over the years. Summary sheets are provided for each block.

7.5.1. History

The consortium began on August 20, 1964, with 11 participating countries. On April 6, 1965, Intelsat's first satellite, the Intelsat I (nicknamed *Early Bird*), was placed in geostationary orbit above the Atlantic Ocean by a Delta D rocket. On July 18, 2001, Intelsat became a private company, 37 years after being formed. In the period prior to

Intelsat's privatization in 2001, ownership and investment in INTELSAT (measured in shares) was distributed among INTELSAT members according to their respective use of services. Investment shares determined each member's percentage of the total contribution needed to finance capital expenditures. The organization's primary source of revenue came from satellite usage fees which, after deduction of operating costs, was redistributed to INTELSAT members in proportion to their shares as repayment of capital and compensation for use of capital. Satellite services were available to any organization (both INTELSAT members and non-members), and all users paid the same rates.

In 1973, the name was changed and there were 80 signatories. Intelsat provides service to over 600 Earth stations in more than 149 countries, territories and dependencies. By 2001, INTELSAT had over 100 members. It was also in this year when INTELSAT privatized and its name changed to Intelsat.

Since its inception, Intelsat has used several versions (blocks) of its dedicated Intelsat satellites. INTELSAT competes each block of spacecraft independently, leading to a variety of contractors over the years. Intelsat's largest spacecraft supplier is Space Systems/Loral, having built 31 spacecraft (as of 2003), or nearly half of the fleet.

Intelsat logo from 1998 to 2006

The network in its early years was not as robust as it is now. A failure of the Atlantic satellite in the spring of 1969 threatened to stop the *Apollo 11* mission; a replacement satellite went into a bad orbit and could not be recovered in time; NASA had to resort to using undersea cable telephone circuits to bring Apollo's communications to NASA during the mission. Fortunately, during the Apollo 11 moonwalk, the moon was over the Pacific Ocean, and so other antennas were used, as well as INTELSAT III, which was in geostationary orbit of the Pacific.

Today, the number of Intelsat satellites, as well as ocean-spanning fibre-optic lines, allows rapid rerouting of traffic when one satellite fails. Modern satellites also are themselves more robust, lasting longer with much larger capacity.

7.5.2. Current Operation

Intelsat was sold for U.S. $3.1bn in January 2005 to four private equity firms: Madison Dearborn Partners, Apax Partners, Permira and Apollo Management. The company acquired PanAmSat on July 3, 2006, and is now the world's largest provider of fixed satellite services, operating a fleet of 52 satellites in prime orbital locations. Intelsat maintains its corporate headquarters in Bermuda, with a majority of staff and satellite functions — administrative headquarters — located at the Intelsat Corporation offices in Washington, DC. A highly international business, Intelsat sources the majority of its revenue from non-U.S. located customers.

Spacecraft operations are controlled through ground stations in Clarksburg, Maryland (USA), Hagerstown, Maryland (USA), Riverside, California (USA), and Fuchsstadt, Germany.

Intelsat was operating Intelsat Americas-7 (known formerly as Telstar 7 and now known as Galaxy 27) which experienced a several-day power failure on 29 November 2004.The satellite returned to service with reduced capacity.

7.6. INMARSAT

Inmarsat is an international telecommunications company founded in 1979, originally as an intergovernmental organization. It operates a fleet of eleven (as of 2005) geosynchronous telecommunications satellites. The company is publicly traded on the London Stock Exchange (LSE: **ISAT**) as of June 2005. It is a constituent of the FTSE 250 Index.

Inmarsat provides telephony and data services to users world-wide, via special digital radios called "terminals". An Inmarsat terminal contacts the satellite and communicates to a ground station through the satellite. It provides reliable communications services to a range of governments, aid agencies, media outlets and businesses needing to communicate in remote regions or where there is no reliable terrestrial network.

The INMARSAT satellites cover the four large region of the globe :

- The Atlantic ocean region, east (AOR-E)
- The Atlantic Ocean region, west (AOR-W)
- The Indian ocean region (IOR)
- The Pacific ocean region (POR)

Figure 7.10 shows the geographical location of INMARSAT satellites around the globe.

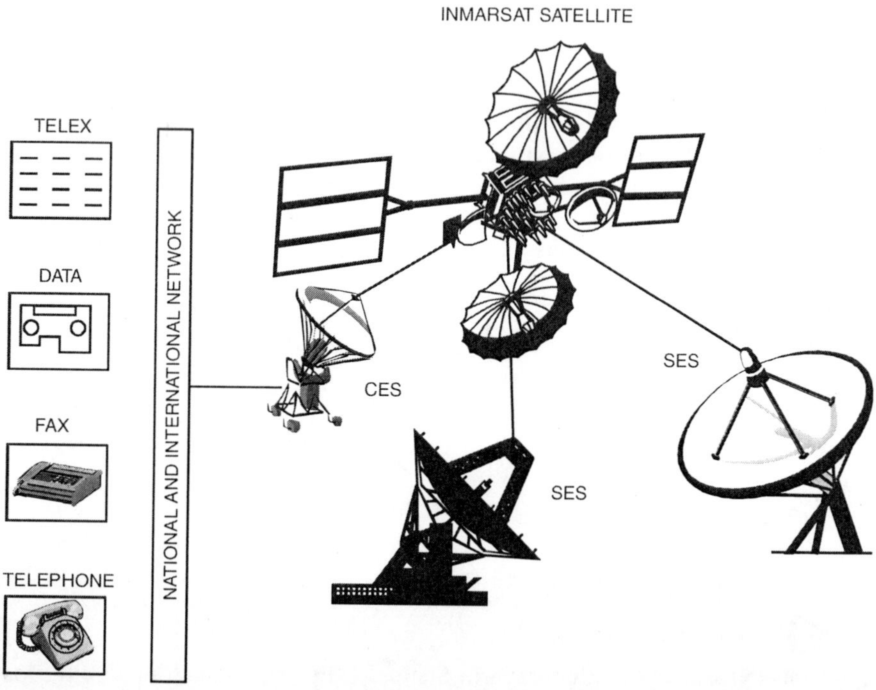

Fig. 7.10. INMARSAT Networking.

The INMARSAT satellite system **comprises of the following essential components :**

- The space segment operating and standby satellites with repeaters.
- Network and land earth stations (LESs) which are owned by some signotories to the INMARSAT organization.
- Subscriber mobile earth station (MESs) or terminals which are either leased or individual owner/operates.
- System control facility which comprises of network. Operation center (NOC) and satellite control center (SCC), which are owned by INMARSAT organization.

Aside from its commercial services, Inmarsat provides global distress and safety services (GMDSS) to ships and aircraft at no charge, as a public service. Services include traditional voice calls, low-level data tracking systems, and high-speed data services as well as distress and safety services. The most recent of these provides GPRS-type services at up to 492 kbit/s via the BGAN IP satellite modem the size of a notebook computer. Other services provide mobile ISDN services used by the media for live reporting on world events via videophone.

Historically expensive, calls via Inmarsat have now dropped to a level where they are comparable, and in many cases favorable, to international roaming costs, or hotel phone calls. Voice call charges are the same wherever in the world the service is used. Newer Inmarsat services using IP technology feature an always-on capability where the users are simply charged for the amount of data they send and receive rather than the length of time they are online. This applies specifically to BGAN, Regional BGAN, and MPDS.

The satellites are digital transponders that receive digital signals, reform the pulses, and then retransmit them to ground stations. The ground stations perform billing and act as gateways to the public switched telephone network and Internet. Inmarsat's most recent series of satellites were launched in June and November 2005. These were the largest commercial telecommunications satellites ever launched. Each satellite is equipped with a global beam, 19 regional spot beams, and 228 narrow spot beams. In addition to her own satellites, Inmarsat has a collaboration agreement with ACeS regarding handheld voice services.

7.6.1. Types of INMARSAT Terminals

INMARSAT-B

- This is successor to INMARSAT-A System, it supports a similar range of services but is cheaper to use because it is based on modern digital communications technology and also digital design makes much better use of satellite power and bandwidth.
- INMARSAT-B Land Mobile Terminals weighs around **15 Kgs.** and antenna size is approximately **1.2 meter x 1.2 meter.**
- VSNL started INMARSAT-B Services from November 30, 1999.

INMARSAT-C

- INMARSAT-C Mobile Earth Station (MES) has a small omni directional antenna, which can be easily mounted on a vehicle or vessel.
- It supports data speed up to 600 bps and provides terrestrial interfaces with telex, fax, X.25, X.400 and Internet.

- Weighs around **3 to 4 Kgs.** and has an omni directional antenna of about **25 centimeter height.**
- VSNL commenced INMARSAT-C Services on 12th November 1993.

INMARSAT-M

- INMARSAT-M System offers voice (4.8 Kbps), FAX (2.4 Kbps) and data (2.4 Kbps) Services.
- INMARSAT-M Terminals are lightweight, portable and available for meritime, land mobile and portable versions.
- Unlike INMARSAT-A Satellite Telephones, INMARSAT-M is a digital system.
- Land Mobile Terminal weighs about **10 Kgs.** and antenna size is approximately **70 cm x 50 cm.**
- VSNL commenced this service in November 30,

INMARSAT-Mini M

- This System offers voice (4.8 Kbps), FAX (2.4 Kbps) and data (2.4 Kbps) Services.
- INMARSAT-Mini M Terminals are lightweight, compact and available in brief case version, vehicle mountable version and maritime version.
- The terminal weighs around **2.5 Kgs.** and the antenna size is approximately **30 cm x 30cm.**
- VSNL commenced this service in May 30, 1997.

INMARSAT-Multimedia Mini M (M4)

- INMARSAT-Multimedia Mini M (M4) is a hybrid system of INMARSAT-B and INMARSAT-Mini M Systems.
- The terminal weighs around 4 Kgs and the antenna size is approximately 100 cm x 70cm.
- It supports high-speed voice, fax and circuit switch data services.
- The terminal shall be low cost, lightweight and would be supporting 64 Kbps data facility with embedded communication applications.

Coverage

There are three types of coverage related to each Inmarsat satellite.

- Global beam coverage. Each satellite is equipped with a single global beam that covers up to one-third of the Earth's surface, apart from the poles. In general, global beam coverage extends from latitudes of –78 to +78 degrees regardless of longitude.
- Wide spot beam coverage. It relates to the overlap of the wide spot beams (a set of narrower beams creating a coverage pattern). Wide spot beam coverage is optimised for covering most areas of interest to Inmarsat's customers and is thus somewhat limited in comparison to global beam coverage. This coverage was introduced with the I-3 satellites. Each I-4 satellite provides 19 wide spot beams.

- Narrow spot beam coverage. It relates to the overlap of the narrow spot beams (a set of very narrow beams creating a coverage pattern). Narrow spot beam coverage is designed to form the backbone of Inmarsat's broadband services, including the Broadband Global Area Network (BGAN). This coverage is introduced with the I-4 satellites. Each I-4 satellite provides 228 narrow spot beams.

Networks

Inmarsat has gradually developed a series of networks providing certain sets of services (most networks support multiple services). They are grouped into two sets, existing and evolved services, and IP-based services. Existing and evolved services are offered through Land Earth Stations which are not owned nor operated by Inmarsat, but through companies which have a commercial agreement with Inmarsat. IP-based services are provided via distribution partners but the satellite gateways are owned and operated by Inmarsat directly.

IP-based shared-carrier services, as follows :

- **Regional BGAN**: RBGAN offers a shared-channel IP packet-switched service of up to 144 kbit/s based on GPRS technology. Coverage is limited to parts of Europe, Asia, Africa & Australia. The new Inmarsat-4 satellite for the Indian Ocean Region (IOR) has taken over RBGAN service from the Thuraya satellite. The new Inmarsat-4 satellite for AOR-West now provides service to South America and the Western edges of Europe and Africa, however no service is offered to North or Central America. Inmarsat have announced the closure of the RBGAN service on December 31, 2008.
- **BGAN**: Broadband Global Area Network (BGAN) benefits from the new I-4 satellites to offer a shared-channel IP packet-switched service of up to 492 kbit/s (uplink and downlink speeds may differ and depend on terminal model) and a streaming-IP service from 32 up to 256 kbit/s (services depend on terminal model). Certain terminals also offer circuit-switched Mobile ISDN services at 64 kbit/s and even low speed (4.8 kbit/s) voice etc services. BGAN service is available on the IOR satellite and AOR-West satellite, with POR service planned for 2007. The BGAN family includes SwiftBroadband, a service planned for aircraft, and FleetBroadband, a service planned for ships and the maritime community.

Historic services, termed "Existing and Evolved", as follows :

- **Inmarsat-A**: provides voice services, telex services, medium speed fax/data services and, optionaly, high speed data services at 56 or 64 kbit/s. This is the only analog network of Inmarsat and is scheduled for phase-out by 31 December 2007. The rest of the Networks are digital and no phase-out date is currently available. The Inmarsat-A service on IOR has been moved to a backup satellite at 109 degrees East longitude due to the use of the new Inmarsat-4 satellite for the primary IOR coverage. The Inmarsat-A service on AOR-West has been moved to a backup satellite at 98 degrees West longitude due to the use of the new Inmarsat-4 satellite for the primary AOR-West coverage.

- **Aeronautical (Classic Aero)**: provides voice/fax/data services for aircraft. Three levels of terminals, Aero-L (Low Gain Antenna) primarily for packet data including ACARS and ADS, Aero-H (High Gain Antenna) for medium quality voice and fax/data at up to 9600 bit/s, and Aero-I (Intermediate Gain Antenna) for low quality voice and fax/data at up to 2400 bit/s. Note, there are also aircraft rated versions of Inmarsat-C and mini-M/M4. The aircraft version of GAN is called Swift64 (below)
- **Inmarsat-B**: provides voice services, telex services, medium speed fax/data services at 9.6 kbit/s and high speed data services at 56, 64 or 128 kbit/s. There is also a 'leased' mode for Inmarsat-B available on the spare Inmarsat satellites.
- **Inmarsat-C**: effectively this is a "satellite telex" terminal with store-and-forward, polling etc capabilities. Certain models of Inmarsat-C terminals are also approved for usage in the GMDSS system, equipped with GPS.
- **Inmarsat-M**: provides voice services at 4.8 kbit/s and medium speed fax/data services at 2.4 kbit/s. It paved the way towards Inmarsat-Mini-M.
- **Mini-M**: provides voice services at 4.8 kbit/s and medium speed fax/data services at 2.4 kbit/s. Uses Inmarsat's zonal ray technology .
- **GAN (Global Area Network)**: provides a selection of low speed services like voice at 4.8 kbit/s, fax & data at 2.4 kbit/s, ISDN like services at 64 kbit/s (called Mobile ISDN) and shared-channel IP packet-switched data services at 64 kbit/s (called Mobile Packet Data Service MPDS, formerly Inmarsat Packet Data Service – IPDS). GAN is also known as "M4".
- **Fleet** : actually a family of networks that includes the Inmarsat-Fleet77, Inmarsat-Fleet55 and Inmarsat-Fleet33 members. Much like GAN, it provides a selection of low speed services like voice at 4.8 kbit/s, fax/data at 2.4 kbit/s, medium speed services like fax/data at 9.6 kbit/s, ISDN like services at 64 kbit/s (called Mobile ISDN) and shared-channel IP packet-switched data services at 64 kbit/s (called Mobile Packet Data Service (MPDS). However, not all these services are available with all members of the family. The latest service to be supported is Mobile ISDN at 128 kbit/s on Inmarsat-Fleet77 terminals.
- **Swift 64**: Similar to GAN, providing voice, low rate fax/data, 64kb/s ISDN, and MPDS services, for private, business, and commercial aircraft. Swift 64 is often sold in a multi-channel version, to support several times 64kb/s.
- **Inmarsat-D/D+**: Inmarsat's version of a pager, although much larger than terrestrial versions. Some units are equipped with GPS. The original Inmarsat-D terminals were one-way (to mobile) pagers. The newer Inmarsat-D+ terminals are the equivalent of a two-way pager. The main use of this technology nowadays is in tracking trucks and buoys and SCADA applications. Inmarsat D+ is the only GPS tracking service available via Inmarsat. Competing systems such as from Skybitz only operate on the MSAT geostationary satellite over North America.
- **Inmarsat-E**: A global maritime distress alerting service using small Emergency Position Indicating Rescue Beacons (EPIRB) that automatically send message

to maritime Rescue Coordination Centres. This service was terminated in 2006 in agreement and coordination with the IMO and IMSO, and is no longer available. Inmarsat moved all users to the 406 MHz system, an equivalent EPIRB working via Cospas-Sarsat satellites, when the Inmarsat-E service was switched off in 2006. The service was terminated due to lack of interest from the maritime community.

- **MPDS (Mobile Packet Data Service)**: Previously known as IPDS, this is an IP-based data service in which several users share a 64kb/s carrier in a manner similar to ADSL. MPDS-specific terminals are not sold; rather, this is a service which comes with most terminals that are designed for GAN, Fleet, and Swift64.

7.7. LANDSAT

LANDSAT is a remote sensing satellite capable of sensing remotely the earth resources and its environment. Remote sensing through earth observation satellite provides :

- Valueable data on natural resource.
- Meteororology and environment leading to better resource management.
- Accelerating national development.

LANDSAT is one of the earth exploration satellite service (EESSS). The EESSS system invovles observation of the earth for various purposes :

- Meterorological (TIROSI, Metostat)
- Geodesy (transit)
- Earth Resources (LANDSAT, SPOT, SEASAT)
- Data collection (ARGOS).

LANDSAT was the first satellite launched in 1972 by NASA for the purpose and since then LANDSAT-2, 3 and 4 were launched later in the year 1975, 1978 and 1982 respectively and LANDSAT-5 in 1984. LANDSAT-6 and 7 are planned to be launched in early nineties.

The working of remote sensing satellite is based on sensors data processing, data analysis and data interpretation for different applications.

- Here physical measurements of reflected, emitted or radiated energy from various objects on the ground are made from distant platforms.
- Since different objects emit radiations in different bands so to recognise them correctly sending is done in discrete bands of the electromagnetic spectrum.
- Remote sensing is carried out in visible, infrared, thermal and microwave bands and each band has its own importance.
- Microwave remote sensing not only gives land observation but also ocean and atmospheric studies and many other new applications.
- Sensors send the data back to earth stations from where it is collected and processed in the form of photographic prints of films, computer compatible taper or floppy disks and stored.

LANDSAT provide world wide high resolution multispectral data of earth's surface. There are three kinds of sensor system in LANDSAT :

- Three camera return beam vidion (RBV).

- Four channel multispectral scanner (MSS).
- Seven band thermatic mapper (TM).

The LANDSAT-4 and LANDSAT-5 satellites are in circular sun synchronous near polar orbit at an altitude of approximately 705 km. They circle the earth every 103 minutes completing 14 orbits per day and view the entire earth in 16 days.

7.8. SEARCH AND RESCUE SATELLITE (SARSAT)

This type of satellite system is used for receiving passengers and crew members of sinking ships and crashed airplanes. This system is also used in USSR under the Russian terminology cospas. Therefore, the SARSAT is also known as **Cosapar-sarsat.** This system is quite different from INMARSAT.

Here the satellite circle the globe at 1000 km altitudes in scanning a wide swath of the earth for distress signals and relaying them to specially built earth stations.

Cospas-Sarsat is useful to private and corporate fliers as well as airline crews, pleasure boaters and small scale commercial fisherman and merchant shippers. When even a plane has to go down in a remote mountain range or ship is floundering in the subarctic Atlantic, satellites direct distress signals and alert rescue teams.

In Cospar-sarsat system a small industry powered transmitter on a ship or aircraft is activated manually or automatically in an accident. After activating the transmitter of beacon emits a low power omnidirectional signal that is picked up by a satellite. Then the satellite relays the distress signals to an earth station which in turn sends the message over conventional communication lines to a rescue center. In future these cospas-sarsat satellites are going to be used with INMARSAT system in a compromise scheme as shown in Fig. 7.11.

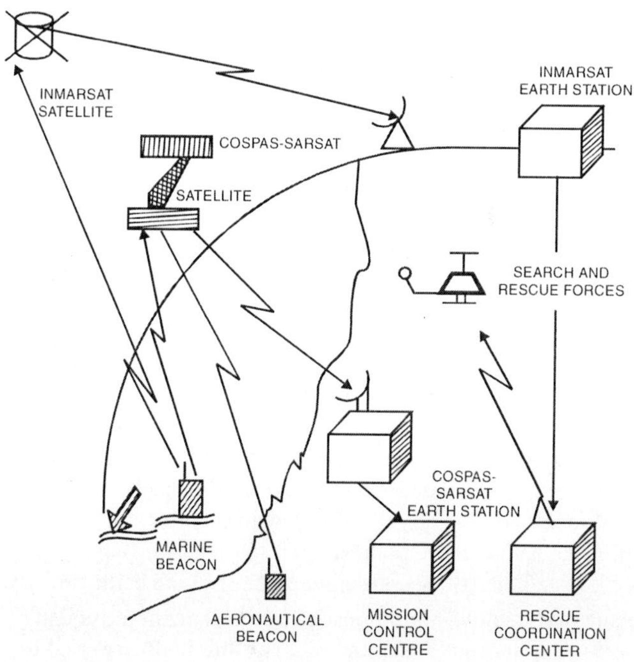

Fig. 7.11. Cospas-sarsat satellite with INMARSAT for maritime services.

7.9. LOW EARTH ORBIT (LEO) AND NON-GEOSTATIONARY

Geostationary satellite has been in use commonly for satellite communication systems as it has proved to be very economical. But there are some specialized applications that require non-geostationary satellites. Surveillance of the earth's surface, for both military data gathering and earth resources applications, requires satellites in low-earth orbit that cover the entire surface of the earth. Satellites providing global navigation, such as GPS constellation, must utilize orbits that place the satellites in widely spaced positions in the sky as seen by the receiver. Most of these satellites must be in inclined orbits with an even distribution over the earth's surface.

A number of LEO systems are being proposed world over to provide global hand held mobile communication. Each proposed system consists of a number of satellites in low earth orbit constellation.

The difference in the systems lie in the proposed :

- Orbital height
- Inclination
- Number of planes
- Number of satellites per plane
- Number of spot beams in mobile bands
- Communication techniques.

The Iridium system of USA was one of the first proponents with a constellation of 66 satellites to provide hand held telephone service using multiple beams TDMA and Inter-satellite links techniques.

Some of the **'big Leo'** systems are Iridium, Global Star, Ellipsat, calling communications, consellation communications, odyssey etc.

ORBCOM of USA is a **'small LEO'** designated to provide message communication only. It has a constellation of 26 satellites at a height of approximately 700 km. The system operates in VHF band (137/149 MHz band).

STARSYS is another **small 'LEO'** system offering services similar to ORBCOM. The advantage of data or message only services is that unlike voice services, the system, can tolerate short breaks in the visibility as repeat requests can be made. Therefore, the number of satellites is much smaller than those required for voice service.

7.9.1. Orbit Considerations

(1) Equatorial Orbits

They lie exactly in the plane of the geographical equator of the earth. That is, the orbital path lies directly above the equator at all times. In order to take advantage of the 0.45 km/s eastward rotational velocity of the earth, most satellites are launched toward the east into a prograde orbit. A westerly directed orbit is called a retrograde orbit. A satellite in an eastwardly directed equatorial orbit will have two periods: a real orbital period that is referenced to the inertial space and an apparent orbital period that is referenced to stationary observer on the surface of the earth.

The plane of a satellite's orbit must be in the plane of the equator for the satellite to be in the equatorial orbit. **The following two methods are used for this.**

(1) The first launch method is to locate the launch site on "the equator and to launch the spacecraft toward the east along the equatorial plane.

(2) The second method is to launch the satellite into an inclined orbit and to execute a maneuver either during the launch trajectory or when the satellite is in the inclined orbit that changes the plane of the initial orbit so that the final orbit is in the plane of the equator.

Figure 7.12(*a*) shows an Equatorial orbit.

(2) Inclined Orbits

Figure 7.12(*b*), the inclined orbit will take the spacecraft, at one time or another, over the earth's entire surface that lies approximately between the latitudes given by +- the orbital inclination. For example, an orbit with an inclination of 30° will cover all regions that lie approximately between latitudes 30° north and 30° south.

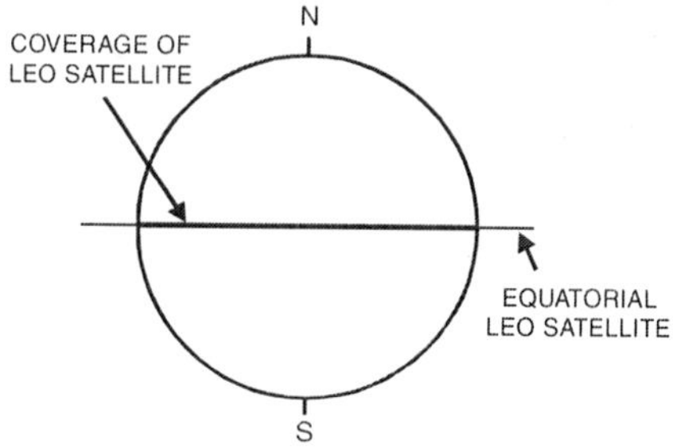

(*a*)

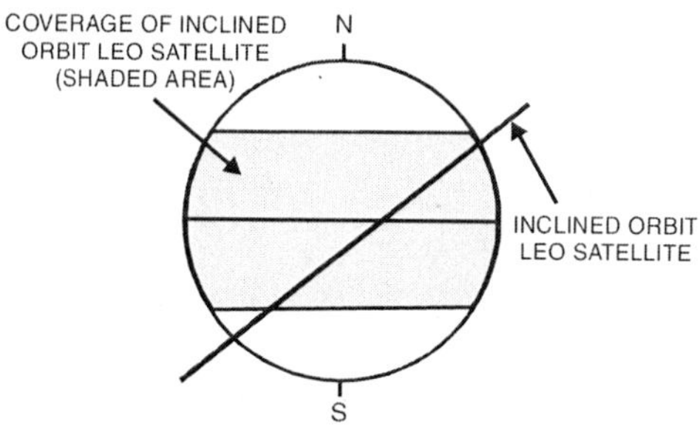

(*b*)

Fig. 7.12.

The **disadvantage** is that the Master Control Station (MCS) will not be able to communicate directly with the satellite on every orbit as with an equatorial orbit satellite. A LEO satellite orbits the earth with a period of 90 to 100 min and, for an inclined orbit satellite; the earth will have rotated the MCS out of the path of the satellite on the next pass over the same side of the earth. The solution to pass data between an inclined orbit satellite and the MCS is to design the satellite to store the data acquired over many orbits and then when it passes within radio range of the MCS to dump the data rapidly to the MCS. This is called store and forward mechanism.

(3) Elliptical Orbits

An elliptical orbit will have a non-zero eccentricity. The orbit eccentricity, e, is determined by the lengths of the semi major axis, a, and the semi ninor axis, b, of the orbit ellipse

$$e^2 = 1 - \left(\frac{b^2}{a^2}\right)$$

(4) Molniya Orbit

The apogee of the Molniya orbit is at an altitude of 39,152 km and the perigee is at an altitude of 500 km. The orbital period is 11h and 38 min and the orbital inclination is 62.9 degrees. This combination of apogee, perigee and inclination ensures that the ground track of the Molniya orbit repeats every other orbit. That is, if the orbit passes exactly over Moscow on orbit one; it will do so again on orbit 3, 5, 7, 9 and so on. This is shown in Fig. 7.13.

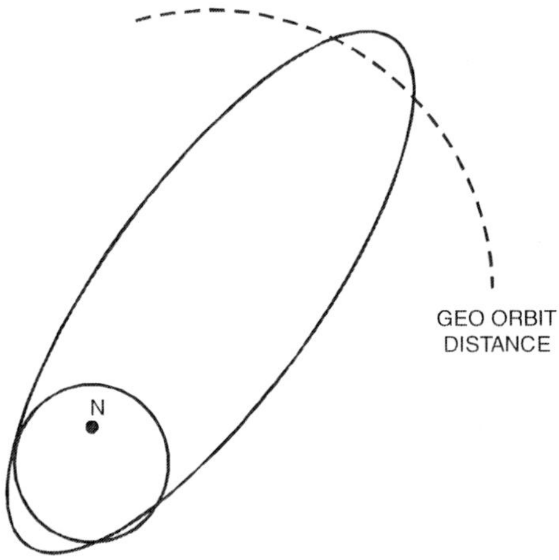

Fig. 7.13. Schematic of a Molniya orbit.

Two Molniya orbits, with the planes of the orbits separated by 180°, will provide coverage over the extreme latitudes of Russia for 24 hours per day using two satellits, correctly phased-one in each of the two Molniya orbits as shown in Fig. 7.14.

The two-satellite, dual Molniya orbit requires that earth station operations be carried out at elevation angles well below 30 degrees for full 24 hrs per day coverage of one region. **The Molniya orbit also suffers from three drawbacks:**

(1) The requirement to track the spacecraft.

(2) The need to switch communications to the other Molniya satellite when the first goes out of coverage as the other comes into coverage.

(3) The radiation environment that the satellite has to pass through four times a day-twice on ascent and twice on descent.

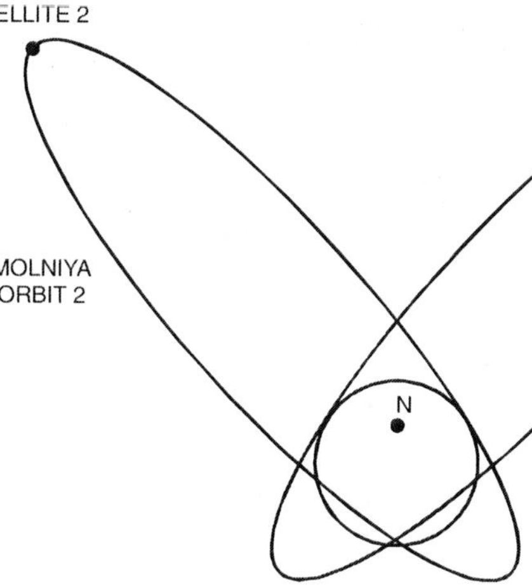

Fig. 7.14. Schematic of an operational Molniya system. Satellite 1 in Molniya orbit 1 is providing service over Russia at close to its apogee while the second satellite is also close to its apogee in Molniya orbit 2.

(5) Sun Synchronous Orbit

It is a special form of low earth orbit where the plane of the orbit maintains a constant aspect angle with the direction to the sun. One example is an earth resources satellite that requires a large amount of direct sunlight to illuminate the region below the satellite so that photographs can be taken. Figure 7.15 below illustrates how the sun synchronous orbit is achieved.

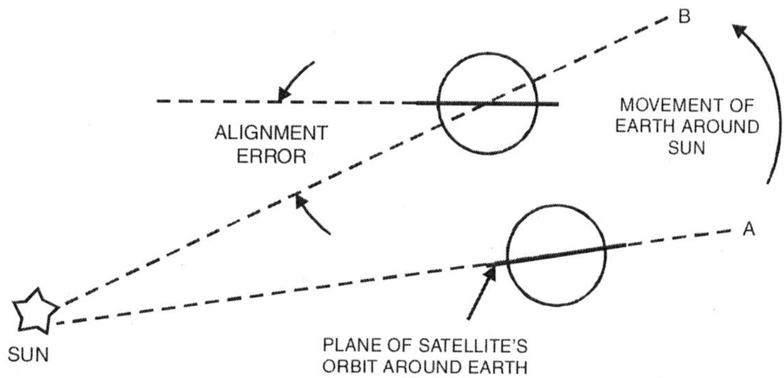

Fig. 7.15.

7.9.2. Coverage and Frequency Considerations

In Figure 7.16 spacecraft orbits at distance *rs* from the center of the earth, *C*. Earth station is located at *E*. The elevation angle to the satellite is θ. Using the sine rule, we have

$$\left[\frac{rs}{\sin(90+\theta)}\right] = \left[\frac{d}{\sin(\gamma)}\right]$$

which gives $\cos\theta = \dfrac{[rs \sin(\gamma)]}{d}$

All 3 parameters in the above equation have key inputs to architecture of the satellite system. The angle y will yield the coverage area on the surface of the earth assuming the satellite has a symmetrical coverage about nadir. The distance d will determine the free space path loss along the propagation path and will be a factor in the link budget design. The elevation angle θ will influence the G/T ratio of the antenna and the likely propagation impairments that will be encountered along the path of the satellite.

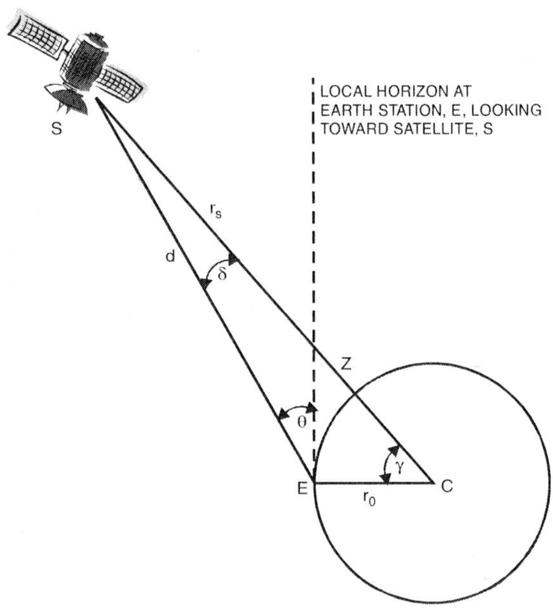

Fig. 7.16

7.9.3. Frequency Band

LEO satellite systems providing data and voice service to mobile users tend to use the lowest available RF frequency. The EIRP required by the satellite transponder to establish a given *C/N* ratio in the mobile receiver is proportional to the square of the RF frequency of the downlink.

The received power at the mobile earth station is given by

$$P_r = \frac{P_t G_t G_r \lambda^2}{(4\pi A)} \text{ Watts}$$

This power at the mobile terminal with an omnidirectional antenna increases as the square of the wavelength, or decreases as the square of the frequency. The lower the RF frequency the greater the received power for any given coverage zone.

Elevation Angle Considerations

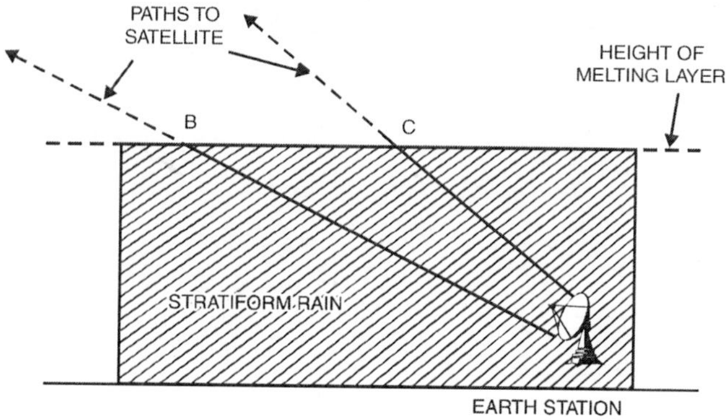

Fig. 7.17. Illustration of the decrease in the path through rain as the elevation angle to the satellite increases.

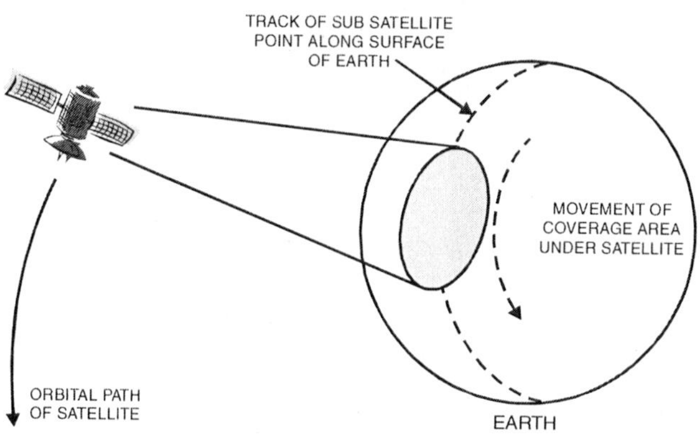

Fig. 7.18. Illustration of coverage area under a satellite.

7.9.4. Off-axis Scanning

The design of a point-to-point wireless communication system requires that the antennas at the either end be directed toward each other for maximum gain advantage. If the transmitting antenna has to communicate with more than one receiving antenna, and these antennas are located in different positions a compromise must be reached between the gain of the transmitting antenna toward the various receiving antennas.

Figure 7.19 illustrates the path loss and the scan angle loss for a phased array. The phased array has its prime axis pointed at nadir. The energy received at nadir from the satellite will be greater than that received at edge of coverage for two reasons. First, the path loss will be less since the nadir distance, d_N, is less than the EOC distance, d_{EOC}. Second there will be a scan loss associated with the antenna reaching out to cover the EOC region.

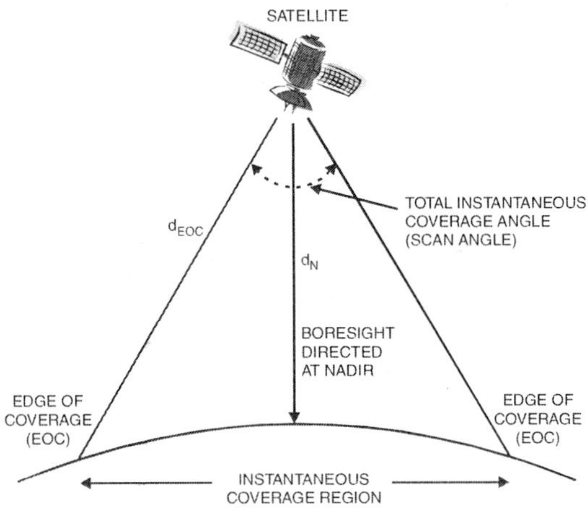

Fig. 7.19

7.9.5. Comparison of Using Satellite in LEO, MEO and GEO for Mobile Satellite Communications

S. No.	*Parameter*	*LEO*	*MEO*	*GEO*
1.	Distance from earth	100 miles to 300 miles	6000 miles to 12,000 miles	19,000 miles to 25,000 miles
2.	Rotation period	1.5 hours	5 to 12 hours	24 hours
3.	Number of satellites required	30 to 60	10 to 20	3 to 6
4.	Life time	5 years	10 to 15 years	10 to 15 years
5.	Launching cost	Low cost for each satellite	Moderate cost for each satellite	High cost for each satellite
6.	Coverage area	Small, therefore hundreds of ground stations are required	Large area, therefore less number of ground stations are required.	Much large area, therefore small number of ground stations are required.
7.	Line of sight time	0.5 hours	2 to 4 hours	24 hours
8.	Ellipse interval	Frequent day-night cycle. Satellite remains in darkness 30% of time.	Infrequent day-night cycle. Satellite remains in darkness 2% of time	Infrequent day-night cycle. Satellite remains in darkness 1-2% of time.
9.	Elevation angle variation	Rapidly varying elevation angle	Slowly varying elevation angle.	No variation in elevation angle.
10.	Tracking	Slow speed tracking.	Slow speed tracking.	Tracking is not required.
11.	Travelling speed	≃ 17,500 mph	≃ 10,000 mph	≃ 6,879 mph
12.	Application	Wheather forecasting	Communication and navigation purpose	Telephony, data/TV distribution, mobile communication, broadcasting

7.10. GLOBALSTAR

Globalstar is a low Earth orbit (LEO) satellite constellation for telephone and low-speed data communications, similar to (and competing with) the Iridium satellite system.

System Architecture

Globalstar differs from Iridium in several important ways :

- Globalstar satellites are simple "bent pipe" repeaters; there is no inter-satellite linking. A network of ground gateway stations provides connectivity from the 40 satellites to the public switched telephone network; users are assigned telephone numbers on the North American Numbering Plan in North America or the appropriate telephone numbering plan for the country that the overseas gateway is located in, except for Brazil, where the official globalstar country code (+8818) is used. Because there is no inter-satellite linking, a satellite must have a gateway station in view to provide service to any users it may see. Because there are no gateway stations to cover certain remote areas (such as oceans far from land), no service can be provided in those areas even though the satellites fly over them.
- Globalstar orbits have an inclination of 52 degrees, compared to the near-polar 86.4-degree orbits used by Iridium. Globalstar also does not cover polar areas due to the lower orbital inclination.
- The Globalstar system uses the Qualcomm CDMA air interface; however, the Ericsson and Telit phones accept standard GSM SIM cards, while the Qualcomm GSP-1600/1700 phones do not have a simcard interface, but use CDMA/IS-41 based authentication. Therefore the Globalstar gateways need to support both the CDMA/IS-41 and the GSM standards, which, however not all gateways do. This results in the lack of coverage for GSM-authentication-standard based phones in the Eastern Asian and Caribbean region.

While for Iridium only Telstra Australia offers GSM-roaming, most Globalstar providers have roaming agreements with local cellular operators, enabling the use of a cellular SIM card with a Globalstar handset and vice versa.

Despite (or because of) the limitations above, Globalstar's operational costs were significantly cheaper and can support somewhat higher data rates than Iridium. However, such cost differences later proved meaningless when both companies shed their multi-billion dollar debts through bankruptcy.

System Deployment

Like Iridium, Globalstar received its U.S. spectrum allocation from the FCC in January 1995, and continued to negotiate with various other sovereign nations for rights to use the same radio frequencies in their countries.

The first satellites were launched in February 1998, but system deployment was delayed through a series of embarrassing and costly launch failures, notably the September 1998 loss of 12 satellites in a launch by the Russian Space Agency. In February 2000, it launched the last of 52 satellites: 48 satellites and four spares (reduced from the original plan of eight spares).

The first call on the Globalstar system was placed on November 1, 1998, from Irwin Jacobs (chairman of Qualcomm) in San Diego to Bernard Schwartz (CEO and chairman of Loral Space and Communications) in New York.

In October 1999, the system began "friendly user" trials with 44 of 48 planned satellites. In December 1999, the system began limited commercial service (200 users) with the full 48 satellites (no spares in orbit). In February 2000, it began full commercial service with its 48 satellites and 4 spares in North America, Europe and Brazil. Initial prices were $1.79/minute vs. $9/minute for Iridium.

In 2005, some of the satellites began to reach the limit of their operational lifetime of 7.5 years. In December of 2005, Globalstar began to move some of its satellites into a graveyard orbit above LEO.

Recent Problems

According to documents filed with the SEC on January 30, 2007, Globalstar's previously identified problems with its S-band amplifiers used on its satellites for two-way communications are occurring at a rate faster than previously believed, possibly eventually leading to total loss of service in 2008.

In the filing, Globalstar made the following statements:

"Based on data recently collected from satellite operations, the Company has concluded that the degradation of the amplifiers is now occurring at a rate that is faster than previously experienced and faster than the Company had previously anticipated."

"Based on its most recent analysis, the Company now believes that, if the degradation of the S-band antenna amplifiers continues at the current rate or further accelerates, and if the Company is unsuccessful in developing additional technical solutions, the quality of two-way communications services will decline, and by some time in 2008 substantially all of the Company's currently in-orbit satellites will cease to be able to support two-way communications services."

Industry analysts speculate the problem is caused by radiation exposure the satellites receive when they pass through the South Atlantic Anomaly in their 800 km altitude orbits.

"The S-band antenna amplifier degradation does not affect adversely the Company's one-way "Simplex" data transmission services, which utilize only the L-band uplink from a subscriber's "Simplex" terminal to the satellites. "

"The Company is working on plans, including new products and services and pricing programs, and exploring the feasibility of accelerating procurement and launch of its second-generation satellite constellation, to attempt to reduce the effects of this problem upon its customers and operations. The Company will be able to forecast the duration of service coverage at any particular location in its service area and intends to make this information available without charge to its service providers, including its wholly owned operating subsidiaries, so that they may work with their subscribers to reduce the impact of the degradation in service quality in their respective service areas. The Company is also reviewing its business plan in light of these developments. "

"The Company's liquidity remains strong. At December 31, 2006, in addition to its credit agreement, the Company had unrestricted cash on hand and undrawn amounts under the Thermo Funding Company irrevocable standby stock purchase agreement of approximately $195 million."

Globalstar's Second-Generation of Satellite Constellation

In December 2006, Globalstar announced that Alcatel Alenia Space has been awarded a • 661 million contract for the Second- Generation 48 satellite constellation.

"The satellites are being designed with a life expectancy of 15 years, which is significantly longer than the design life of Globalstar's first generation constellation."

In addition, Globalstar announced on April, 03, 2007 that it has signed a EUR 9 million (approximately US$12 million) agreement with Alcatel Alenia Space to upgrade the Globalstar satellite constellation, including necessary hardware and software upgrades to Globalstar's satellite control network facilities.

"This Second-Generation of satellites are expected to provide service to Globalstar customers with satellite voice and data services until at least 2025."

Summary

1. Intially satellite communication was used for transmission of telephone and television signals but now a number of applications are there : Fixed satellite communication systems, broadcast satellite communication system, land mobile satellite communication, aeronautical satellite communication system etc.
2. Basically satellite TV configurations is based on simply point to multipoint connectivity of the satellite communication.
3. Number of countries have their own domestic satellite systems to be used for direct broadcast purpose. INSAT series of satellites are multipurpose in nature and also have direct broadcasting capability.
4. Some of the popular and more recent satellite systems development for earth observations are LANDSAT, ERS-1, IRS etc.
5. For cartography the satellites are equipped with high resolution cameras, that have been used to correct and update certain features of existing maps in different countries.
6. The first wheather satellite, TIROS-1, was launched by USA in 1960 and was used by a large number of countries to monitor the behaviour of atmosphere over a large area of the earth.
7. IRAS (Infrared Astronomical Satellite) was launched in 1983. It was the first satellite to carry an all-sky survey to look for astronmical objects emitting infrared radiation.
8. NAVSTAR is a satellite to provide navigation and they are capable of providing data to enable position fixing with an accuracy of a few meters.
9. IMMARSAT is an international organisation that controls satellite systems for communication between ships and coast so that emergency life saving services may be provided.
10. There are 6 INTELSAT satellite systems which have been launched from time to time. These are probably known as INTELSAT I, II, III, IV, V, VI.
11. VSAT (Very Small Aperture Terminal) have become popular because they were successfully used with Ku band (12-14 GHz) satellite systems.
12. The MSAT (Mobile satellite communication) systems provide direct satellite service to mobile phones. It gives greater improvement in rural communications.
13. In MSAT direct communication links can be established between a satellite and a variety of stations : mobile units, portable units, gate way earth stations, base earth stations and a control earth station.

14. Search and rescue satellite (SARSAT) system is used for receiving passengers and chew members of sinking ships or crashed airplanes.
15. LANDSAT is one of the earth exploration satellite service and it provide world wide high resolution multispectral data of earth's surface.
16. In VSAT (Very small aperture terminal) the earth station antenna size is typically less than 2.4 in diameter.
17. VSAT system operate in star configuration *i.e.* the connection of one VSAT to another must be made through the hub.
18. The major shortcoming of present day VSAT systems are
    - High initial costs
    - Tendency towards optimizing systems for large networks
    - Lack of direct VSAT to VSAT links.
19. In the ALOHA concept there are three variations
    - Pure ALOHA
    - Slotted ALOHA.
    - Reservation ALOHA
20. Dynamic accessing schemes may be classified into two categories :
    - Contention or random access schemes
    - Reservation schemes
21. Radarset is an earth resource remote sensing satellite which is a part of the canadian space program.
22. Radarset is a large spacecraft, the total man is the orbit is about 3100 kg and it works at a carrier frequency of 5.3 GHz.
23. Mobile stations use two types of antennas
    - Steered antennas
    - Nonsteered antennas
24. There are two different kind of DGPS
    - Post processing DGPS
    - Inverted DGPS
25. GPS can also be used to determine precise time, time intervals and frequency.

Review Questions

1. List the various application of satellites.
2. Describe the operation of a typical VSAT system. State briefly where VSAT systems find widest application.
3. With the help of suitable diagram explain the satellite communication system architecture using VSAT. Also write the applications of VSAT. **[UPTU 2011-12]**
4. Explain the important features of a RADARSAT.
5. Discuss the various application areas of VSAT and data broadcast satellite.
6. Explain the various applications of Radarset and MSAT.
7. Draw the network structure of mobile satellite communication (MSAT) communication. How many types of communications are possible through this network ?

8. How many types of antennas can be used in MSAT system ? Explain briefly.
9. What is the need of mobile services ?
10. What is the role of INMARSAT ? Mention its area of coverage also.
11. Mention some of the typical applications of VSAT.
12. With the help of block diagram of INMARSAT network, explain its operation.
13. Mention the reason of development of INMARSAT series of satellites.
14. What is LANDSAT ? Mention the services provided by LANDSAT.
15. Describe the main features of RADARSAT. Explain what is meant by a 'down to dusk' orbit and why the RADARSAT follows such an orbit.
16. Explain the role of ORBCOMM and mention its different segments.
17. Name the other satellites like LANDSAT, which are used for study of earth surface and still in operation.
18. Describe the main features and services offered of INTELSAT satellite system. How do these services compare with services offered by low earth orbit satellites, used for communication.

❑❑❑

8 CHAPTER

SATELLITE APPLICATIONS

INSIDE THIS CHAPTER

8.1. INTRODUCTION

In the last chapters we have studied different aspects of satellite and communication through satellite. In the field of satellite communication applications, satellite television, telephone and data communication are the major application areas. Satellite can be used as navigational aids by vehicles on land, in the air and on the sea. Some of the metrological applications like wheather forecasting, flood forecast etc. are also possible by the use of satellite.

In this chapter, we shall discuss the different applications of satellites.

A. Satellite for Earth Observation

With the TIROS wheather satellite, the satellites can be used for a variety of earth observation applciations. Some of the most popular satellite system developed for this role internationally include LANDSAT, ERS-1, IRS etc. Some of the applications are :

(1) Cartography : One of the early applications for earth resources is cartography *i.e.* map-making. Satellite images can been used to produce related maps for the construction of roads, railways and irrigation channels. Satellite equipped with high resolution cameras have been used to correct and update certain features in the exciting maps. LANDSAT has played a vital role in mapping. Space imagery is most cost effective as compared to the other methods because of the speed with which space surveys can be made.

(2) Oceanography : It is possible to make the pictures of ocean waves, ice fields, ice bergs, ice leads and sea conditions along the coastal lines with the help of satellite. Satellite data can be used to determine sea conditions such as sea surface temperature, wind speed, wind direction, local air temperature and moisture contents etc.

Department of ocean development is planning to start a program for remote sensing program in ocean related applications. From satellite data analysis, sea surface temperature (SST) charts are being generated. Potential fishing zones are being identified by the SST data.

(3) Ice Reconnaissance : Remote sensing satellite data can be used in forecasting of ice and ice bergs, it is very important for navigation of ship in cold water. To do the forecasting, it is necessary to have the knowledge of air and sea temperatures and wind direction etc. Data related to the eye thickness can be obtained by infrared sensors.

(4) Monitoring Oil Pollution : Satellite can be used for monitoring oil slicks in the sea. Sometimes it is very difficult to see such leakage from aircraft, and are restricted in vision to narrowbands of ocean because of their low altitudes. On the other hand, the oil slicks can be spotted very effectively by the satellites on global scale except in the places of persistent low loads. Space borns sensors measure the amount of sunlight reflected from the sea surface. Spilled out means it is clearly from the ordinary sea water in the near ultraviolet region around 0.38 micron and in near red around 0.6 micron.

(5) Oil and Mineral Exploration : Satellite data can be used for oil and mineral exploration. The satellite data can be used by geologists and it indicates the trans current faults, and used to make searches from unknown deposits to searches from unknown deposits of oil. Rock colour and topographic form is used to identify the minerals.

Remote sensing satellites IRS has been effectively used for identifying location for gold mineralisation.

(6) Monitoring Agriculture and Forestry : Remote sensing data from the earth observation satellite is used for several applications such as crop identification, agricultural crop aereage and field estimation, flood mapping, drought monitoring and assessment, landuse and land cover mapping and forest resources management etc. Present day remote sensing satellite equipped with the state of the art high resolution camera are capable of monitoring characteristics like crop identification, Crop field, harvest time and land fertility etc.

B. Satellite for Weather Forecast

Satellite data can be used for wheather, TIROS-1 was launched by U.S.A. in 1960 and was used by a large number of countries to monitor the behaviour of atmosphere. In present days, some instruments are also provided with satellite and they give the information regarding loud formation, cloud type and position intensity of frontal depression, thunderstorm, typhoons etc. Sometimes wheather satellites may be useful to find the details of snow cover, sea ice, sea surface temperatures, cloud top temperatures and measurement of humidity through the depth of atmosphere.

The accuracy of the wheather forecasting depends upon the quality and quantity of observational data from the surface of earth and surrounding atmosphere. Basically there are two categories of operational wheather satellite depending upon the type of orbit :

(1) Polar orbit (2) Geostationary orbit

Polar orbiting satellite can provide the global coverage for every twelve hours while geostationary orbiting satellite can monitor, wheather determining atmosphere parameters over a very large area almost continuous.

Radiometer is an important device, that is equipped with wheather satellite and it measures the amount of radiation. Measurement in the visible part of spectrum gives pictures about clouds, deserts etc. Infrared measurements give information both during the day as well as during the night and it show the relative temperature of the different features. Usually pictures are obtained in black and white, black areas indicates high temperature and white areas low temperature.

In India, INSAT–series have been used for metrological data. The metrological data utilization centre (MDVC) of Indian Metrological Department (IMD) processes and disseminates the INSAT metrological images.

C. Satellites for Scientific Studies

Satellite can also be used in a variety of scientific experimental studies including studying the characteristics of upper region of the atmosphere, probing other planets and stars etc. Satellites equipped with scientific payloads offers a very good and effective solution to carryout atmospheric and space studies.

In India, Indian Space Research Organisation (ISRO) is pursuing the programme of using space borne vehicles for carrying out scientific experiments. IRAS (Infrared Astronomical Satellite) was the first satellite, launched in 1983, to carry an sky survey to look for astronomical objects emitting infrared radiation.

D. Satellites for Military Applications

Satellites have played a major role in military applications, both during the war time and peace time. The earlier methods of reconnaissance, communication, navigation, remote sensing etc. had some contrains because of the land based on needing the services of specialised aircraft or ships. Some of the more commonly used satellite applications are :

(1) Command and Communication : Today satellites can provide instant communication between the operational forces and their command authorities irrespective the distance or time of the day.

Defence Satellite Communication System (DSCS) launches courier 1B in 1960, and military communication satellites have been started. Till the mid- eighties DSCS-1, DSCS-2, DSCS-3 series launches. For naval applications, most popular military satellite include SEASAT.

(2) Navigation Satellites : NAVSTAR is the satellite that provides navigation. It gives data to enable position fixing with an accuracy and it gives a position accuracy of 15 m. NAVSTAR is a replacement for the US Navy's TRANSIT series. TRANSIT-series is a series of navigation satellites operated by the US Navy but also used by civilians, especially sailors. Receivers fix position through a combination of Doppler effect and accurate orbit details transmitted by satellites. In most of the world, position accuracy of about 100 m is possible.

(3) Early Warning Satellites : Basically early warning satellite warns to the owners of a missile attack. The Integrated Missile Early Warning Satellite (IMEWS) first launched in 1970 and, these satellites carry infrared telescope and television cameras to transmit photos of detected ICBMS to prevent false alarm that pluged MIDAS. IMEWS series satellites are capable of detecting ICBMS within seconds of ignition. This programme was renamed as Defence Support Programme (DSP).

(4) Metrological Functions : Satellite data can be used for accurate wheather forecast. Wheather satellites provide long range forecast for military planners. They give advance information to the mission planners of reconnaissance satellites on weather the target area will be clear of cloud cover or not. National Oceanographic and Atmospheric Administration (NOAA) made available data to the civilians. These satellite provide infrared and visual imagery, temperature moisture surroundings, auroral detection and upper atmosphere surroundings.

(5) Nuclear Detection : USA originates the idea of nuclear detection using satellites. Twelve satellites of vela-series (vela-1 and vela-2) were launched during time period 1963-1970, for nuclear detection. Then IMEWS early warning satellite has taken over.

(6) Reconnaissance and Intelligence gathering functions : It is widely used satellite application for military and is used to varify the extent and composition of military forces. This is achieved by means of variety of electronic and photographic. Some other surveillance and reconnaissance related military applications of satellite include :

(*a*) Ocean surveillance

(*b*) Electromagnetic reconnaissance dealing in the location and cataloging of military ratio and radar stations.

8.2. SATELLITE TELEVISION

As well as the number of users are increase, the conventional methods for coverage of TV by microwaves space communication and coaxial cable links are become expensive. Thus the geostationery communication satellites are used for TV applications. The satellites are launched into synchronous orbits around the earth and it enables the national as well as international programmes to be relayed between a number of ground stations around the world. Three artificial satellites are placed in an equatorial orbits at 120° from each other so that it can cover the whole area of the world.

The frequency band recommended for satellite broadcasting are 620 to 790 MHz. 2.5 to 2.69 GHz and 11.7 to 12.2 GHz on a shared basis with other fixed and mobile services.

The size of the satellite antenna and RF power depend on the frequency of operation. Space erectable antennas are used for the 620-790 MHz band with the size limited to about 15 meters while rigid antennas are used for 2.5 and 20 GHz bands and their size are limited to about 3 meters. For ground terminals, beam width and frequency limits the maximum diameter of the antenna. The coast and complexity in the receiver circuit increases with the increase in frequency.

The Fig. 8.1 shows the block diagram of the direct reception of signal from the satellite to the TV receiver.

- High power and highly directive transmitters transmit signals to the geostationary satellite. Each microwave channel can accommodate many TV signals and thousands of telephone channels or suitable combination of these. The width of microwave channel is several tens of megahertz.

- A satellite is positioned at a height of 36,000 km in a geostationary synchronous orbit with sub-satellite longitude of 33° east was used for TV programmes over most parts of the country.
- The satellites usually powered by solar batteries, it receivers the signal transmitted by the transmitter, demodulate and amplify it and redemodulate it on a different carrier before transmitting again.
- By using the suitable reflector in the transmitting antenna on the satellite, radiated beam can be directed to a narrow region on the earth. Since high power satellites can provide large power flux densities, so smaller size antennas can be used for reception.
- For national distribution, the transmission is downwards from a wide angle antenna so that the whole national area is illuminated by transmission.
- For international distribution, the transmission is toward the one or two satellites from highly directive antennas. The demodulation amplification remodulation transmission process is repeated in the second satellite. Finally the down channel transmission is received by large cress section antenna and processed in low noise receivers and finally reradiated from the regular TV transmitters. The processing of the received signal from the satellites to the loudspeaker and picture tube is shown in Fig. 8.1.

Cost of satellite communication : The cost of satellite communication depends the life of satellite. The life of satellite is limited because a good synchronous satellite uses a high gain antenna and it requires the control of both, its position and attitude in the orbit. The position and attitude control rockets require fuel that has to put on a for all before launch. Therefore for a given payload, for longer life satellite will be heavier and cost will be high.

Satellite TV for Rebroadcast

TV programme can be broadcast to different places through normal terrestrical transmitters connected to earth stations which receive signal at the microwave frequency for a satellite. Figure 8.2 shows the block diagram of TV broadcasters satellite system.

- In this system, the baseband signal from the studio is modulated by a carrier at intermediate frequency (70 MHz) using frequency modulator. Then it is up converted to a microwave frequency, 6 GHz and amplified to TWT power amplifier to the desired valve.
- The uplink consists of two parabolic reflectors, one at the earth for the transmission and other at the satellite for receiving. FM signal goes to the satellite transponder through uplink. Then this incoming signal is amplified and down converted to 4 GHz.
- Similarly the downlink consists of one high gain parabolic reflector at the satellite for transmission and another parabolic dish at the earth for receiving.
- At the earth receiving station the frequency modulated signal is amplified by LNA and then converted to 70 MHz IF using a mixer and local oscillator. FM signal is demodulated using FM detector and the original baseband signals are recovered.

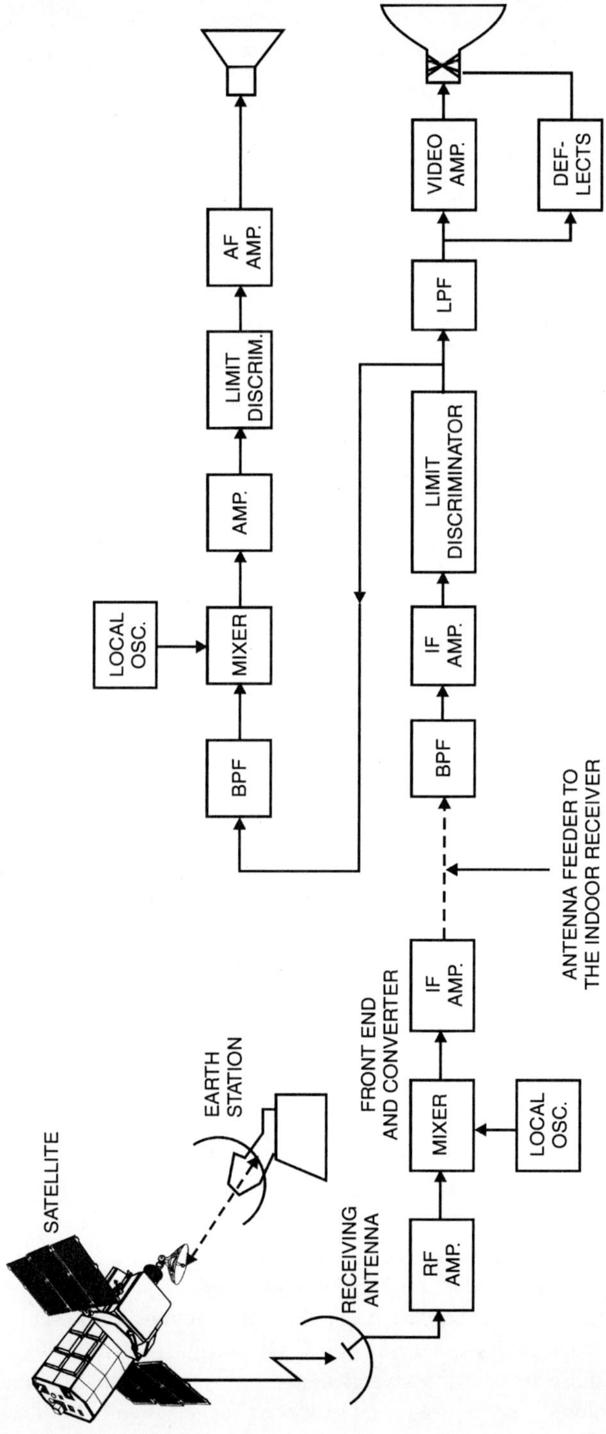
SATELLITE
EARTH STATION
RECEIVING ANTENNA
FRONT END AND CONVERTER
RF AMP.
MIXER
LOCAL OSC.
IF AMP.
ANTENNA FEEDER TO THE INDOOR RECEIVER
BPF
IF AMP.
LIMIT DISCRIMINATOR
LPF
VIDEO AMP.
DEF-LECTS
BPF
MIXER
LOCAL OSC.
AMP.
LIMIT DISCRIM.
AF AMP.

Fig. 8.1.

- These baseband signal modulate a conventional RF carrier frequency, UHF and VHF using AMVSB method of modulation for video and FM for audio. Then the modulated signal is amplified and broadcast through the turnslite antenna system mounted on a high tower.

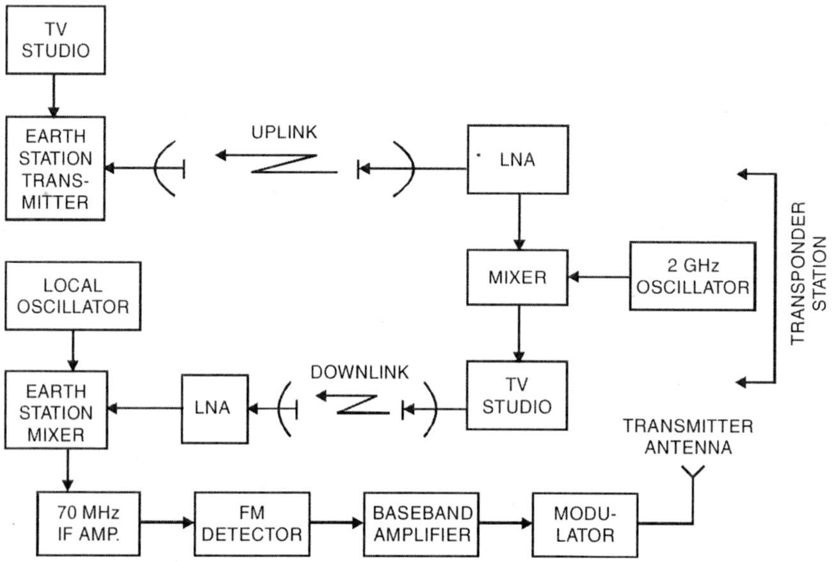

Fig. 8.2. Block diagram of TV broadcast satellite system.

Cable Television (CATV)

Cable Television is in extension of master antenna TV system. CATV system is a cable system which distributes good quality television signals to a very large number of receivers throughout an entire community. It covers the TV receivers located in shadow zone or fringe areas as well as those which are in good line of sight and gives the excellent signal strength. There are several additional features in CATV and these are :

1. The CATV service provider may receive signals from satellites and make a large number of channels available to subscribers.
2. Own generated TV programme by the cable operator, can be send on cables.
3. Internet services through cable TV has also been introduced.
4. A CATV system may have many more active VHF and UHF channels than a receiver tuner can directly select.

Previously CATV system were only employed in far-fringe areas or in valley but now-a-days it is being used in a big cities where signal level become weak and cause ghosts due to multipath reflections, because of tall buildings. If the signals are coming from the different stations, several high gain and properly oriented antennas are employed. In those areas where several signals are coming from the same direction, a log periodic antenna may be used to cover those channels. Figure 8.3 shows the typical plan of a CATV system.

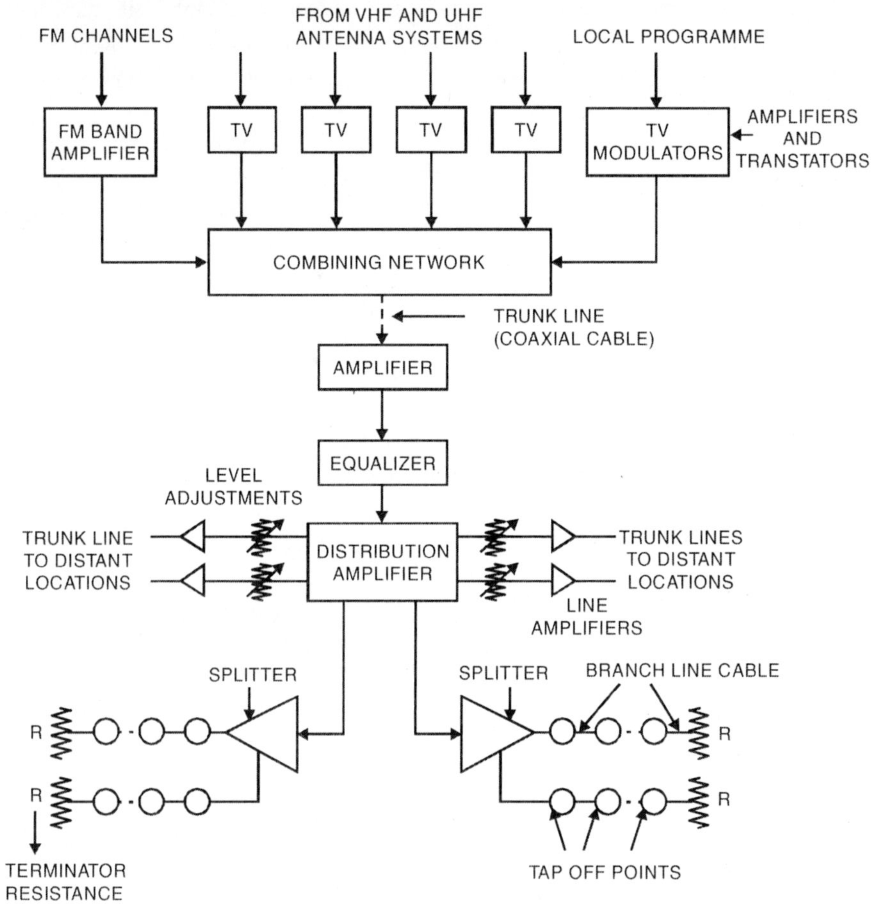

Fig. 8.3. Simplified plan of a CATV.

Head end : The broadcasted signals are picked up by the respective antennas are processed. They are amplified by a low noise amplifier and adjusted for correct level. When UHF reception is provided in addition to VHF, a frequency translator is used to convert the UHF frequencies to VHF frequencies. Translation is important because transmission loss through the cable is more at UHF than at VHF frequencies.

Various inputs, including those coming from the translater are combined in a suitable mixer. This complete set up from the antennas to this combiner is called a head end. The signal from the head end then goes to the trunk line.

Trunk line cables : Usually it is a coaxial cable which is an efficient wibeband line. It has the advantage of shielding. For the trunk line, a good coaxial is about 19 mm in diameter. Optical fibres can also be used but they are not used by the cable operators. In future optical fibre may be used with reduced cost because of the advantage of very high capacity, low loss and no electromagnetic interference.

Amplifier : These are provided at several points along the line to overcome progressive signal attenuation which occurs due to attenuation losses. These losses increase in proportion to the square root of the frequency. At high VHF, these losses may be doubled in comparison to the losses at low VHF. Thus the trunk amplifiers with a gain 20 to 30 dB are placed at suitable interval along the route. This keeps the signal level between 1 and 3 mV.

Equalizer : An equalizer consists of a bandpass filter arrangement with an adjustable frequency response. It is operating by the introduction of a relative low frequency loss so that outputs from the splitters should have uniform relative amplitude response across the entire VHF band.

Distribution amplifier : It takes care of the impedence mismatch caused by the connection with the trunk line and compensates the loss in the trunk line upto the point of connection, gain of this amplifies is 20 to 30 dB.

Line amplifiers : Even the branch lines are shorter in length, they require amplifier of 20 to 30 dB gain at suitable intervals. It is also known as the line extender because it becomes necessary when a branch line is to be extended

Directional coupler : To ensure that each branch line coming out of the trunk line extracts only a small energy from the trunk line, the device used is known as the directional coupler. It is a three terminal device Fig. 8.4 shows a directional coupler and its symbol.

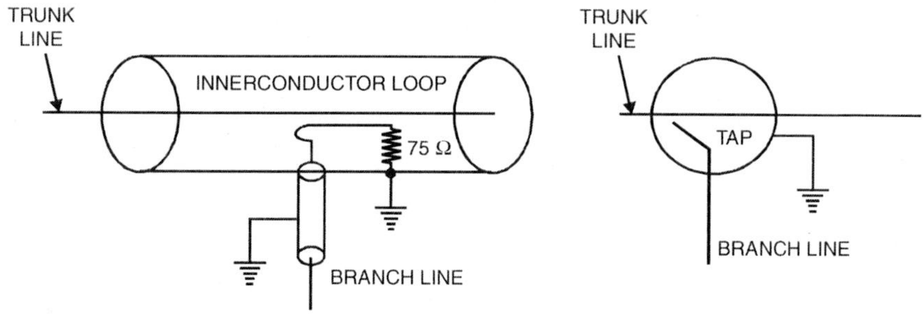

(a) Directional coupler *(b) Symbol of the directional coupler*

Fig. 8.4.

The inner conductor loop works as a capacitance as well as inductance and is internally terminated in a 75 Ω resistance. The directional couples has a very small insertion loss *i.e.* about 1 dB at 300 MHz and it ignores the reflected energy. The tap loss is about 13 dB and is compensated by the distributing amplifier.

Channels for Cable TV. In earlier days of cable TV signals on cables were sent on VHF channels and it is used for the TV broadcast. Early models of televisions had eleven channels in VHF band.

Ist channel ⟶ 41–47 MHz (not used for TV)

2nd-4th channel ⟶ 47–68 MHz

5th-12th channel ⟶ 174–230 MHz

Cable operators can send many more channels through cable lines when they receive signals from the satellites. These signal are converted into VHF/UHF frequencies and multiplexed at the head end before transmitting them through trunk lines. At the receiver, a converter is used to select the desired channel. This converter is known as set-top converter.

The details of cable channels available to CATV operator are given in Table 8.1.

Table 8.1.

Band	Channel name	Frequency range	Purpose
VHF band I	1st channel 2nd - 4th channel	41–47 MHz 47–68 MHz	Not used for TV Used for TV broadcast
VHF band II	X channel *Y* channel Z channel 2 + 1 channel 2 + 2 channel (S_1 to S_{10}) 10 channels	68–75 MHz 75–82 MHz 82–89 MHz 89–96 MHz 96–103 MHz 104–174 MHz	Not used for TV broadcast but used in defence, police and FM broadcast for ATV network
VHF band III	E_5 to E_{12} 8 channels	174–230 MHz	Allows for TV broadcast.
Superband	S_{11} - S_{12} channels 10 channels	230–300 MHz	Used for CATV
Hyperband	S_{21} - S_{41} 21 channels	302–470 MHz	Cable TV
UHF band IV	E_{13} - E_{28} 16 channels	470–502 MHz	Used for cable TV
VHF band V	S_{47} - S_{71} 33 channels	606–870 MHz	Allocated for TV broadcast.

8.3. TELEPHONE SERVICES VIA SATELLITE

Satellite has a very important role in :

- point to point services.
- long distance trunk.

Now-a-days about 80% of International telephone traffic is handled by satellite. Satellite link serves like a repeater station. Satellite link is advantageous when :

(1) The distance involved is more than 1000 km.

(2) The region to be covered is sparsely populated or has difficult geographical terrain.

Figure 8.5 shows a setup of international telephone service via satellite.

In earlier days of development, communication via satellite suffered from echo problem but with the development of device technology and digital techniques this problem is overcome.

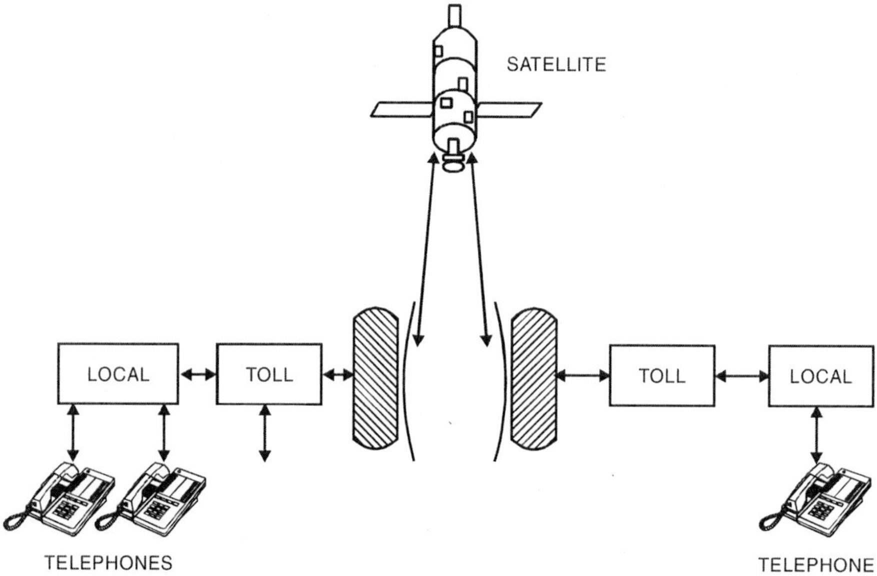

Fig. 8.5.

8.4. DATA COMMUNICATION SERVICES

Data communication services are involved when the advances made in the field of computers and telecommunications. The function of data communication is to link computers and other data processing facilities with other remotely located computers by a communication channel. The telecommunication channel could be :

- terrestrial based
- satellite based.

Terrestrial links for data communication have :

- low data transfer speed
- high cost
- less realisable
- throughput capability.

While satellite links are attractive for data communication because :

- high data transfer rate
- high reliability
- high performance to cost ratio.

8.4.1. Data Broadcasting Using Satellite

Data processing based on point to multipoint configuration is commonly used data communication application using a satellite. Figure 8.6 shows its basic setup.

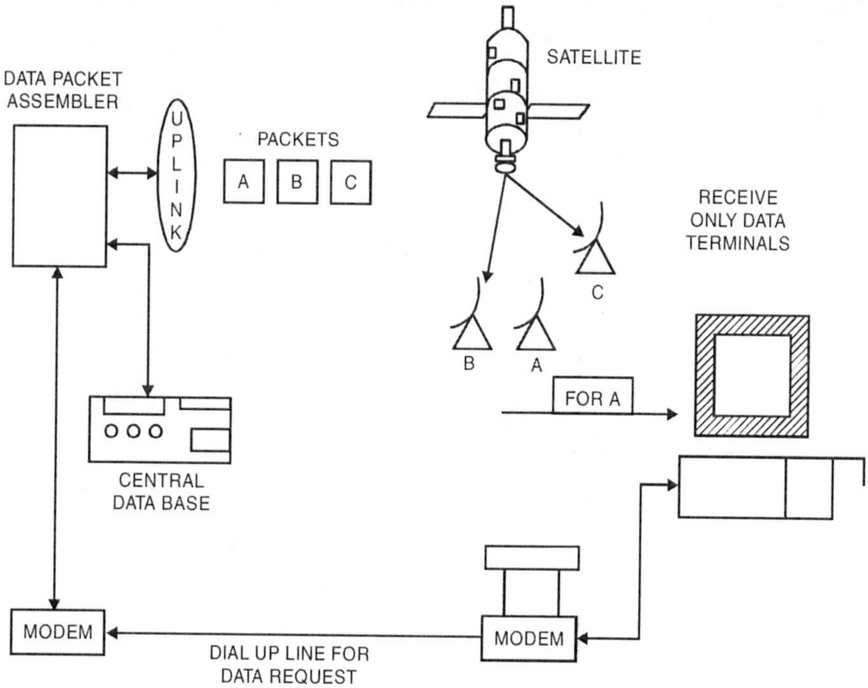

Fig. 8.6.

Data is routed to the earth station from a central facility with the up-linking capability via data packet assembler. The *data packet assembler* is a digital processing device that receives the data to be broadcast, organizes it into packets and places appropriate address bits at the beginning of each packet.

Each data packet is self contained with source and destination address thus it can be routed in any way over satellite and finally through a terrestrial packet switched network. Every data packet is coded also so that only authorised users can have access to the information. In case of news service, full access to all data is provided. The data beamed up at the satellite is retransmitted towards earth to be received by Receive only terminals.

The *Receive only terminals* have digital processing capabilities to identify data packet addressed to then. The respective RO terminal receives the data meant for user terminals linked to them for subsequent delivery.

8.4.2. Interactive Data Communication

In an interactive data communication network, the remotely located user terminals can also exchange information between one another. Such type of network can be best implemented using VSATs and each VSAT supporting large number of user terminals. Figure 8.7 shows an arrangement where the remote stations can transmit information to the central facility of the same satellite.

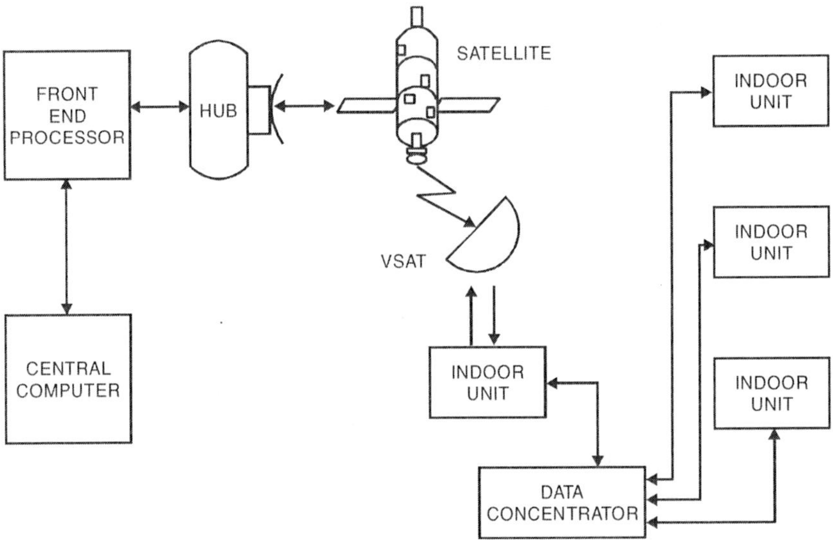

Fig. 8.7.

The data concentrator multiplexes the data flows from the different user terminals and sends a single stream of bits to the VSAT indoor electronic unit. It also demultiplexes the data broadcast received from the satellite for delivery to prospective user terminals.

8.5. DIRECT BROADCAST SATELLITE TELEVISION AND RADIO

Definition : Direct broadcast satellite (DBS) refers to satellite television (TV) systems in which the subscribers, or end users, receive signals directly from geostationary satellites. Signals are broadcast in digital format at microwave frequencies. DBS is the descendant of direct-to-home (DTH) satellite services.

A DBS subscriber installation consists of a dish antenna two to three feet (60 to 90 centimeters) in diameter, a conventional TV set, a signal converter placed next to the TV set, and a length of coaxial cable between the dish and the converter. The dish intercepts microwave signals directly from the satellite. The converter produces output that can be viewed on the TV receiver.

A number of companies provide DBS and DTH service throughout the world. In the United States, DirecTV and EchoStar are the main vendors.

8.5.1. Digital DBS TV

Digital TV signals are transmitted as a 20 Mbps QPSK signal occupying about 27 MHz of transponder bandwidth. The 40 Mbps signal has a data rate between 23 and 27 Mbps with the remaining bits used for error control and system operation. DBS-TV signals make extensive use of error correction and error detection techniques in the form of a double layer of error control coding with interleaving.

Each DBS-TV satellite carries up to 32 high power transponders covering part of the 12.2 to 12.7 GHz broadcast satellite band (BSS), and the satellites at each orbit location transmit in opposite hands of circular polarization (CP). Signals with opposite hands of circular polarization are orthogonal, and a suitably designed earth station antenna can separate two signals with opposite hands of circular polarization. In the simplest DBS- TV receiving terminal, an electronically controlled polarizer is used immediately behind the antenna feed. The polarizer can be set to receive LHCP (left-hand circular polarization) or RHCP by changing the voltage supplied to the low noise block converter (LNB) unit at the antenna. The polarizer converts the circularly received polarization signal to a linearly polarized signal in a section of waveguide, and a linear probe in the waveguide converts the signals to currents that drive the LNA input.

Customers wanting to receive signals from more than one orbital location need an antenna with two feeds.

DBS-TV receiving antennas are typically an offset parabolic reflector design with the feed below the antenna aperture. The offset feed design eliminates blockage of aperture by the feed which occurs in symmetrical reflector antenna designs, and improves the aperture efficiency of the antenna, and therefore increases its gain. Offset fed parabolic reflectors have a beam squint effect in the plane of symmetry when operated in opposite hands of circular polarizations.

8.5.2. DBS TV Receiver

Figure 8.8 shows a block diagram of a DBS-TV receiver. The entire front end of the receiver is located at the antenna feed in the form of an LNB to minimize loss of

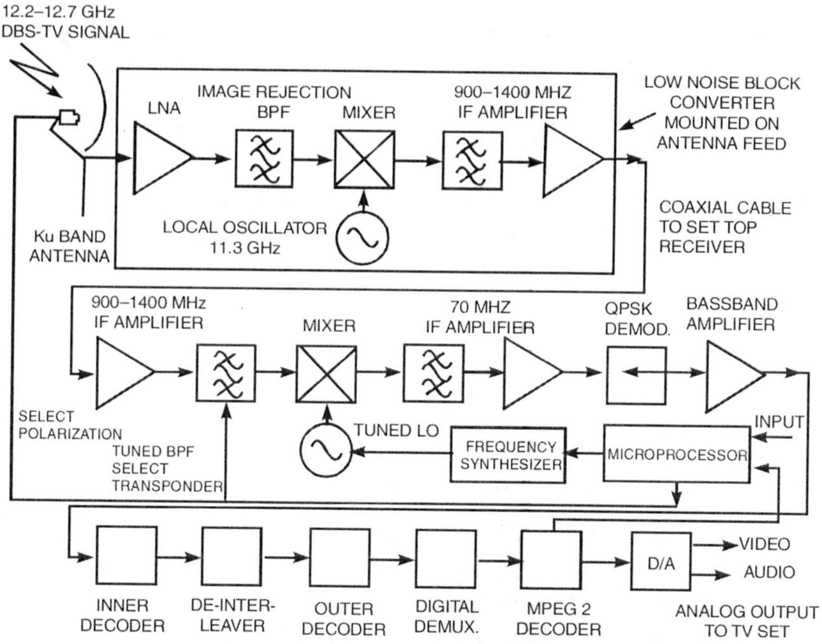

Fig. 8.8.

signal and hence to maintain the lowest possible system noise temperature. The electronic polarizer is switched by changing the voltage supplied to the LNB via the cable that interconnects the antenna and set-top receiver. The entire 12.2 to 12.7 GHz band is down converted by the LNB to the 900-1400 MHz band, where cable losses are much lower than at Ku band. The down-converter consists of a dielectric resonator local oscillator and mixer, followed by an IF amplifier and band-pass filter. The high gain LNB can drive 100 m of coaxial cable without any reduction in signal quality. Where longer cable runs are needed, amplifiers for the 900-1400 MHz band can be used to boost the signal strength. The set-top box accepts the entire 500 MHz band and separates the individual transponder frequencies. Any one of these frequencies can be selected on demand by the user.

The user enters a desired channel number into the set-top box using an IR remote control, for example, channel 362, which is converted via a stored look-up table in the receiver to an RF-channel frequency and polarization. The signal from the required transponder is then selected by the receiver by setting the correct polarization at the antenna and tuning the set-up local oscillator to the appropriate IF channel frequency. The QPSK signal is then demodulated. The result is a multiplexed bit stream, typically at a rate up to 40 Mbps, which contains the bits for channel 362 and several other video signals. The bit stream is encrypted and contains error control coding bits and data bits. The bit stream is processed to correct and detect errors, de-interleaved, and decrypted. A digital demultiplexer then extracts the bits for the wanted channel, 362 in this example, sends them to a MPEG 2 decoder, and finally generates analog video and audio signals with D/A converters to drive the TV set.

8.5.3. Satellite Radio Broadcasting

Satellite Radio broadcasting commenced in 2001 from three Sirius satellites in elliptical orbits and two XM satellites in GEO. The signals are transmitted in S band at 2.3 GHz and are aimed primarily at automobiles, which is where most people listen to the radio. Repeaters are used in the city areas to overcome signal blockage by tall buildings.

Its advantages are :

- DTH acts as a string alternative to cable TV.
- Availability of satellite broadcast in rural and semi-urban areas where cable is difficult to install.
- Due to digital compression techniques, many more niche channels are available than can be provided by cable operators.
- The services are more reliable vis-a-vis repeated breakdowns that can occur in cable transmissions as against satellite broadcast.
- DTH also offers digital quality signals which do not degrade the picture or sound quality.
- One of the great advantages of the cable industry has been the ability to provide local channels, but this handicap has been overcome by many DTH providers using other local channels or local feeds.

8.6. GLOBAL POSITIONING SYSTEM (GPS)

The global positioning system (GPS) is a satellite based navigation system made up of a network of satellite placed into orbit by the U.S. department of defense. GPS was originally intended for military application, but in 1900s, the government made the system available for civilian use. GPS works in any wheather conditions, anywhere in the world, 24 hours a day.

Originally designated the NAVSTAR Navigation system with timing and ranging global positioning system, GPS was developed by the US department of defence to provide round the clock navigation capabilities for military, ground, sea and airforces. Since its implementation, GPS has also become an integral asset in numerous civilian applications and industries around the globe, including recreational uses, corporate vehicle fleet tracking and surveying. GPS employs 24 spacecraft is 20,200 km circular orbits inclined at 55°. These spacecraft are placed in 6 orbit planes with four operatinal satellites in each plane.

8.7. GPS POSITION LOCATION PRINCIPLE

The global positioning system is comprised of three segments : Satellite constellation ground control/monitoring network and user receiving equipment. Formal GPS joint program office (JPO) programmatic terms for these components are space, operational control and user equipment segments, respectively.

1. The satellite consetellation contains the satellites in orbit that provide the ranging signals and data messages to the user equipment.
2. The operational control segment (OCS) tracks and maintains the satellites in space. The OCS monitors satellite health and signal integrity and maintains the orbital configuration of the satellite. Furthermore, the OCS updates the satellite clock corrections and ephemerides as well as numerous other parameters essential to determining user position, velocity and time (PVT).
3. The user receiver equipment performs the navigation, timing or other related notation.

GPS Satellite Constellation

The satellite constellation consists of the nominal 24-satellite constellation (the first was launched in 1978 and the 24th in 1994). They transmit signals (at 1575.42 MHz) that can be detected by receivers on the ground. The satellites are positioned in six Earth-centred orbital planes with four satellites in each plane. This means that signals from six of them can be received 100 percent of the time at any point on earth. The nominal orbital period of a GPS satellite is one half of a sidereal day or 1 1 hr 58 min. The orbits are nearly circular and equally spaced about the equator at a 60° separation with an inclination relative to the equator of nominally 55°. The orbital radius is approximately 26,600 km (*i.e.*, distance from satellite to centre of mass of the earth).

GPS satellites transmit two low power radio signals, designated LI and L2. Civilian GPS uses the L1 frequency of 1575.42 MHz in the UHF band. A GPS signal contains three different bits of information — a pseudo-random code, ephemeris data and almanac data. The pseudo-random code is simply an I.D. code that identifies which satellite is transmitting information.

Several different notations are used to refer to the satellites in their orbits. One particular notation assigns a letter to each orbital plane (*i.e.*, *A*, *B*, *C*, *D*, *E* and *F*) with each satellite within a plane assigned a number from 1 to 4. Thus, a satellite referenced as B3 refers to satellite number 3 in orbital plane *B*. A second notation used is a NAVSTAR satellite number assigned by the U.S. Air Force. This notation is in the form of a space vehicle number (SVN) 1 1 to refer to NAVSTAR satellite 11.

Operational Control Segment (OCS)

The OCS has responsibility for maintaining the satellites and their proper functioning. This includes maintaining the satellites in their proper orbital positions (called station keeping) and monitoring satellite subsystem health and status. The OCS also monitors the satellite solar arrays, battery power levels, and propellant levels used for manoeuvres and activates spare satellites. The overall structure of the operational ground/control segment is as follows: Remote monitor stations constantly track and gather C/A and P(Y) code from the satellites and transmit this data to the Master Control Station, which is located at Falcon Air Force Base, Colorado Springs. There is also the ground uplink antenna facility, which provides the means of commanding and controlling the satellites and uploading the navigation messages and other data. The unmanned ground monitor stations are located in Hawaii, Kwajalein in the Pacific Ocean, Diego Garcia in the Indian Ocean, Ascension Island in the Atlantic and Colorado Springs, Continental United States. Ground antennas are located in these areas also. These locations have been selected to maximise satellite coverage.

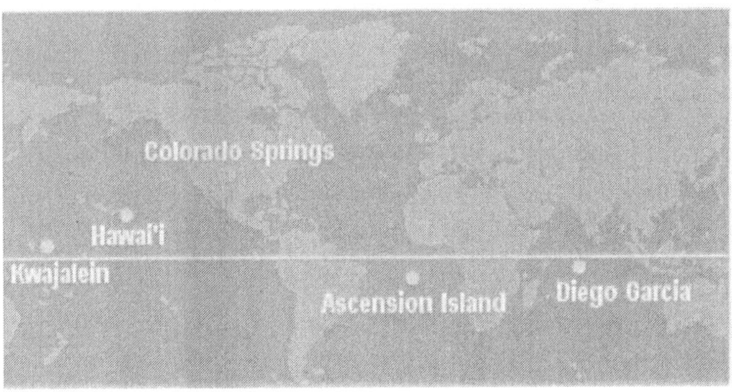

Fig. 8.9. Location of GPS ground stations.

User Receiving Equipment

The user receiving equipment, typically referred to as a GPS receiver processes the L-band signals transmitted from the satellites to determine PVT. There has been a significant evolution in the technology of GPS receiving sets since they were initially manufactured in the mid-70's. Initially, they were large, bulky and heavy analog devices primarily used for military purposes. With today's technology, a GPS receiver of comparable or more capability typically weighs a few pounds or ounces, and occupies a small volume. The smallest of today's are those of a wrist watch size, while the largest is a naval shipboard unit (weighing about 32 kgs). The basic structure of a receiver is the antenna, the receiver and processor, the display and a regulated dc-power supply. These receivers can be mounted in ships, planes and cars, and provide exact position information, regardless of weather conditions.

8.8. GPS CODES

GPS satellite transmits using pseudo random sequence (PN) codes. The original GPS design contains two ranging codes:

- *Coarse/Acquisition* (C/A) code, which is freely available to the public.
- Restricted *Precision* (P) code, which is usually reserved for military applications.

Coarse/Acquisition code

The C/A code is a 1,023 bit which deterministic sequence called pseudorandom noise (also known as **pseudorandom binary sequence**) (PN or PRN code) which, when transmitted at 1.023 megabits per second (M bit/s), repeats every millisecond. These sequences only match up, or strongly correlate, when they are exactly aligned. Each satellite transmits a unique PRN code, which does not correlate well with any other satellite's PRN code. In other words, the PRN codes are highly orthogonal to one another. This is a form of **code division multiple access (CDMA),** which allows the receiver to recognize multiple satellites on the same frequency.

Precision Code

The P-code is also a PRN; however, each satellite's P-code PRN code is 6.1871×10^{12} bits long (6,187,100,000,000 bits, ~720.213 gigabytes) and only repeats once a week (it is transmitted at 10.23 Mbit/s). The extreme length of the P-code increases its correlation gain and eliminates any range ambiguity within the Solar System. However, the code is so long and complex it was believed that a receiver could not directly acquire and synchronize with this signal alone.

First the receiver would lock onto the relatively simple C/A code and then, after obtaining the current time and approximate position, synchronize with the P-code. Whereas the C/A PRNs are unique for each satellite, the P-code PRN is actually a small segment of a master P-code approximately 2.35×10^{14} bits in length (235,000,000,000,000 bits, ~26.716 terabytes) and each satellite repeatedly transmits its assigned segment of the master code.

To prevent unauthorized users from using or potentially interfering with the military signal through a process called spoofing, it was decided to encrypt the P-code. To that end the P-code was modulated with the W-code, a special encryption sequence, to generate the Y-code. The Y-code is what the satellites have been transmitting since the anti-spoofing module was set to the "on" state. The encrypted signal is referred to as the P(Y)-code. The details of the W-code are kept secret, but it is known that it is applied to the P-code at approximately 500 kHz, which is a slower rate than that of the P-code itself by a factor of approximately 20.

8.9. SATELLITE SIGNAL ACQUISITION

GPS signal acquisition is a search process. The main purpose of acquisition is to identify the visible satellites in the data and then find the beginning point of the C/A code and the Doppler frequency by correlating the incoming signal with receiver generated signal.

The GPS signal is combination of carrier wave, C/A code and navigation message. In order to extract the navigation message from the GPS signal, it is necessary to remove the carrier wave (demodulate) and the C/A code (de-spread). After, demodulation and de-spreading, we get only the navigation message and noise. These navigation messages are then interpreted based on the definitions or coding scheme provided in the interface control document (ICD) of GPS.

Demodulation and data recovery of a direct sequence spread spectrum signal involves several operations:

- Carrier synchronization;
- Chip synchronization;
- Coarse despreading signal acquisition;
- Code synchronization code tracking;
- Data demodulation.

Signal acquisition is described below. Since there are several strategies and system configurations that can be used for acquisition, this explanation is based on the block diagram in Figure 8.10. We make the following assumptions:

- The spreading code period is relatively short.
- Demodulation is coherent, that is, the local oscillator is phased locked to the carrier, or the IF, if down conversion is done in stages.
- Transmitter and receiver chip rates are equal.
- The code is an m-sequence, generated in a linear feedback shift register.
- No data is transmitted during acquisition.
- There is no noise or interference.
- Code modulation is BPSK.

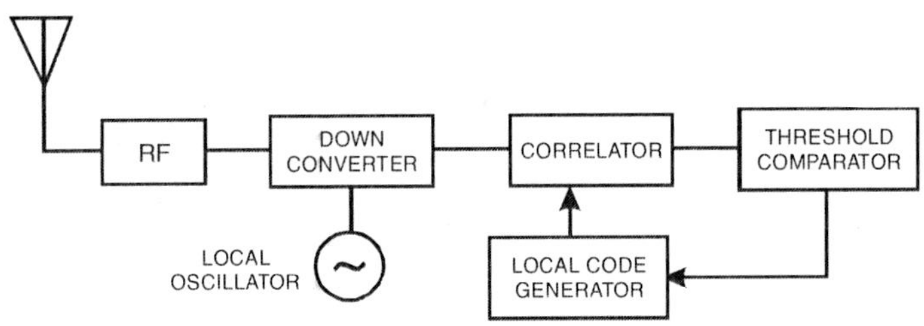

Fig. 8.10. Receiver configuration for signal acquisition.

This search process, like the tracking process, requires replication of both the code and the carrier of the SV to acquire the SV signal (i.e., the signal match for success is two dimensional). The range dimension is associated with the replica code. The initial search process is always a C/A code search for C/A code receivers and usually begins with a C/A code search for P(Y) code receivers. The initial C/A code search usually involves replicating all 1,023 C/A code phase states in the range dimension. The criteria for direct C/A code and direct P(Y) code acquisitions were discussed in the previous section.

If the range and Doppler uncertainty are known, then the search pattern should cover the 3-sigma values of the uncertainty. If the uncertainty is large in either or both dimensions, the search pattern is correspondingly large, and the expected search time increases. Some criteria must be established to determine when to terminate the search process for a given SV and select another candidate SV. Fortunately, the range dimension for C/A code search is bounded by the ambiguity of C/A code to only 1,023 chips total range uncertainty, but it is essentially unbounded for direct P(Y) code search.

8.10. GPS NAVIGATION MESSAGE

The GPS satellites transmit two microwave carrier signals:

- The L1 frequency (1575.42 MHz) carries the navigation message and the SPS code signals.
- The L2 frequency (1227.60 MHz) is used to measure the ionosphere delay by PPS equipped receivers.

The positioning information from GPS satellites in sent in the form of repeating codes which identify the sending satellite, give locations of the other satellites in the system, and give the navigation data. The codes which form the GPS signal structure are superimposed upon two carrier waves in the L-band, a frequency range set aside for satellite communication. Both the carrier frequencies and the signal frequencies are derived directly from the onboard atomic clock oscillator frequency of 10.23 MHz. The coded signals are repeated regularly in epochs on the order of 15 seconds. The codes used are known as "pseudo-random codes. The codes are well suited to decoding a message embedded in noise signals which may be orders of magnitude larger than the signal itself. Such techniques were valuable in picking up radar echoes off the moon and asteroids. Use of these codes facilitates the sorting out of signals which arrive simultaneously from several GPS satellites.

The Navigation Message is a 50 Hz signal consisting of data bits that describe:

- GPS satellite orbits
- Clock corrections
- Other system parameters.

Structure of the GPS Navigation Data

Structure of GPS navigation message (data) is shown in Fig. 8.11. The user has to decode the data signal to get access to the navigation data. For on line navigation purposes, the internal processor within the receiver does the decoding. Most of the manufacturers of GPS receiver provide decoding software for post processing purposes.

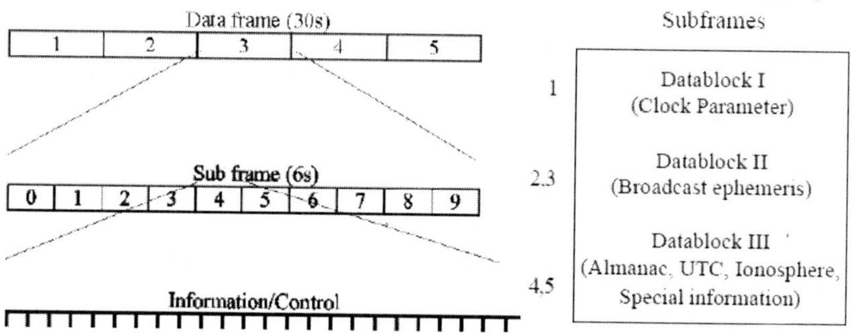

Fig. 8.11. Structure of navigation data.

With a bit rate of 50 bps and a cycle time of 30 seconds, the total information content of a navigation data set is 1500 bits. The complete data frame is subdivided into five sub frames of six-second duration comprising 300 bits of information. Each sub frame contains the data words of 30 bits each. Six of these are control bits. The first two words of each sub frame are the Telemetry Work (TLM) and the C/A-P-Code

Hand over Work (HOW). The TLM work contains a synchronization pattern, which facilitates the access to the navigation data.

The navigation data record is divided into three data blocks:

Data Block I : appears in the first sub frame and contains the clock coefficient/ bias.

Data Block II : appears in the second and third sub frame and contains all necessary parameters for the computation of the satellite coordinates.

Data Block III : appears in the fourth and fifth sub frames and contains the almanac data with clock and ephemeris parameter for all available satellite of the GPS system. This data block includes also ionosphere correction parameters and particular alphanumeric information for authorized users.

Unlike the first two blocks, the sub frames four and five are not repeated every 30 seconds.

8.11. GPS RECEIVER

Components of GPS Receiver : The main components of a GPS receiver are shown in Fig. 8.12.

- **Antenna : Sensitive antenna** of the GPS receiver detects the electromagnetic wave signal transmitted by GPS satellite and converts the wave energy to electric current, amplifies the signal strength and sends them to receiver electronics. Commonly used GPS antennas are : Monopole or dipole, spiral helix, quadrifilar helix, microstrip (patch) and choke ring.
- **RF section with signal identification :** The incoming GPS signals are down converted to a lower frequency in the RF section and processed with one or more channels. Receiver channel is the primary electronic unit of a GPS receiver. A receiver may have one or more channels. In the parallel channel concepts each channel is continuously tracking one particular satellite. A minimum of four parallel channels is requried to determine position and time. Modern receivers contain upto 12 channels for each frequency. Continuous tracking parallel channels are cheap and give good overall performance.

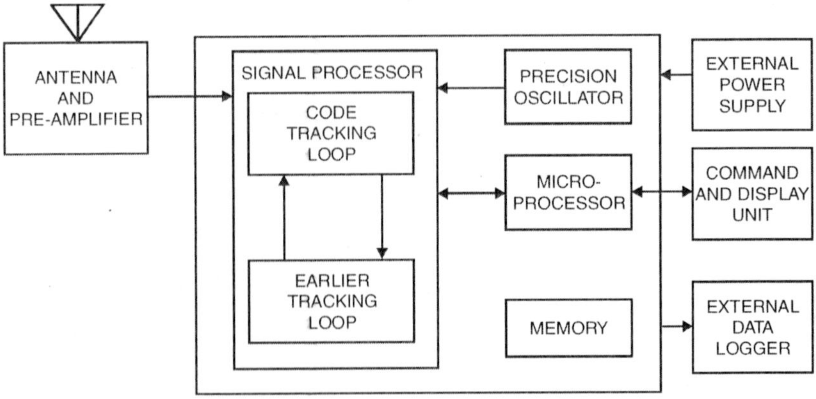

Fig. 8.12. Major components of GPS receiver

- **Microprocessor :** To control the operation of a GPS receiver, a microprocessor is essential for acquiring the signals, processing of the signal and the decoding of the broadcast message.
- **Precision oscillator** : A reference frequency in the receiver is generated by precision oscillator. Usually less expensive, low performance quarter oscillator is used in receiver since the precision clock information is obtained from the GPS satellites. The user clock error can be eliminated through double differencing technique.
- **Power supply :** First generation GPS receivers consumed very high power but modern receivers are design to consume as little energy as possible. Most receivers have an internal rechargeable Nickel-cadmium battery in addition to an external power input.
- **Memory capacity :** For port processing purposes all data have to be stored on internal or external memory devices. Post processing is essential for multistation techniques applicable to geodatic and surveying problems. GPS observation for pseudoranges, phase data, time and navigation message data have to be recorded. Modern receivers have internal memories of 5 M bytes or more. Some receivers some data on magnetic tape or on hard disk using external microcomputer connected through RS-232 port.

GPS Receiver Block Diagram : In any satellite receiver since the input power is of the order of picowatts and the C/N to be maintained to get demodulated S/N above the threshold fixed for error free processing. Figure 8.13 shows the configuration of GPS receiver.

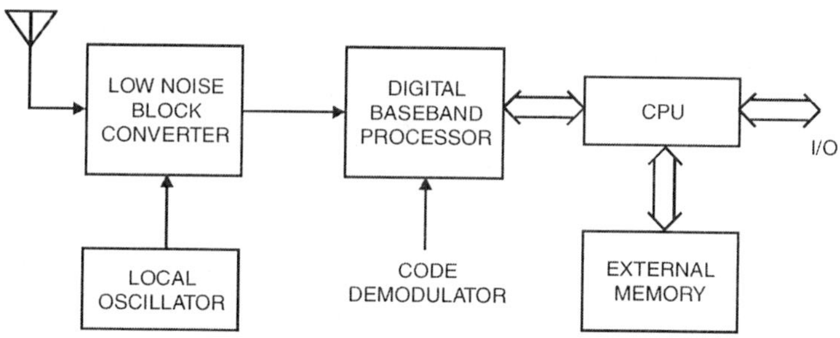

Fig. 8.13. GPS Receiver block diagram.

To recover the baseband cost's loop is used with phase synchronization of local oscillator or else the signal cannot be detected.

The required PN code is locally generated to detect the chip of the spread spectrum received signal. This requires as many correlators as the visible satellites. Most of the receivers have 12 correlators. Diversity reception is used and the strongest signals are used for processing.

The CPU uses software to calculate *XY* and *Z*-coordinates and hence find latitude, longitude and altitude.

8.12. ERROR CORRECTION

A variety of different errors can occur within the system, some of which are natural, while others are artificial. First of all, a basic assumption, the speed of light, is not constant as this value changes as the satellite signals travel through the atmosphere. As a GPS signal passes through the charged particles of the ionosphere and then through the water vapour of the troposphere it gets slowed down, and this creates the same kind of error as bad clocks. This problem is tackled by attempting to use modelling of the atmospheric conditions of the day, and using dual-frequency measurement, *i.e.* comparing the relative speeds of two different signals. Another problem is multipath error, this is when the signal may bounce off various local obstructions before it gets to our receiver. Sophisticated signal rejection techniques are used to minimize this problem.

There are also potential problems at the satellites. Minute time differences can occur within the on-board atomic clocks, and sometimes position errors can occur. These other errors can be magnified by a high GDOP **"Geometric Dilution of Precision"**.

8.13. DIFFERENTIAL GPS

Using a modified form of GPS called **Differential GPS** (originally initiated by the U.S. Coast Guard to counter the accuracy degradation caused by Selective Availability) can significantly reduce the above errors. Even with SA eliminated, DGPS continues to be a key tool for highly precise navigation on land and sea. DGPS can yield measurements accurate to a couple of meters in moving applications and even better in stationary situations.

- Differential GPS involves the co-operation of two receivers, one that's stationary and another that's revolves around making position measurements.
- As each GPS receivers use timing signals from at least four satellites to establish a position then each of those timing signals is going to have some error or delay depending on what sort of problems have occurred it on its journey down to Earth. Since each of the timing signals that go into a position calculation has some error, that calculation is going to be a compounding of those errors.
- However if two receivers are fairly close to each other, say within a few hundred kilometres, the signals that reach both of them will have travelled through virtually the same slice of atmosphere, and so will have virtually the same errors.
- This means that you could use have one receiver to measure the timing errors and then provide correction information to the other receivers that are moving around. This allows virtually all errors to be eliminated from the system.
- The reference station operates by receiving the same GPS signals as the moving receiver but instead of working like a normal GPS receiver it uses its known position to calculate timing, rather than using timing signals to calculate position.
- Essentially determining what the travel time of the GPS signals should be, and compares it with what they actually are. The difference is an "error correction" factor. The receiver then transmits this error information to the roving receiver so it can use it to correct its measurements.
- Since the reference receiver has no way of knowing which of the many available satellites a roving receiver might be using to calculate its position, the reference

receiver quickly runs through all the visible satellites and computes each of their errors. Then it encodes this information into a standard format and transmits it to the roving receivers.

- The roving receivers can then apply the corrections for particular satellites they are using. The United States Coast Guard and other international agencies are establishing reference stations all over the place, especially around busy harbours and waterways.

There are also different kinds of DGPS, for use when users don't need precise positioning immediately. This is termed **Post Processing DGPS,** and is used when the roving receiver just needs to record all of its measured positions and the exact time it made each measurement. Then later, this data can be merged with corrections recorded at a reference receiver for a final clean-up of the data, meaning you don't need the radio link required in real-time systems. Another form of DGPS, called **Inverted DGPS,** which is used to save money when operating a large fleet of users. With an inverted DGPS system the users would be equipped with standard GPS receivers and a transmitter and would transmit their standard GPS positions back to the tracking station (the main office). Then at the tracking station the corrections would be applied to the received positions.

8.14. CARRIER-PHASE GPS

This is a new version of GPS that can eliminate errors even better than other forms. Recall that a GPS receiver determines the travel time of a signal from a satellite by comparing the pseudo random code it's generating, with an identical code in the signal from the satellite. The receiver slides its code later and later in time until it syncs up with the satellite's code. The amount it has to slide the code is equal to the signal's travel time. The problem is that the bits (or cycles) of the pseudo-random code are so wide that when the signals sync up there is room for error. Survey receivers are better as they start with the pseudo-random code and then move on to measurements based on the carrier frequency for that code. This carrier frequency is much higher so its pulses are much closer together and therefore more accurate. At the speed of light the 1.57 GHz GPS signal has a wavelength of roughly twenty centimetres, so the carrier signal can act as a much more accurate reference than the pseudo random code by itself.

- In essence this method is counting the exact number of carrier cycles between the satellite and the receiver.
- The problem is that the carrier frequency is hard to count because it's so uniform. Every cycle looks like every other.
- The pseudo-random code on the other hand is intentionally complex to make it easier to know which cycle you're looking at. But Carrier-phase GPS tackles this problem by using code-phase techniques to get close.
- If the code measurement can be made accurate to say, a meter, then we only have a few wavelengths of carrier to consider as we try to determine which cycle really marks the edge of our timing pulse.
- Resolving this carrier phase ambiguity for just a few cycles is a much more tractable problem and as the computers inside the receivers increase in

processing power and functionality it's becoming possible to make this kind of measurement without all the steps that survey receivers go through.

8.15. APPLICATIONS OF GPS TECHNOLOGY

GPS technology has matured into a resource that goes far beyond its original design goals. These days people from a plethora of professions are using GPS in ways that make their work more productive, safer, and sometimes even easier.

There are five main uses of GPS today:

1. Location- determining a basic position.
2. Navigation - getting from one location to another.
3. Tracking - monitoring the movement of people and things.
4. Mapping- creating maps.
5. Timing - providing precise timing.

1. Location

The first and most obvious application of any Location Based Service_such as GPS is the simple determination of a "position" or location. GPS was the first positioning system to offer highly precise location data for any point on the planet, in any weather. Knowing the precise location of something, or someone, is especially critical when the consequences of inaccurate data are measured in human terms.

2. Navigation

GPS helps you determine exactly where you are, but sometimes it is more necessary to know how to get somewhere else. Recall that GPS was originally designed to provide navigation information for ships and planes. So it's no surprise that while this technology is appropriate for navigating on water, it's also very useful in the air and on the land.

3. Tracking

GPS used in conjunction with communication links and computers can provide the backbone for systems tailored to applications in agriculture, mass transit, urban delivery, public safety, and vessel and vehicle tracking. Therefore, more and more police, ambulance and fire departments are adopting systems like GPS-based AVL (Automatic Vehicle Location) Manager to pinpoint both the location of the emergency and the location of the nearest response vehicle on a computer map. With this kind of clear visual picture of the situation, dispatchers can react immediately and effectively.

4. Mapping

Mapping the planet has never been an easy task, but GPS today is being used to survey and map it precisely, saving time and money in this most stringent of all applications. GPS can help generate maps and models of everything in the world, mountains, sea, rivers, cities and help manage endangered animals, archaeological treasures, precious minerals and all sorts of resources, as well as accurately managing the effect of damage and disasters.

For example peace brokers used GPS generated maps to determine the partitions of Bosnia under the Dayton Peace Accord, and GPS was used to map Cambodia accurately for the UN stabilisation force introduced after the civil war ended in 1993.

In fact large tracks of Cambodia had never been mapped at all, due to inhospitable terrain and low population density. GPS solved this in weeks, whereas it would have taken months prior to the introduction of GPS.

5. Timing

GPS can also be used to determine precise time, time intervals, and frequency. There are three fundamental ways we use time:

- As a universal marker,
- As a way to synchronize
- To provide an accurate, unambiguous sense of duration.

Summary

1. Remote sensing satellite working is based on sensors data processing, data analysis and data interpretation for different applications.
2. Global positioning system consists of three segments
   - Satellite contellation.
   - Ground control/monitoring network
   - User receiving equipment.
3. The orbital period of a GPS satellite is one half of a side real day of 11 hr 58 min.
4. The modified form of GPS satellite is known as differential GPS.
5. There are five main uses of GPS location, navigation, tracking, mapping and timing.
6. Satellite can be used as navigational aids by vehicles on land, in the air or on the sea.
7. Cartography *i.e.* map making in which satellite images can be used to produce related maps for the construction of roads, railways and irrigation channels.
8. In oceanography it is possible to make the pictures of ocean waves, ice fields, ice bergs, ice lands and sea conditions along the coastal lines with the help of satellite.
9. Present day remote sensing satellite equipped with the state of the art high resolution camera are capable of monitoring characteristics like crop identification.
10. There are two basic categories of operational wheather satellite depending upon the type of orbit :
    - Polar orbit
    - Geostationary orbit.
11. Radiometer is an important device that is equipped with wheather satellite and and it measures the amount of radiation.
12. The first early warning satellite IMEWS was launched in 1970. It warns to the owners of a missible attack.
13. USA generates the ideal of nuclear detection using satellites. Twelve satellites of vela series (Vela-1 and Vela-2) were launched during time period 1963-1970 for nuclear detection.
14. In direct broadcast satellite service, planned broadcasting directly to home TV receivers takes place in the KU band *i.e.* 12 GHz.
15. The idea of direct broadcasting of TV from a satellite was introduced around 1986 with the existing communication satellites.
16. For DBS TV transmission the downlink frequency is about 12 GHz and uplink frequency is about 14.5–14.8 GHz.

17. Major difference between the DBS TV and conventional TV is that with DBS, frequency modulation is used whereas with conventional TV, amplitude modulation in the form of vestigial single sideband is used.
18. In VSAT (Very small aperture terminal) the earth station antenna size is typically less than 2.4 in diameter.
19. VSAT system operate in star configuration *i.e.* the connection of one VSAT to another must be made through the hub.
20. The major shortcoming of present day VSAT systems are
    - High initial costs
    - Tendency towards optimizing systems for large networks
    - Lack of direct VSAT to VSAT links.
21. In the ALOHA concept there are three variations
    - Pure ALOHA
    - Slotted ALOHA.
    - Reservation ALOHA
22. Dynamic accessing schemes may be classified into two categories :
    - Contention or random access schemes
    - Reservation schemes
23. Radarset is an earth resource remote sensing satellite which is a part of the canadian space program.
24. Radarset is a large spacecraft, the total man is the orbit is about 3100 kg and it works at a carrier frequency of 5.3 GHz.
25. Mobile stations use two types of antennas
    - Steered antennas
    - Nonsteered antennas
26. There are two different kind of DGPS
    - Post processing DGPS
    - Inverted DGPS
27. GPS can also be used to determine precise time, time intervals and frequency.

Review Questions

1. How many types of orbits can be used for satellites in wheather forecasting ?
2. Discuss the direct broadcast satellite (DBS) frequency plan with orbit location, allocated to India.
3. List the various advantages to use Ku band for direct broadcast satellite service.
4. Draw the block diagram of outdoor unit of a DBS home receiver. Explain the function of each block briefly.
5. State and explain the different segments of a global positioning satellite system.
6. Explain why a minimum of four satellites must be visible at an earth location utilizing the GPS system for position determination.
7. What do you meant by differential GPS ? Explain briefly.
8. Discuss the various applications of GPS (global partitioning satellite) system.
9. Draw and discuss the block diagram of a DBS system.

Objective Type Questions

1. One of the following satellite system is used for wheather forecast application
(*a*) GORIZONT (*b*) TIROS-N
(*c*) COMSAT (*d*) SPOT

2. The satellite system used for earth's resources survey is
(*a*) INTELSAT (*b*) SPOT
(*c*) EUTELSAT (*d*) TIROS-N

3. The satellite series developed by ISRO for measurement of earth observation application is the
(*a*) INSAT-2 series (*b*) INSAT-1 series
(*c*) IRS series (*d*) SROSS-C series

4. In India, which series have been used for metrological data
(*a*) INSET series (*b*) IRS series
(*c*) SROSS-C series (*d*) None of these.

5. Which of the organisation is pursuing the programme of using space borne vehicles for carrying out scientific experiments
(*a*) ISRO (*b*) DRSO
(*c*) NASA (*d*) None of these

6. For DBS TV transmission the downlink frequency is about
(*a*) 14 GHz (*b*) 10 GHz
(*c*) 6 GHz (*d*) 12 GHz

7. The number of channels, that can be accommodated in the downlink frequency band of DBS system
(*a*) 32 (*b*) 64
(*c*) 24 (*d*) 23

8. The uplink frequency in DBS TV transmission is
(*a*) 13–14.5 GHz (*b*) 14.5–14.8 GHz
(*c*) 12.5–13.5 GHz (*d*) 15.5–15.8 GHz

9. Type of modulation used in direct broadcast satellite
(*a*) Frequency modulation (*b*) Amplitude modulation
(*c*) Pulse code modulation (*d*) None of these

10. VSAT stands for
(*a*) Very small aperture terminal
(*b*) Very small antenna terminal
(*c*) Very small antenna transistor
(*d*) None of these

11. The type of configuration used in VSAT system is
(*a*) Hybrid (*b*) Data
(*c*) Mesh (*d*) Star

12. Time period of a RADARSET is
(*a*) 90.3 min (*b*) 100.3 min
(*c*) 100.7 min (*d*) 110.7 min

13. Number of satellites in global positioning system.

(*a*) 12 (*b*) 24

(*c*) 3 (*d*) 6

14. Generally the VSAT network is operated in

(*a*) FDMA mode (*b*) TDMA mode

(*c*) CDMA mode (*d*) None of these

15. Infrared Astronomical Satellite (IRAS) was launched in

(*a*) 1983 (*b*) 1973

(*c*) 1960 (*d*) 1970

ANSWERS

1. (*b*)	**2.** (*b*)	**3.** (*c*)	**4.** (*a*)	**5.** (*a*)
6. (*d*)	**7.** (*a*)	**8.** (*b*)	**9.** (*a*)	**10.** (*a*)
11. (*d*)	**12.** (*c*)	**13.** (*b*)	**14.** (*a*)	**15.** (*a*)

❑❑❑

9

CHAPTER

Global Mobile Satellite Systems

INSIDE THIS CHAPTER

9.1. INTRODUCTION

Initially mobile communication systems were conceived for the transmission and reception of telegraphy and telephony signals via antenna from ships, cars, trains and aircrafts.

- The consideration of antenna transmission is inevitable in GMSC systems because their propagation characteristics are very much affected by different and changable local environments during movement of mobile and mainly they differ from those observed in fixed satellite systems.
- To design antenna hardware for mobile GMSC systems, to realize full mechanical and transmission potentials and transmission potentials, engineers have to consider all related factors.
- This chapter introduces the antenna characteristics basic relationship of antenna system and their requirements, antenna classification and antenna mounting and tracking systems.

9.2. EVOLUTION OF ANTENNA SYSTEMS FOR MOBILE RADIO COMMUNICATION

- In 1895, the "Russian Professor of Physics Propov" designed first world's radio receiver with an antenna in the shape of wire mounted on a ballon in the air and transmitter with lightning conductor as an antenna including a metal filings coherer and a detector element with telegraph relay and a bell.
- After that Marconi started commercially to deploy radio and antenna equipment on board different merchant ships. He also establish his own company for the production of maritime radio and antenna equipment.

- Initially mobile radio was used for long distance wireless communications at LF.
- First shipboard antenna were all made of haphazard lengths of wire string as high as possible above the ships topside because the thinking was that the longer and higher wire gives better result.
- After that various kinds of wire and whip antennas were developed for MF/HF/VHF maritime and aeronautical applications.

9.2.1. Development of Antenna for MSC

The MSC systems introduced new complexities into the design of shipboard antennas.

- The direct line of sight between antenna and satellite require the antenna to see from horizon to over head in elevation and 360° in azimuth angle with total hemispherical coverage. In case of transceiver antenna it is fulfilled through the use of tracking rotatable, high gain antenna often installed in pairs on board ship to attain full coverage, besides blockage in the form of funnel, musts, stacks and other deck objects.
- Ship platform imposes even some more stringment requirements. The MSC antenna's narrow radiation beam must be pointed accurately from any position on the high sea in spite of constant vibrating, pitching, rolling and yawing during bad weather condition. **This situation is less complicated in land or aero antenna**.
- In 1970's the idea came to virtually replace HF radio in Navy with the new flatsatcom military mobile system.
- Till this time several UHF transceiving antennas developed for us naval shipboard services were used. Example : **Oldest crossed dipole array**. Its improved version **wash tube** similar to an **SBF antenna** and one type of SHF parabolic dish antenna.
- The first real global MSC system was the Marisat system which employed SES and L-band antenna systems similar to current Inmarsat-A and B terminals.

9.2.2. Classification and Types of Mobile Satellite Antennas (MSA)

The general classification of MSA is based on service operator/provided. They are of four major types :

- Satellite communication antennas.
- Satellite broadband (multimedia) antennas.
- Satellite navigation antennas.
- Military satellite antennas.

But fundamentals classification of MSA according to gain values and characteristics are of three types :

- Low gain omnidirectional antenna.
- Medium gain directional antenna.
- High gain directional operator antenna.

9.3. ANTENNA REQUIREMENTS AND TECHNICAL CHARACTERISTICS

The general requirements for mobile antenna solutions used in GMSC systems for maritime, land and aeronautical application, including antennas for personal handheld terminals are discussed in this section.

The important characteristics of mobile antennas are :

- Compact
- Flexible
- Light weight
- Perform with good mechanical and electrical characteristics.

These characteristics are very important for :

- heavy mobiles such as ships and aircraft.
- Owing to the special conditions of installation.
- Influence of changeable environmental conditions.

9.3.1. Mechanical Characteristics

Mobile antenna have to satisfy the requirements of mechanical characteristics such as :

- Construction strength
- Easy installation
- Compactness
- Light weight

Easy installation and appropriate physical shape are very important. The installation requirements are not severe in the case of ship borne antennas as compared to aircraft and cars because there is a comfortable space to install an antenna set even in small ships. But the only problem is because all types of ships satellite antennas are sometimes :

- Understress from vibration.
- Sloping caused by strong winds
- Ship's rolling and pitch
- Corrosion by sea salt

Because of these problem, a **ship's antenna** has to be well protected by plastic radome and properly mounted on a strong mast, especially designed for a certain size of antenna. In case of **road vehicles**, especially in small cars low profile and light weight equipment is required. Almost same requirements are there for **space craft** although more severe conditions are required to satisfy avionic standards. Vehicle and aircraft utilize smaller and more aerodynamic sizes of antenna. **Low air drag** is one of the most important requirements for aircraft antennas.

9.3.2. Electrical Characteristics

Sometimes the mechanical construction of antenna is perfect because of functional or electrical characteristics. But antenna designers should understand that the compact design of antenna has two major disadvantages in electrical characteristics :

- Low gain
- Wide beam coverage

Theoretically the gain of antenna is determined by its physical dimensions, so reducing size of antenna means decreasing its gain. Because of low gain and limited electric power supply, it is very difficult for mobile antennas to have enough receiving capability (G/T) and transmission power (EIRP). **These disadvantage of MES can be compensated by** a satellite that has large antenna and HPA with enough electrical power.

Another disadvantage is that wide beam antenna may transmit undesired signals to receive them from an undesired direction, which may cause interference with other systems wide beam antenna is also responsible for several fading effects like sea surface reflections in MMSC and AMSC and multipath fading in LMSC. Therefore a compact mobile antenna system is required to prevent fading and interference.

Because of small physical dimensions of mobile antenna, it is difficult to achieve good performance like gain, radiation power and receiving capability. Without considering this, the requirement of transmitting and receiving performance of mobile antennas mainly depend on satellite transmission capability.

- The first and second generation of inmarsat satellites have a global beam. The third generation has spot beam and global beams to provide global coverage.
- The region or domestic MSS such as AMSC, MSAT and optus have spot beams.
- The spot beam gives higher satellite capacity than global beams.

9.3.3. Basic Relationship of Antennas

The basic relationships of antenna system are very important to understand the mode of antenna functions in two way satellite transmission systems, example : MES transceiving antennas. These characteristics of antenna systems are also required for :

- Link budget calculations
- Good satellite up and downlink designs.

They can provide reliable and acceptable quality satellite communications.

9.3.3.1. Frequency and Bandwidth

Almost all present and upcoming MSS using GEO satellites. The L-band 1.6/1.5 GHz is used for a link between the satellite and MES. The frequency bandwidth required in L-band MSS is about 8% to cover transmission and receiving channels. To use narrowband antennas like omnidirectional patch antenna, some efforts have to be made to increase bandwidth.

The S and L-band are allocated in WARC-92 for big LEO Iridium and Gloabalstar systems. They require frequency bandwidth of about 5%.

9.3.3.2. Gain and Directivity

Gain is a parameter which measures the degree of directivity of the antenna radiation patterns. The required antenna gain is determined by link budget, which is calculated by considering the required channel quality and the satellite capability. The channels are expressed as C/N_0 and depends on :

- G/T value
- EIHF value
- MES

Low, medium and high gain antenna systems can be differentiated by the gain quantum, shape of the antenna and type of service. But In present and upcoming L-band GMSC applications, classification of L-band MES antenna system can be done by their receiving capabilities as discussed in Table 9.1.

Table 9.1. Classification of L-band Antenna System in GMSC.

S. No.	Type of Antenna	Gain Class	Typical Gain [dB]	Typical G/T [dBK]	Typical Antenna (Dimension)	Typical GMSC Services
1	Omnidir-ectional	Low	0–4	–27 to –23	Quadrifilar Drooping -dipole Patch	LSD (Message) Ship (Inmarsat-C) Vehicles & Aircraft
2	Semidir-ectional (Only in Azimuth)	Medium	4–8 8–16	–23 to –18 –18 to –10	Array (2–4 elements) Helical, Patch SBF (0, 4m ϕ Phased Array (20 elements)	Voice/HSD Ship (Inmarsat-M) Vehicles Aircraft (Inmarsat-Aero)
3	Direc-tional	High	17–20 20–24	–8 to –6 –4	Dish (0, 8m ϕ) Dish (1 m ϕ)	Voice/HSD Ship (Inmarsat-A, B)

Ideal Gain Antenna : It is defined with an isotropic antenna which has an isotropic radiation pattern without any losses. Therefore it radiates power uniform with intensities in all directions.

The power flux density per unit area (P_{id}) at distance (r) from the antenna is given by

$$P_{id} = \frac{P_{in}}{4\pi r^2} \text{ [w/m}^2\text{]} \qquad ...(9.1)$$

where P_{in} = Input power put into an isotropic antenna.

If radiated power density is $\frac{P(\theta, \phi)}{r^2}$ in direction at distance (r) from the antenna under elevation, the gain of antenna is given by

$$G(\theta, \phi) = \frac{P(\theta, \phi)}{r^2} \Big/ P_{id}$$

$$= \frac{P(\theta, \phi)}{r^2} \Big/ \frac{P_{in}}{4\pi r^2}$$

$$= \frac{4\pi P(\theta, \phi)}{P_{in}}$$

$$G(\theta, \phi) = \frac{P(\theta, \phi)}{P_{in}} \Big/ 4\pi (dBi) \qquad ...(9.2)$$

where, θ = Angle between the considered direction and the one in which maximum power is radiated. It is known as **boresight**.

ϕ = Phase

The above defined gain is known as **absolute gain or directive gain**. It is determined by directivity of antenna only without considering any losses such as impedance mismatch loss or spill over loss. If direction is not specified and gain is not given a function of (θ, ϕ), it is assumed to be maximum gain.

A **general relationship** between absolute gain and the physical dimensions of antenna is given by :

$$G = \left(\frac{4\pi}{\lambda^2}\right)\eta a \quad \text{...(9.3)}$$

where, η = Aperture efficiency

a = Physical aperture *i.e.* effective aperture of the antenna.

Above equation describes that compact antennas with small apertures must have low gain. If an antenna aperture is a dish, known diameter (*d*) above eqn. in dB is

$$G = \left(\frac{\pi d}{\lambda}\right)^2 \eta$$

$$G = 10 \log \eta \left(\frac{\pi d}{\lambda}\right)^2 \; [dBi]$$

Therefore it can be calculated that the gain in the Inmarsat shipborne antenna with a diameter of $d = 1$ m operated at 1.5 GHz is about 21 dBi.

Antenna Directivity : It is defined as the ratio of $P(\theta, \phi)$ to the power per unit solid angle from an isotropic antenna radiation *i.e.* P_r. It is denoted by $D(\theta, \phi)$ and does not include dissipative losses. It is defined as :

$$D(\theta, \phi) = \frac{P(\theta, \phi)}{\dfrac{P_r}{4\pi}} \quad \text{...(9.4)}$$

Radiation Efficiency of Antenna : It is defined as the ratio of $G(\theta, \phi)$ to $D(\theta, \phi)$.

9.3.3.3. Radiation Pattern, Beam Width and Sidelobes

In principle, radiation calculation is possible if EM field can be described quantitatively at all points of the antenna surface whose boundaries are those of the apertures. Since aperture type antenna has been generally used in MSC especially in MMSC], we consider radiation pattern from a circular aperture.

If an antenna generates a single focused beam, the basic parameter affecting the antenna radiation pattern $E(\theta, \phi)$ is the aperture illumination distribution $E_a(r, \psi)$ after the aperture size (*a*). **Aperture illumination distribution** is defined as the amplitude of the far field radiation pattern (*E*) at point (θ, ϕ).

Fourier transform of illumination distribution is given by

$$E(\theta, \phi) = \frac{1}{\pi a^2}\int_0^{2\pi}\int_0^{a} E_a(r, \psi) \exp\left[-jkr \sin\theta, \cos(\phi - \psi)\right] r \, dr \, d\psi \quad \text{...(9.5)}$$

Example : Let us consider an antenna employed in MSC, which utilizes a circular aperture. For circular symmetric aperture above equation reduces to :

$$E(\theta) = \frac{2}{a^2} \int_0^a E_a(r)\, J_0\, [Kr \sin\theta \cos(\phi - \psi)]\, r\, dr \qquad ...(9.6)$$

where, $a = \frac{d}{2}$, radius of antenna aperture.

J_0 = First kind and zero order bassel function.

$K = \frac{2\pi}{2}$ = Denotes wave number.

Figure 9.1 shows the other notations that denote distance and angles in coordinates are define in geometry.

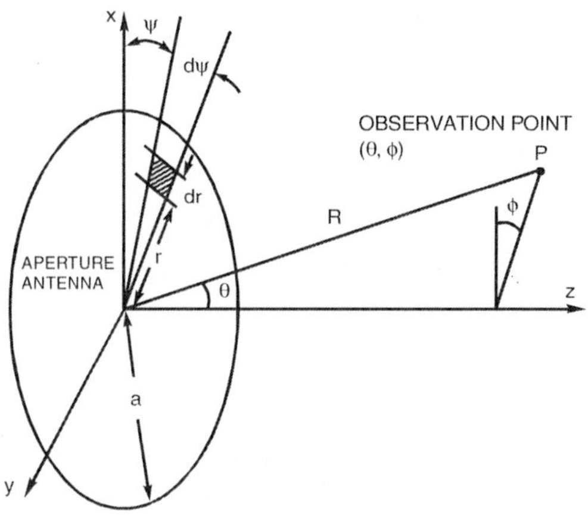

Fig. 9.1 Geometrical Parameters of Antenna Pattern.

The antenna radiation pattern is three dimensional in nature so it should be represented from the point of view of single axis plot. The characteristics of MES antenna radiation pattern affect interference levels directly.

To improve antenna pattern we can either increase antenna diameter or with a constant diameter use specific techniques for reducing the sidelobes. This method is applicable when the MSC network is in initial stages of development.

Gain is expressed in [dBi] where i denote the fact that it is relative to the isotropic gain. The matter of moment is dual polarization frequency reuse satellite communication system is polarization discrimination between copolar and crosspolar signal especially in the antenna main beam region, as shown in Fig. 9.2.

Major parameter used in an antenna's specification is the beam width evaluated by Half power beam width (HPBW) $2Q_{HP}$, where Q_{HP} is the half power angle when radiated power becomes half the maximum level (–3 dB). The HPBW (Q_0) is given by

$$\boxed{Q_0 = 65\left(\frac{\lambda}{d}\right)} \qquad ...(9.7)$$

Here it is possible to realize that the half power bandwidth is inversely proportional to

- Operating frequency
- Antenna diameter.

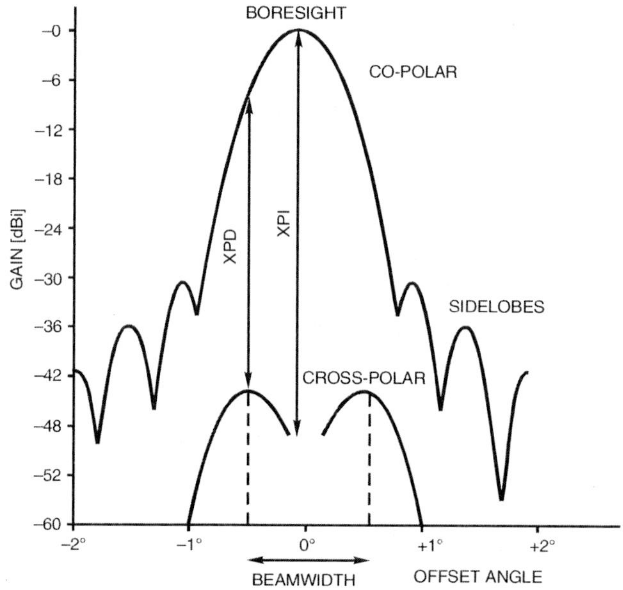

Fig. 9.2. Gain Characteristics.

Example : A 1 m receiving antenna operating in the C band (4 GHz) has a 3 dB bandwidth of roughly 4.9°, while the same antenna operating in Ku band (11 GHz) has a 3 dB bandwidth of approximately 1.8°.

The level of antenna pattern's sidelobe is also very important because it represents gain in an unwanted direction.

- For a transmitting gain, it leads to the transmission of unwanted power which results in interference to other systems.
- In case of receiving antenna, it leads to the reception of unwanted signal or noise.

The sidelobe characteristics of MES is major factor in determining :

- Minimum spacing between satellites.
- Satellite orbit.
- Spectrum utilization efficiency.

The ITU-RS.465-5 recommendation gives a reference radiation diagram for use in coordination and interference assessment, which is defined by

$$G = \begin{cases} 32 - 25 \log \phi \text{ for } \phi_{min} \le \phi \le 48^\circ \\ 10 \text{ dBi} \qquad \text{for } 48^\circ \le \phi \le 180^\circ \end{cases} \qquad ...(9.8)$$

where, G = Gain relative to an isotropic antenna.

ϕ = Off axis angle referred to main lobe axis

$$\phi_{min} = 1 \text{ or } \frac{100\,\lambda}{d} \text{ degree whichever is greater.}$$

Sidelobes may be because of antenna defects which can be minimized by proper design.

9.3.3.4. Polarization and Axial Ratio

The antenna and EM field transmitted or received have polarization properties. The polarization of EM wave describes the shape and orientation of the locus of the extremities of the field vectors as a function of time.

A wave may be described as :

- **Linearly Polarized :** In linear polarization, electric *E*-field is oriented at a constant angle as it is propagated which can be either vertically or horizontally. If a plane wave is propagated along the *Z*-axis and electric field *E* is an $(X - Z)$ or $(Y - Z)$ planes, linear vertical and horizontal polarization can be written as :

$$E_x = E_a \,.\, e^{j(\omega t - KZ + \phi a)}$$

and $$E_y = E_b \,.\, e^{j(\omega t - KZ + \phi b)} \quad ...(9.9)$$

where, E_a, E_b = Maximum amplitude of electric field

w = Angular frequency

K = Wave number

ϕ_a, ϕ_b = Initial phase of wave

- **Circular Polarization :** It is the super position of two orthogonal linear polarization i.e. vertical and horizontal with a 90° phase difference. The tip of resultant *E*-field may be imagined to rotate as it propagates in a helical path. There is a Left Hand Circularly Polarized (LHCP) wave with anticlockwise rotation and a Right Hand Circulary Polarized (RHCP) wave with clockwise rotation.
- **Elliptically Polarized :** It is the result of either two linearly or two circularly polarized waves with opposite directions. In this type of polarization, the amplitude and the phase difference between the two waves are not equal $\left(\frac{\pi}{2}\right)$.

We have already discussed that signals field can contain copolar and cross polar components. If co-polar component is vertical then cross polar component is horizontal. Circular cross polarization is opposite to the desired principle or reference polarization. Impure cicular polarization is elliptical. The level of impuring is measured by the elliptical and is known as **Axial Ratio (AR).**

The axial ratio can be defined as the ratio of the major axis electric components to that of minor axis and is defined as :

$$\boxed{|AR| = \left|\frac{E_1}{E_2}\right| \left(1 \le |AR| \le \infty\right)} \quad ...(9.10)$$

The signal for *AR* denotes the direction of rotation. Usually an absolute value is used to evaluate circular polarized radiated waves. In dB it is expressed as :

$$|AR| = 20 \log \left| \frac{E_1}{E_2} \right| \text{[dB]} \quad \text{for } \left(0 \leq |AR| \leq \infty\right) \quad ...(9.11)$$

AR is determined by the performance of the antenna, so it is one of the most important parameters of circular polarized antennas. Circular polarized waves are used to eliminate the need for polarization tracking.

9.4. CLASSIFICATION OF MOBILE SATELLITE ANTENNAS (MSA)

Currently available mobile satellite antennas for MSC applications form the weakest link of the system. If the mobile antenna has the high gain then it has to track satellite following both mobile and satellite orbital motions. Sometimes it is difficult and expensive to synchronize. Therefore if the vehicle antenna has low gain, it has no need to perform tracking but capacity of communications link is limited.

Depends on transmission detection, **generally there are three types of MSA :**

(*i*) **Transmitting and receiving** (i.e. transceiving) as a part of all three types of MES.

(*ii*) **Only receiving** is the part of special Inmarsat EGC receiver.

(*iii*) **Only transmitting** is built in satellite because antennas for maritime, land and aeronautical applications.

The Inmarsat, Eutelsat, ESA, Cospas-sarsat, Iredium, Global star, ICO and other GEO and non GEO current and upcoming mobile satellite providers have conducted research on all network segments including different types of MSA and, their future developments and improvement. While Engineering Test Satellite-V (ETS/V) experiments conducted in Japan for the transmission of voice, video and different data rate digital communications between ships, land vehicle and aircraft were successful.

9.4.1. Shipborne MSA

The various kinds of shipborne satellite antenna systems were developed for installation on :

- Board ocean going ships
- Inland sailing vessels.
- Sea platforms.
- Offstore infrastructures.

Generally these antennas must have strong and rugged constructions with corresponding mechanical and electrical particulars.

- The Imarsat-A SES is the inheritor of the first generation of Marisat. It is **similar to SES** and it was the first Inmarsat operating standard of MMSC.
- In 1982 analog standard was operated. It uses Inmarsat standard-A transceiving antenna system known as **Above Deck Equipment (ADU).**
- Meanwhile inmarsat-C and ECG were developed with small omnidirectional antenna.
- Using second generation of Inmarsat satellites, at the end of 1993, Inmarsat-B digital standard started to be in service on ships. This standard is compatible with inmarsat-A and uses the same antenna specifications.

- After employing more power full satellite constellations, the Inmarsat system developed Inmarsat-M, mini- M and D+ standards with special shipborne antenna systems.
- Above all mentioned MSA are transceiving antennas except for the Inmarsat-EGC receiver, which can also use a receiving antenna.
- The Cospao-Sarsat system developed shipborne EPIRB with small built in VHF and UHF antennas.
- Therefore rugged MSA with value added system functions compatible with shipborne operation for Inmarsat, Iridium, Globalstar and other systems. Compact antenna for small lifeboat operation are major concern for antenna designed in the future.

9.4.2. Vehicleborne MSA

Vehicleborne MSA were developed for installation on road and rail vehicles such as buses, cars, trucks, trailers and trains.

- The transceiving Inmarsat–C, D, M, Mini-M and other MSA are used on board land vehicles.
- Like maritime application, the cospas-sarsat system has developed personal or vehicleborne PLB handheld satellite because with small built in VHF and UHF band transmitting antennas.

9.4.3. Airborne MSA

The airborne MSA were developed for installation on board aircraft such as airplanes and helicopters.

- Several types of low and high gain airborne satellite transceiving antennas are there. They are used for Inmarsat–C, H, H+, Z, L, Mini-M and other MSS.
- Parallel to the maritime and land distress applications, the cospas-sarsat system developed airborne ELT satellite beacons with small built in VHF and UHF band transmitting antenna.
- Since the speed of aircraft is high, the aerodynamic constraints are more significant. Antenna for both radio and satellite systems must confirm to minimum drag and reliability requirements.

9.4.4. Transportable MSA

Inmarsat developed special transceiving satellite antenna systems. They can be integrated into mobile transportable earth stations (TES) of Inmarsat-B, C, M and mini terminals.

9.4.5. MSA for Personal Satellite Terminals

The new GEO and non GEO MSC systems developed small satellite antennas for GMPSC terminals.

Example : Handheld antennas, antennas for mobiles and, roof antennas for in door and outdoor satellite terminals. But some of these systems provide mobile tracking, so tracking antenna has to be combined or integrated with the communication antenna.

9.4.6. Other Types of MSA

The other types of MSA are receiving broadcast radio and TV MSA, receiving GNSS MSA, transceiving broadband MSA and antennas for military satellite communication and navigating equipment.

9.5. LOW GAIN OMNIDIRECTIONAL ANTENNA

The antenna system for MSC are classified into :

- Omnidirectional
- Directional

The omnidirectional antenna has low gain, generally from 0 to 4 dB L-band and they do not require the capability of satellite tracking. The low gain omnidirectional antennas are very attractive for all mobile applications because of :

(*i*) Small size

(*ii*) Light weight

(*iii*) Circular polarization properties.

These antennas are also used as elements of directional antennas for special configurations. There are three types of low gain omnidirectional antennas.

9.5.1. Quadrifilar Helix Antenna (QHA)

The QHA low gain model consists of four identical helixes would equally spaced on a cylindrical surface. The helix elements are fed with equal amplitude and a relative phase of 0°, 90°, 180° and 270°. This antenna can easily generate circular polarized waves without a balloon or a 3 dB power divider and they are required to excite a balanced fed dipole and circular polarized cross dipoles. Since it is a travelling wave type antenna, it can also be operated on a wide frequency bandwidth of up to 200%. Fig. 9.3 show the QHA.

The components of QHA are :

- Ground plane (g)
- Pitch (P)
- Pitch angle (a)
- Length (l)
- Diameter (d)

Usually the diameter of ground plane is selected to be larger than one wavelength and the number of turns is $N = 1/P$.

The value of parameter a is about 12 to 15 and the circumference of helix (πd) is about 0.75 to 1.25 wavelengths. The helical antenna gain depends on the number of turns (N) and typical gain. Half power beam width are about 8 dBi and 50°. This antenna is also a component of AMSC, MSAT, MSAT-X, Iridium, Globalstar, Mobilesat ETS-D, upcoming ICO and other MSC terminals.

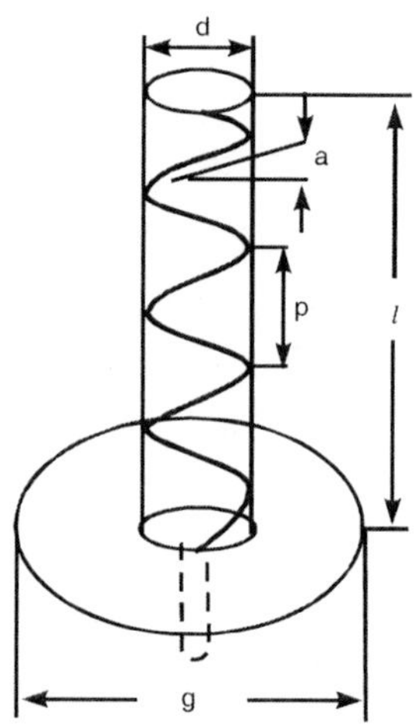

Fig. 9.3. Quadrifilar Helix Antenna (QHA).

In general **QHA as mobile antenna has two advantages :**

- An increase in bandwidth, it can generate axial mode circular polarized waves in the frequency range from 0.4 to 2.0 wavelength of the helix circumference.
- Lowered frequency for axial mode operation.

The **major disadvantage** is an increase in the complexity of the field systems. Usually the area of ground plane is about 3 times the diameter of the helix.

9.5.2. Crossed-Drooping Dipole Antenna (CDDA)

Widely used dipole antenna has half wavelength $\left(\frac{\lambda}{2}\right)$. It is most popular and have been used in antenna systems such as the parabolic antenna for MSC. A half wavelength dipole is a linear antenna whose current amplitude varies one half of a sine wave with a maximum at the center. Since dipole antenna radiates linearly polarized waves, two crossed dipole antenna have been used to generate circular polarized waves. Two dipoles are geometrically orthogonal and equal amplitude signals are fed to them with $\frac{\pi}{2}$ in phase difference.

To optimize radiation pattern, a set of dipole antennas is bent toward the ground so it is known as **drooping dipole antenna**. Fig. 9.4 shows the CDDA.

The CDDA serves as a transceiver antenna for L-band Inmarsat-C SES and VES applications mounted inside a radome. The CDDA is most interesting for LMSC, where required angular coverage is narrow in elevation and is almost constant in azimuth angle.

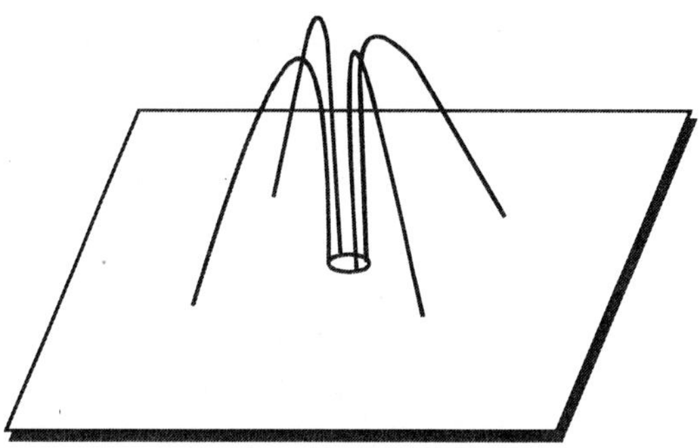

Fig. 9.4. Crossed Drooping Dipole Antenna (CDDA).

The **general characteristics** of this antenna are :

- Gain : 4 dBi minimum
- Axial ratio : 6 dB maximum
- Antenna height : about 15 cm.

This antenna has maximum gain in boresight direction.

9.5.3. Microstrip Patch Antenna (MPA)

A microstrip patch (disc) antenna has a very low profile and has mechanical strength. It is considered as the best type for mobiles like cars and especially in aircraft at the hybrid L to Ku-band. They requires low air drag.

Generally circular disk antenna element has a circular metallic disc supported by a dielectric substrate material and printed on a thin dielectric substrate with a ground plane. To produce a circular polarized wave, a patch antenna is excited at two points orthogonal to each other and feel with signal in equal amplitude and a relative phase of 0° and 90°.

Therefore higher mode patch antenna can be designed to have similar radiation pattern to drooping dipole. The antenna is excited at higher order modes to produce conical radiation patterns suitable for land mobile satellite applications. The configuration of circular patch antenna is shown in Fig. 9.5.

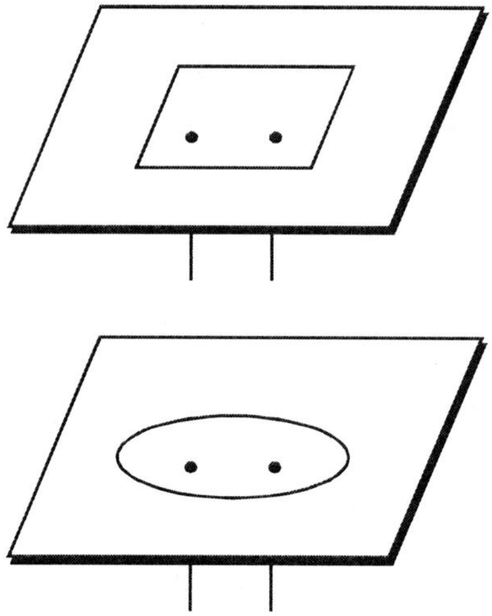

Fig. 9.5. Microstrip Patch Antenna (MPA).

It has two feed points to generate circular polarized waves. The resonant frequency excited by basic mode is given by :

$$f = \frac{\frac{184C}{2\pi a}}{\sqrt{\varepsilon_r}}$$

where, a = Radius of circular disc

C = Velocity of light in free space

ε_r = Relative dielectric constant of the substrate.

The area of a higher mode circular MPA is about 1.7 times larger in radius on the gain is about 6 to 8 dBi. The circular patch is also suitable as a satellite receiving antenna for GPS receivers.

9.6. MEDIUM GAIN DIRECTIONAL ANTENNAS

The medium gain directional MSA gives solution with a typical gain between 12 to 15 dBi. Some of the antennas can have even bigger gains. These MSC antenna system can provide voice, Fax and HSD for Inmarsat-M shipborne and vehicleborne applications.

9.6.1. Aperture Reflector Antennas

The aperture reflector antennas are good solutions with medium gain characteristics used in MSC. Basically **there are three types of aperture reflector antennas :**

- SBF (Short Backfire)
- Modified SBF
- Improved SBF

The main characteristics of these three antennas are discussed in Table 9.2.

Table 9.2. Particular of Aperture Types of Antennas.

Characteristics	SBF Antenna	Modified SBF	Improved SBF Antenna Antenna
Effective Gain	14.5 dB	15 dB	15 dB
Half-Power Bandwidth	34°	34°	34°
Directive Gain	14.8 dB	15.5 dB	15.5 dB
First Sidelobe Level	–21 dB	–22 dB	–22.5 dB
Axial Ratio	–1.3 dB	–1.1 dB	–1.1 dB
Aperture Efficiency Effective-Directive Gain	65% – 75%	75% – 80%	76% – 85%
RF VSWR Bandwidth under 1.5	3%	7%	9%
Diameter of Large Reflector (D_R) :			
Bigger (D_{R1})	40 cm (2.05‰)	40 cm (2.05‰)	40 cm (2.05‰)
Smaller (D_{R2})	—	27 cm (1.38‰)	—
Diameter of Small Reflector (D_r) :			
Bigger (D_{r1})	9 cm (0.46‰)	9.5 cm (0.48‰)	9 cm (0.46‰)
Smaller (D_{r2})	—	8.5 cm (0.43‰)	8 cm (0.41‰)
Width of a Rim	4.9 (0.25‰)	—	4.9 (0.25‰)
Distance Between (D_R) and (D_r)	9.7 cm (0.49‰)	19.5 cm (0.99‰)	12.9 cm (0.66‰)
Distance Between Exciter and D_r	4.9 cm (0.25‰)	—	5.7 cm (0.29‰)
Distance Between (D_{r1}) & (D_{r2})	—	—	1.8 cm (0.09‰)
Slanting Angle of a (D_R)	0°	—	15°

Figure 9.6 shows the three basic representative of aperture reflector antennas.

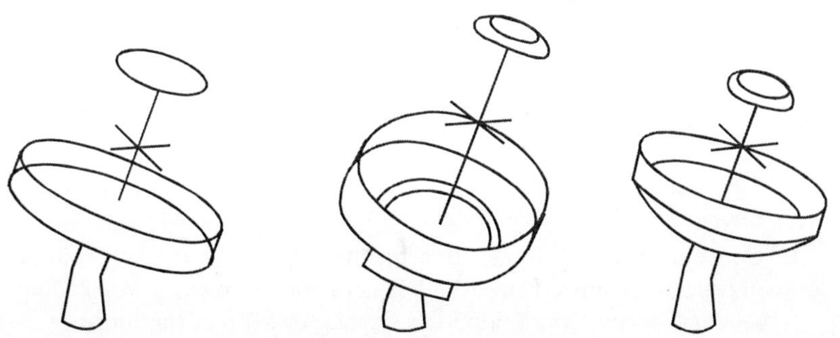

(a) (b) (c)

Fig. 9.6. Types of Aperture Antennas.

The SBF antennas consists of :

- Stabilized platform with two gyroscopes for azimuth and elevation angle.
- Diplexer
- HPA
- LNA

To stabilize antenna, two gyro wheels are rotated in opposite direction on a platform. Because of excellent radiation characteristics of SBF antennas, all three types of aperture antennas with half power beam width of about 34° have been used for shipborne antenna of Inmarsat-M.

9.6.1.1. Short Back Fire (SBF) Plane Reflector Antenna

Experimentally the SBF plane reflector antenna was developed by H.W Ehrenspeck in 1960s. It is well known as a highly efficient antenna of very simple and compact construction. Since it has high directivity and low sidelobe characteristics, a single antenna with high, even values is applicable to MSC, tracking and telemetry. An SBF antenna is very attractive for gains in the order of 8 to 15 dBi peak RHCP.

This type of antenna consist of two circular planar reflectors of different diameter, separated of about one half wavelength, forming a shallow leaky resonator with a radiation beam normal to small reflector. The antenna is fed by a dipole at around the midpoint between two reflectors and it has a quarter wavelength rim on the larger reflector. **It has the problem of** narrow bandwidth of about 3% because of its leaky cavity operation.

The **basic configuration** of SBF antenna consist of :

- Cross dipole element, required to generate a circular polarized wave.
- Large and small reflector.
- Circular metallic rim

This **antenna has many beneficial factors :**

- Efficiency
- Easy to construct
- Can be considered a good option for a compact and high efficiency shipboard antenna.

9.6.1.2. Modified SBF Plane Reflector Antenna

It is different from the conventional SBF. Difference is that there is either an additional step on the large reflector or the change in the shape of large reflector from a circular to conical plate, to improve gain characteristics and frequency bandwidth of VSWR.

- The dual reflector improves the input impedance characteristics covering the frequency range between transmitting and receiving sides.
- Conventional SBF model is a resonant type antenna. It produces input impedance characteristics which is narrow in bandwidth. Wide bandwidth is required to cover the 1.6/1.5 GHz range for MES of the Inmarsat system.
- Input impedance improvement mainly depends on the size and separation of the small reflectors.

9.6.1.3. Improved SBF Conical Reflector Antenna

The electrical characteristics of a simple SBF antenna can be improved by changing its main reflector from a flat disk to a conical or step plate and by adding a second small reflector.

- Gain is improved by 1 dB without changing sidelobe levels.
- Antenna stabilization is obtained by two-axis stabilized method and satellite pointing is carried out by a tracking program which uses output signals from the ships gyroscope.

9.6.2. Wire Antenna

Wire antennas are monosyllabic construction or combination of elements, example : different shapes of wire spirals and helixes, dipoles and patches. These antennas have :

- Very simple construction
- With any specified reflector for medium gain directional antenna.
- With some modification they responds well to the demands of MSC applications.

9.6.2.1. Helical Wire Antennas

An axial mode helical antenna has good circular polarization characteristics over wide frequency range, so practically it can be used as a single wire antenna or as an array element with respect to the structure, this antenna can be considered as a compromise between the dipole and the loop antennas and the **radiation mode varies with :**

- Pitch angle
- Helix circumference.

Particularly a helix with a pitch angle of 12° to 15° and a circumference of about 1‰ has a sharp directivity towards the axial direction of antenna. This radiation mode is known as axial mode and it is very important for helical antennas.

Fig. 9.7. Conical Helix Antenna.

Conical Helix Antenna

Conical helix antenna is shown in Fig. 9.7. A conical helix is important for L-band MSS enabling HPBW in the order of 100° and circular polarization without hybrid gain of 4 to 7 dBi. This type of antenna is more suitable for wide beam width applications with good efficiency. This type of antenna can also achieve an input VSWR of 1.5 : or better than 5% frequency bandwidth.

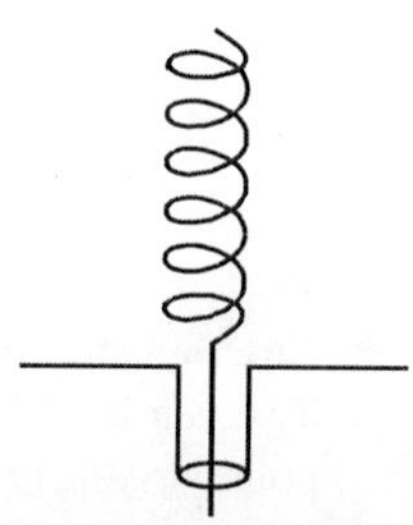

Fig. 9.8. Cylindrical Helix Antenna.

Cylindrical Helix Antenna

Figure 9.8 shows the cylindrical helix antenna. Cylindrical antennas can be monofilar or multifiear. It is suitable for L-band MSS. In a short cut cylindrical helix antenna, rapid fluctuating travelling wave will be dominant, especially in two turn (N = 2) helical antenna.

Two Turn Cylindrical Helix Antenna

This antenna has two turns of wires, forming a simple helical antenna solutions with reflector, it is shown in Fig. 9.9.

Fig. 9.9. Two Turn Cylindrical Helical Antenna.

This antenna has relatively high gain and excellent polarization characteristics for its size. **Radiation pattern characteristics are :**

- Calculated w.r.t. (E_0) and (E_ϕ) planes.
- Antenna gain is 9 dBi.
- Axial ratio is 1 dB
- Reflector diameter (d) is about 1 λ.

Besides their small size and easy construction, these antennas have comparatively high performance.

Five Turn Cylindrical Helix Antenna

Figure 9.10 shows five turn cylindrical helix antenna.

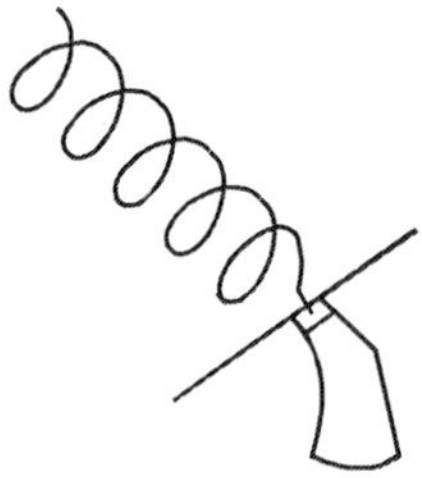

Fig. 9.10. Five Turn Cylindrical Helix Antenna.

The **main electrical characteristics are :**

- Gain : 12.5 dBi of peak RHCP for Tx and 11.5 dBi for Rx.
- Sidelobe level : –13 dB
- Axial ratio : 3 dB
- Beam width : 3 dB has an angle of –47°
- G/T terminal : –16 dBk
- EIRP terminal : 29 dBw

This antenna is designed and developed by the European research Institution ESTEC.

9.6.2.2. Inverted V-Form Cross Dipole Antenna

It is an advanced circularly polarized antenna with thick V-elements. Fig. 9.11 shows this type of antenna.

- Antenna resonance is obtained when its length is somewhat shorter than a free space half wavelength. Therefore when thickness increases, resonant length decreases.
- Circular polarization can be produced by a pair of orthogonally positioned dipoles driven in quadrature phase with equal amplitudes.

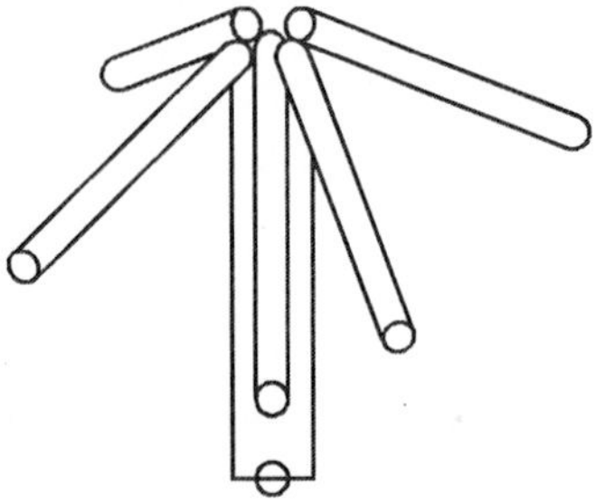

Fig. 9.11. Inverted V-Form Cross Dipole Antenna.

- The crossed dipole antenna arrangement can not provide a good axial ratio off boresight because the radiation patterns for the straight dipole are different in both planes *i.e.*, E & H-planes.

This drawback can be improved by modifying straight dipole to nonstraight version i.e. U an V-forms.

- The improved dipoles are known as V and U-type dipoles. According to some conducted measurements, U-type gives better electrical performance than V-type though V-type is simpler in mechanical structure and is less complex.
- The crossed dipole can also produce circular polarization without using any external circuits like hybrid component.
- Two of elements are at 0° phase angle and other two at 180° phase angle. The desired 90° phase difference is obtained by designing the orthogonal elements such that one is larger relative to making it inductive but other is smaller to make it capacitive.
- This type of antenna is very useful for KU band aeronautical satellite communications.

9.6.2.3. Crossed Slot Antenna

Crossed slot antennas are useful for L-band aeronautical satellite communications on high speed aircraft. They are suitable for flush mounting applications and they are low profile in structure. Crossed slot antenna is shown in Fig. 9.12.

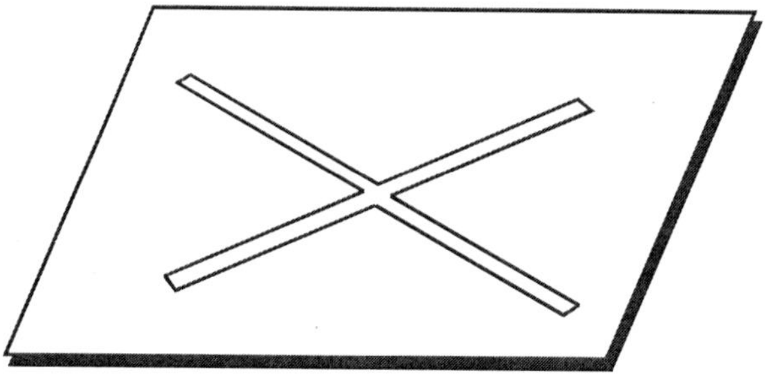

Fig. 9.12. Crossed Slot Antenna.

- The slot antenna is circularly polarized and is complementary with the corresponding dipole antenna, so that the radiation pattern is same as that of horizontal dipole.
- **Two differences are there :** First the property of electric and magnetic fields are interchanged and second is that slot electric field normal to the perfectly conducting sheet is discontinuous from one side of sheet to other because the direction of field reverses.
- This antenna can be complementary with corresponding crossed dipole antenna but the feeding method of circular polarization is more complicated.
- Therefore on a model of this antenna known as **cavity backed**, it needs one 90° hybrid to produce the circular polarization.

9.6.2.4. Conical Spiral Antenna

This type of antenna has spiral wire elements on a cone with circular polarization and is shown in Fig. 9.13.

Conical spiral antenna is suitable for L-band LMSC and GPS applications while bifilar version is also used in KU-band satellite communications.

- As compared to conical helix antenna, it gives better performance and is more versatile although the geometry is quite complex.
- This antenna is independent of frequency and its geometry can be presented mathematically in spherical coordinates (r, θ, ϕ) as :

$$\boxed{r = e^{a\theta}\, g(\theta)} \qquad \text{...(9.12)}$$

Fig. 9.13. Conical Spiral Antenna.

where, a = Arbitrary constant

$g(\theta)$ = Angular function

- Its radiation mechanism can be understood by considering two spirals as a transmission line.
- When two conductor arms are fed in antiphase at cone apex, waves travel out from the feed point and propagate along the spirals without radiating until a resonant length has been traversed.

9.6.2.5. Planar Spiral Antennas

These types of antennas are very suitable for flux mounting on aircraft for L to KU-band satellite communications. Generally this antenna is fed by using external balun but it can also be fed at center point, or apex, from a coaxial cable bonded to one of the arm. Without any external balun like conical spiral antenna. Mainly there are three categories of this type of antenna :

- Equriangular
- Logarithmic
- Archimedean

Equiangular Spiral Antenna

This antenna geometry is a special case of conical spiral antenna, bifilar with logarithm period, cavity backed and can be obtained by substituting $\frac{\pi}{2}$ into θ_0, we get

$$r_1 = r_0\, e^{a\phi}$$
$$r_2 = r_0\, e^{a(\phi - \Delta)}$$
$$r_3 = r_0\, e^{a(\phi + \pi)}$$
$$r_4 = r_0\, e^{a(\phi - \Delta + \pi)} \qquad \text{...(9.13)}$$

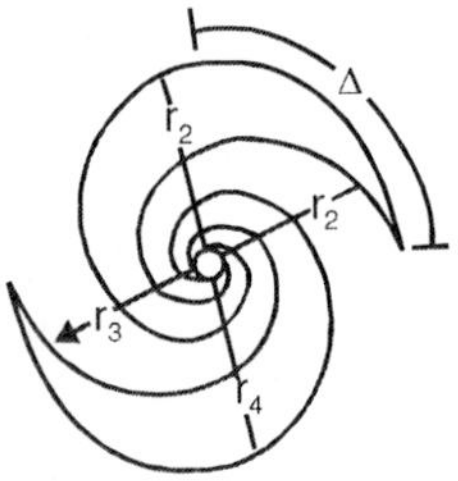

Fig. 9.14. Equiangular Spiral Antenna.

Figure 9.14 shows the antenna pattern and it needs no external hybrid circuit to produce circular polarization.

- When the pair of spirals is excited in antiphase at the centre, this antenna can radiate LHCP waves outward from the page and RHCP waves into the page.
- According to experimental measurements, the axial ratio is near unity and HPBW is in the order of 90° over a decade bandwidth or even more.
- This structure is defined as self complementary like conical spiral antenna.

Logarithmic Spiral Antenna

This is a bifilar antenna design with logarithmic period and cavity backed, as shown in Fig. 9.15.

Mathematically it can be represented as :

$$r_1 = a^{\phi}$$
$$r_2 = a^{(\phi - \Delta)}$$
$$r_3 = a^{(\phi + \pi)}$$
$$r_4 = a^{(\phi - \Delta + \pi)} \qquad ...(9.14)$$

When the pair of spirals is excited with an antiphase at center, this antenna can radiate RHCP waves outward from the page and LHCP waves into the page without any external hybrid circuits.

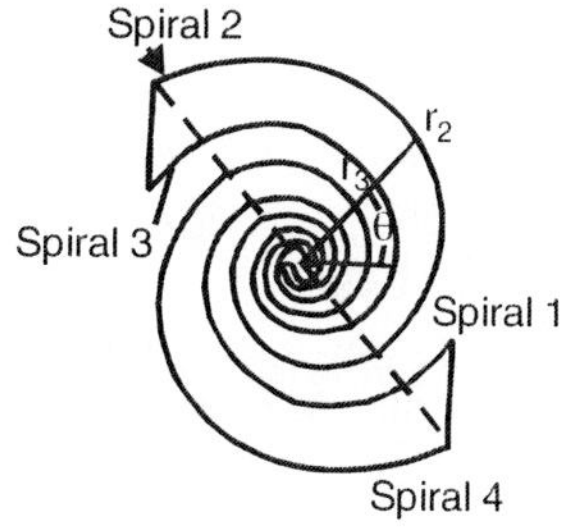

Fig. 9.15. Logarithmic Spiral Antenna.

Archimedean Spiral Antenna

It is the another geometry of the planar spiral. Archimedean spiral thin wire bifilar cavity backed antenna is shown in Fig. 9.16.

- This antenna consists of a pair of thin wire arms and it has superior bandwidth properties. Its geometry can be presented as :

$$r_1 = r_0\phi$$
$$r_2 = r_0(\phi - \pi) \qquad ...(9.15)$$

- To produce circular polarization, this antenna needs no external hybrid circuits. It can radiate RHCP waves outward from the page and LHCP waves into the page.

Fig. 9.16. Archimedean Spiral Antenna.

- It is a broadband antenna and has similar properties to that standard planar spiral antenna.
- When placed in a quarter wave cavity, this antenna can achieve near octave bandwidth even when the cavity consists in a metal based cylinder without any absorber.
- This structure has many mechanical advantages : it is compact and very simple to construct, and the spiral arms can be easily fed using a suitable impedance transforming belun.

9.6.3. Array Antenna

Various types of antenna can be arranged as array in space to make a directional pattern or one with desired radiation pattern. This type of integrated and combined antenna is known as **array antenna** and it consists of more than two elements like microstrip, cross-slot, cross dipole, helixes or other wire elements and is suitable for MSC. Every element of array antenna is excited by equal amplitude and phase and it radiation pattern is fixed.

9.6.3.1. Microstrip Array Antenna

Figure 9.16 shows Microstrip Array Antenna (MAA). It is a nine element flat antenna, disposed in three spaced at 94 mm about half wavelength at 1.6/ 1.5 GHz and its volume is about 300 × 300 × 10 mm.

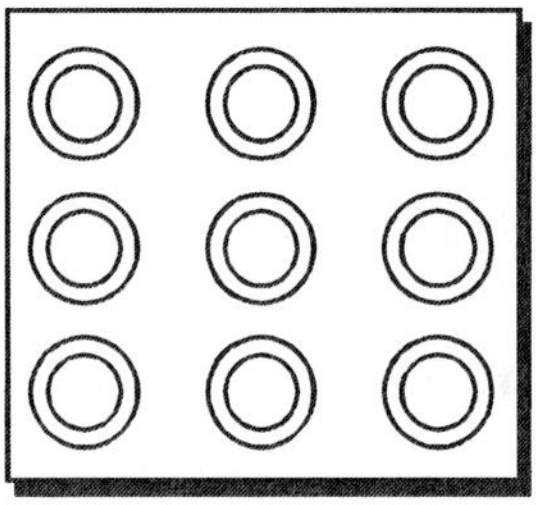

Fig. 9.16. Microstrip Array Antenna.

The elements of MAA are arranged in 3 × 3 rows square array to obtain similar radiation pattern in different cut planes. The microstrip array antenna beam scanning is performed by controlling four bit variable phase shifter attached to each antenna element.

Application of MAA is for MES Inmarsat-M and Inmarsat-Aero standard.

9.6.3.2. Cross Slot Array Antenna

Figure 9.17 shows the cross slot array antenna (XSA). It has 16 elements solution with 97 mm spacing and their volume is about 560 × 560 × 20 mm.

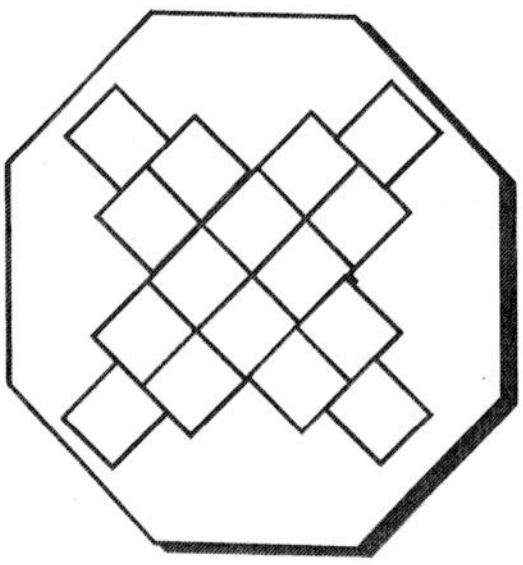

Fig. 9.17. Cross Slot Array Antenna.

To obtain similar radiation pattern in different cut planes, the elements arrangement of XSA is modified to 4 × 4 square array. The XSA antenna beam scanning is being done to control four-bit variable phase shifters associated to each antenna element.

XSA antenna application is suitable for the MES Inmarsat-M maritime and land including Inmarsat-Aero solutions.

9.6.3.3. Cross Dipole Array Antenna

Figure 9.18 shows cross dipole array antenna. It consists of 16 crossed dipoles fed in phase with a peak gain of 17 dBi and a feeding circuit which is behind the radiating aperture.

Fig. 9.18. Cross Dipole Array Antenna.

The major electrical characteristics of array antenna are :

- **Gain :** 15 to 17 dBi with peak RHCP transmit
- **Axial ratio :** 0.7 dB
- **Beam width :** 3 dB is –34°
- **G/T terminal :** –9.5 dBk
- **EIRP terminal :** 32 dBw

Otherwise antenna system consists in a stabilization mechanism for tracking, flat antenna array, diplexer, HPA and LNA. All of these components are protected by a plastic radome.

Antenna stabilization is obtained by single wheel gyroscope when azimuth pointing is controlled by the output from the ship's gyrocompass.

This antenna is also suitable for maritime and aeronautical applications like previous two models.

9.6.3.4. Four Element Array Antenna

Several four element antenna models have been developed such as Yogi Uda, Quad Helix, and four elements SBF array.

Yogi-Uda Crossed Dipole Array Antenna

Figure 9.18 shows this type of antenna and it has been developed for use on board ships and is protected with a radome.

This antenna feeder is a simple formation of four in line crossed dipoles fixed in middle of the reflector. The gain of end-fire array is between 8 and 15 dBi and it has circular polarization.

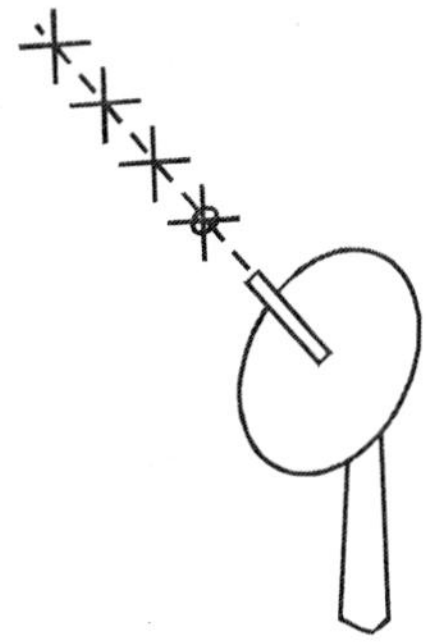

Fig. 9.18. Yogi-Uda Crossed Dipole Array Antenna.

Quad Helix Array Antenna

This antenna is composed of four identical two turn helical wire antenna in square shape. The orientation of elements of quad helix array antenna is shown in Fig. 9.19.

- Single helical antenna has axial ratio of 1 dB but in case of array antenna with an array spacing of 0.7λ, this value is degraded to about 4.5 dB.
- **Best properties of antenna gain and axial ratio** can be obtained at a rim height of about 0.25λ.
- The antenna gain is improved by 0.4 dB and axial ratio is improved by 3.5 dB compared to that of quad helix array antenna with out lines.
- This helix integrated antenna is more suitable for shipborne Inmarsat M standards.

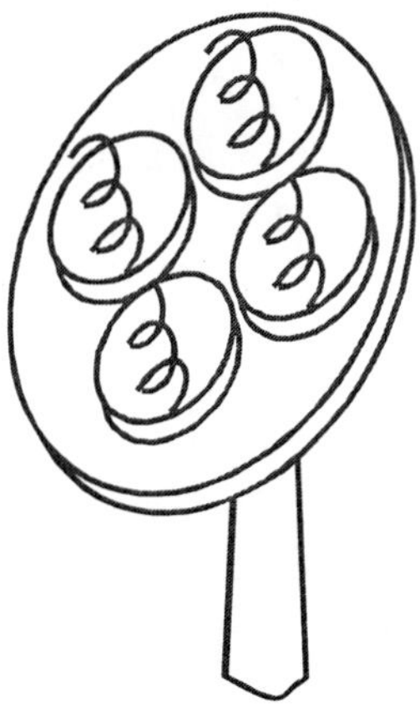

Fig. 9.19. Quad Helix Array Antenna.

Four Element SBF Array

Figure 9.20 shows four element SBF array antenna and it is developed on the basis of a conventional SBF antenna as an integrated array with four SBF element.

This antenna provides :

- High aperture efficiency.

- Circular polarization.
- High performance gain between 18 and 20 dBi.

Four element SBF array antenna is more suitable for maritime applications as a shipborne antenna because of its high gain characteristics.

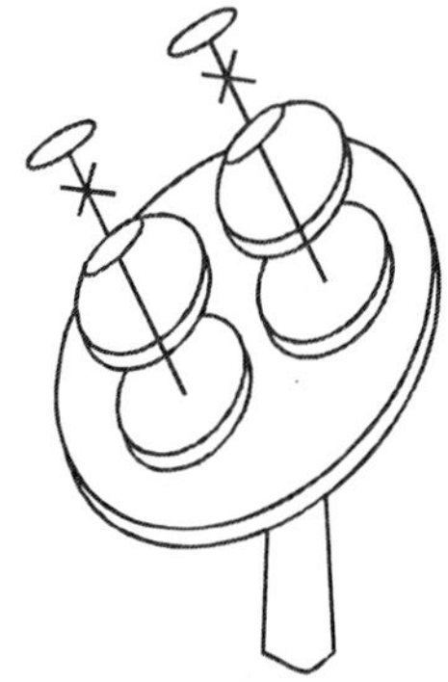

Fig. 9.20. Four Element

9.6.3.5. Spiral Array Antenna

Directional antennas for LMSC provide voice and HSD links for long have tracks and for private cars also. Cost is a major factor which should be considered in designing antenna systems.

- In earlier LMSC, a mechanical steering antenna was considered best for vehicles. But it will be replaced by phased array antenna in near future because of **many advantages** : low profile, low cost and high speed tracking.
- Figure 9.21 shows a mechanical steering antenna with eight spiral elements. It uses closed loop tracking method and gives about 15 dBi system gain.
- The antenna radius is 30 cm, height 35 cm and weight 1.5 kg. The array consists of 2 × 4 spiral elements and it form a fan beam with HPBW of 21° in azimuth and 39° in elevation plane at L-band.
- It has a peak gain of about 15 dBi including feeder losses. It is suitable to track the satellite for MSC.
- The antenna beam direction can be shifted in two azimuth direction from E or W side by switching the pin diode phase shifters.
- The difference between the received signals in both directions is used to drive the antenna system towards the satellite.

Fig. 9.21. Spiral Array Antenna.

9.6.3.6. Patch Array Antennas

The major feature of future MSC is portability. Portability means a person can directly access the satellite to establish a link using a very small TES transceiver with antenna system. In the present Inmarsat L band system, great efforts have been made to develop transportable and portable terminals with corresponding antennas.

Two Patch Array Antenna

Figure 9.22 shows two patch array antenna. It has two microstrip patch elements *i.e.*, one for Rx and other for Tx.

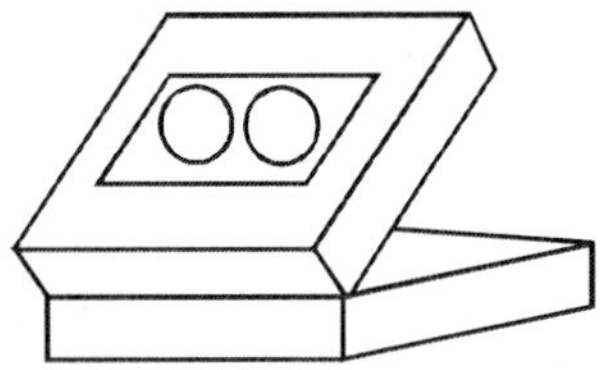

Fig. 9.22. Two Patch Array Antenna.

Separate Tx and Rx antennas is used to eliminate a diplexer. Diplexer is too large and heavy for a compact and light weight terminal. The major characteristics of this antenna are :

- Gain : 6 dBi
- EIRP : 6 dBw
- G/T : –21 dBk

Two microstrip patch array antennas mounted on the lid of the briefcase TES transceiver for low speed data transmission server the Inmarsat and ETS-V TES terminals.

Four Patch Array Antenna

Several four element patch array transportable antennas were developed for universal Inmarsat-C and other TES terminals. JPL and NASA designed similar L-band types of mobile antennas especially for LMSC and regional utilities in USA. Fig. 9.23 shows four patch array antenna.

This antenna has a mechanically steered tilted 1 × 4 patch array and two electrically steered planar phase array antennas. On the other hand mechanically steered four square patch array can be fixed in one line like two patch array. Otherwise it can have the shape of four circular patch array manually steered antenna, arrangement is shown in above Fig. 9.23.

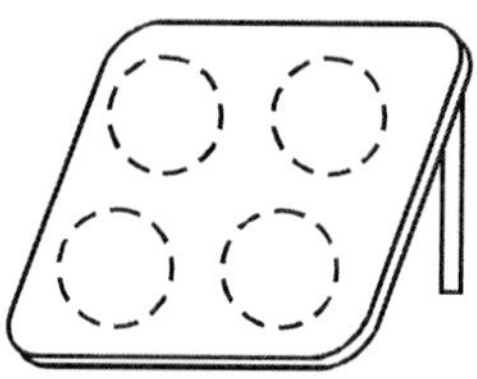

Fig. 9.23. Four Patch Array Antenna.

This antenna characteristics are :

- Gain : 9 to 12 dBi
- Reject multipath signals outside their beam pattern
- Allow two satellites separated by 30° in GEO.

9.6.3.7. Phased Array Antenna

These antennas were developed for maritime, land and aeronautical applications. They provide a design for thin antenna and can be installed on land vehicles and aircrafts.

Radiation pattern of an array antenna is fixed. In space, radiation pattern can be scanned by controlling the phase of the exciting current in each element of the array. This type of antenna is known as **phased array antenna**. In terms of **MSC applications, it has many advantages :**

- Compactness
- Light weight
- Low cost
- High speed tracking performance

Phased Array antenna for AMSC

The key technology in AMSC is a directional medium gain antenna. This antenna was developed and tested by CRL for the ETS-V program.

The microstrip phased array antenna is mounted on the top of the fuselage and has two planes with 16 circular patch elements. Out of which 2 are in elevation and 8 in azimuth, as shown in Fig. 9.24.

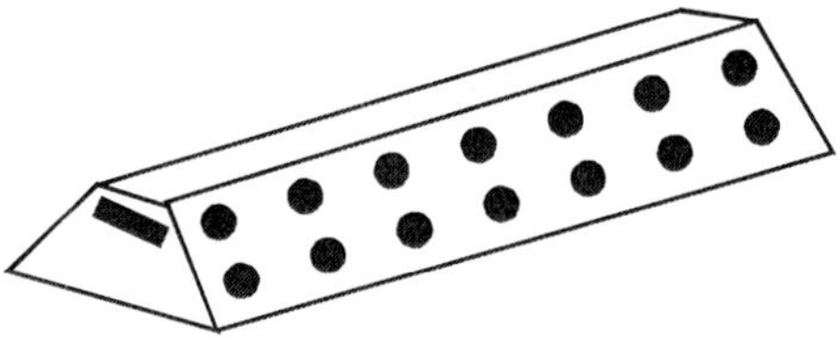

Fig. 9.24. Array Antenna for AMSC.

Practically the required coverage angle is narrow i.e. 20° so beams are not steered in elevation directions and array plane is set at 65° to the horizon to optimize the beam coverage on the flight routes. Therefore by controlling eight 4-bit digital phase shifters, the beam scans in a 4° step with +60° w.r.t. a line perpendicular to the axis of aircraft. A step track method is used to track the satellite.

This **array antenna has the following characteristics :**

- 2 L-band Tx and Rx frequencies.
- Gain : 14.7 dBi for Tx and 13.5 dBi for Rx
- EIRP : 30 dBw
- G/T : –10.8 dBk
- Axial Ratio : 2 dB
- Weight : 18 kg
- Volume : 760 (l) × 320 (w) × 180 (h)
- Polarization in LHCP.

Phased Array Antenna for MMSC

New shipborne phased array antenna has been developed for utilization on board big ocean going vessels. This unit is designed to serve as ADE of the Nera F77 transceiver or other brand SES for new Inmarsat Fleet 77 service. Figure 9.25 shows the Neera F77 antenna.

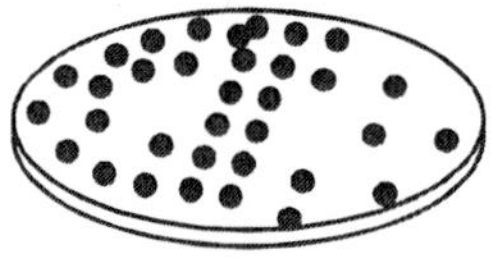

Fig. 9.25. Array Antenna for MMSC.

It is a mechanically steered circular disc with 32 low profile radiating elements arranged in two circles with 16 and 12 elements, and 4 elements are located in the middle of the disc. The ADE unit can be mounted on a mast or directly on the deck and is covered by a radome.

- The dimensions of antenna mast mount are 180 × 132 cm with a weight of 65 kg.
- The dimension of deck mounted antennas are 108 × 91 cm and 50 kg in weight.
- The antenna gain is about 15 dBi for both transmission satellite links.

Breadboard Phased Array Antenna for LMSC

This type of antenna has been developed in USA to meet the L-band frequency requirements :

- 20° to 60° elevation.
- Full 360° azimuth angles of special coverage.
- 10 dBi gain above 30° E and θ dBi above 20°E

- HPBW of 25° in A and 35° in E.
- Beam pointing accuracy of +5°.
- Intersatellite signal isolation of 20 dB between two GEO separated by circa 35°.

Figure 9.26 shows two phased array antennas developed for LMSC : Ball and Teledyne designs.

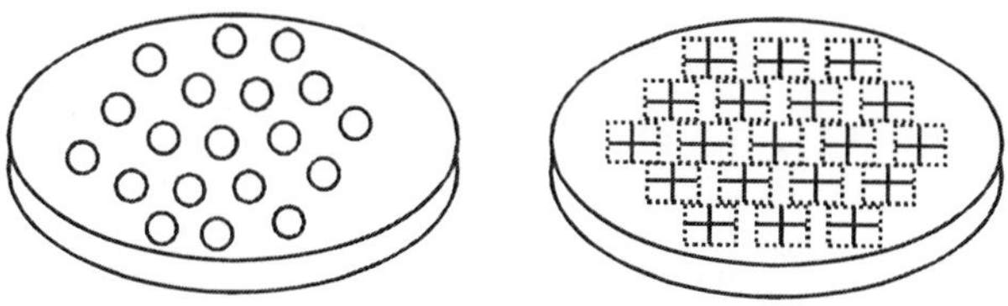

Fig. 9.26. Phased Array Antenna for MMSC.

Each antenna consists of 19 low profile radiating elements with 18 3-bit diode phase shifters. The Ball model uses dual resonant stacked circular microstrip elements to cover both the Tx and Rx bonds. Teledyne model consists of stripling crossed slot radiations. This antenna characteristics are :

- Gain : 10 and 18 dBi for elevation of 30° and 90° respectively.
- System temperature : 200 K.
- Axial ratio : 4 dB for 30° E
- Weight : 5 kg
- Volume : 60 × 4 cm.

9.7. HIGH GAIN DIRECTIONAL APERTURE ANTENNAS

High gain directional aperture antennas are very powerful transmission reflector and panels. They are used for Inmarsat maritime and transportable applications. Typical value of these antenna gain is more than 20 dBi, EIRP is a maximum of 33 dBw and G/T is about –4 dBk.

Two basic types of directional parabolic antennas are there :

- Dish
- Umbrella

9.7.1. Parabolic Dish Antenna

The first generating of parabolic dish antenna uses a reflector in diameter of max 1.2 m but new models reduces the size approximately 0.7 to 0.8 m. Fig. 9.27 shows the parabolic dish antenna.

- Since antenna produces a large proportion of Rx signal gain and Tx EIRP, the area of dish can be reduced if power transmitted from satellite transponder is increased.

Fig. 9.27. Parabolic Dish Antenna.

- Mostly parabolic reflector is used for high directivity of radio signals travelling in straight line same as light rays.
- Like light rays, they can be focused or reflected, a microwave source can be placed at focal point of antenna.
- When all parts of field are reflected from the parabolic surface, they travel to the space line in the same amount of time.
- This antenna is a large microwave parabolic consists of : reflector of dish shape, feeder structure, waveguide assembly, serve and drive system, and protective radome.

Stable Platform : This is antenna support assembly and it must remain perfectly stable when ship is pitching and rolling in extremely bad weather conditions.

The platform consists in a large solid bed mounted in such a way that four gyro compasses are able to sense movement and correct detected errors, holding the platform level. Practically it is a form of electronic gimbal.

Tracking System : With the help of stepping motors antenna is controlled in A/E angles. Electronically it is controlled in a simple feedback system.

The electromechanical antenna arrangement enables the dish to maintain a lock on a satellite. Therefore when ship changes course, both A and E control corrections must be made automatically.

Computer Control : The antenna unit processor controls all ADE functions which include satellite tracking and electronic control.

RF Electronics : It consists of :

- Transmitter HPA
- Receiver RF front-end LNA stages
- Critical band pass signal filter stage.

Multiplexer Unit : It is important to reduce the number of cables between ADE and BDE. It can be done by multiplexing up/down signals or commands between ADE and BDE onto the one coaxial feeder.

9.7.2. Parabolic Umbrella Antenna

Figure 9.28 shows a parabolic umbrella antenna.

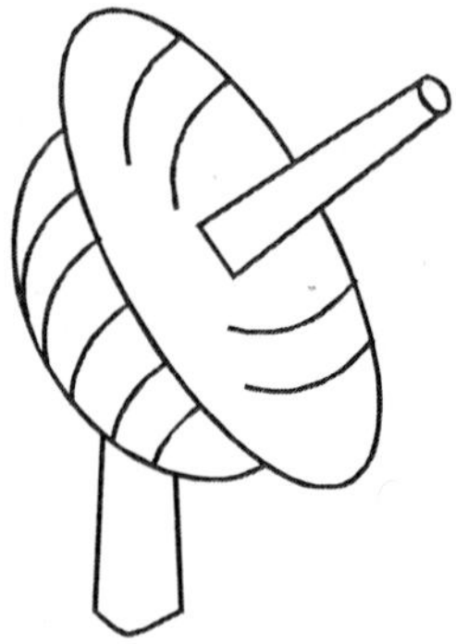

Fig. 9.28. Parabolic Umbrella Antenna.

This antenna is a deployable compact and lightweight. It is suitable for transportable Inmarsat-A and B transceivers. This antenna has almost the same characteristics as a parabolic dish antenna.

9.8. WIRE QUADRIFILAR HELIX ANTENNA (WQHA) FOR HANDHELD TERMINAL

This antenna is invented by "Kilgus" in 1970. The WQHA is highly resonant antenna and it consists of four helices placed at 90° to each other. Radial parts connect the four helical elements and they are feel in 90° phase difference (0°, 90°, 180° and 270°).

Analysis by Kilgus

- For resonant RHA (RQHA) it is assume that the QHA consists of four helical and four radial parts.
- The current distribution on helical element is assumed to be sinusoidal with magnitude maxima at the feed and the distant end for each 1/2 length.
- Uniform distribution gives the approximation of the current in the radius.

Another Analysis Approach

- It is based on the assumption that a QHA consists of two Bifilar Helices (BH) placed at 90° angular distance and fed in phase quadrature.
- The radials in the distant end can be shorted or open circuited gives only the change in input impedance.

The **major disadvantage** of the QHA is that it require complex feed network.

Basically there are two modes in which QHA can operate :

- Satellite mode (Hemisphere Pattern)
- Terrestrial Mode (Toroidal Pattern)

There are three possible ways to accommodate a QHA on handheld terminals :

- Bottom fed, end fire QHA with shorted end radials and normal QHA. (Fig. 9.29)

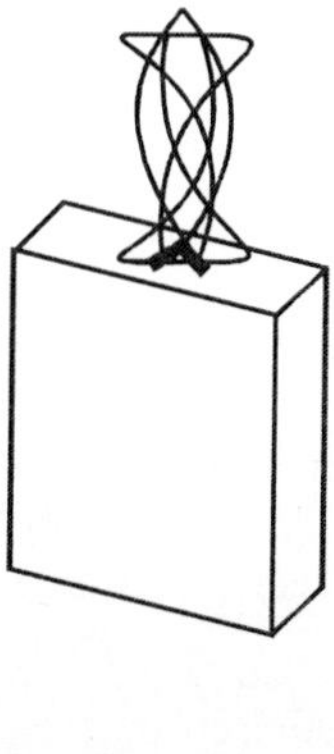

Fig. 9.29.

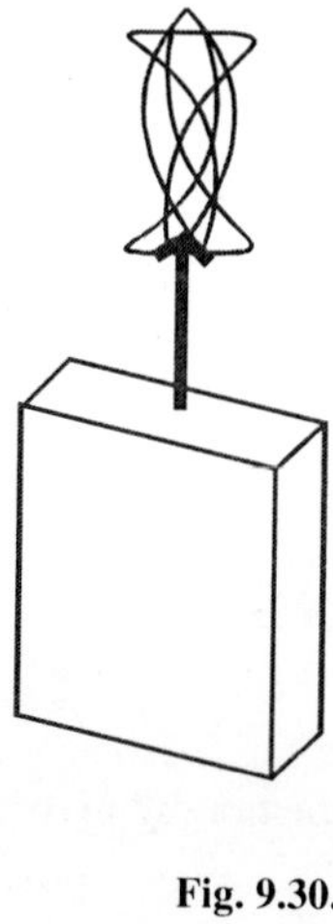

Fig. 9.30.

Fig. 9.31.

- It is know as **moved-up QHA** because the same antenna is uplifted to the head. (Fig. 9.30)
- It is top-fed backfire QHA with good open bottom radials i.e. with open radial QHA. (Fig. 9.31)

9.9. ANTENNA SYSTEMS FOR MOBILE SATELLITE BROADCASTING (MSB)

Satellite broadcasting at 12 GHz band becomes very popular and day by day the need for TV reception by relatively large mobile like ships, trucks, buses, trains and aircraft is rapidly growing. This service is attractive for commercial as well as military applications.

Digital TV and audio distribution by satellite is becoming standard for TV reception in many countries like US, Japan, and European communities.

In USA currently many services are performed by : Direct TV, USSB, Dish network and Alphastar Broadcast emissions.

Since satellite broadcasting area covers a wide area, **it is more suitable for two new areas :**

- Direct Audio Broadcasting (DAB)
- Direct Broadcasting Satellite (DBS)

The reception by mobiles has advantages in propagation conditions compared to its terrestrial TV broadcast counter parts. To receive good quality satellite broadcasts following electrical, mechanical and structural requirements are imposed on mobile antennas for MSB.

■ Accurate Beam Pointing

An accurate beam pointing capability is required all existing MSB systems required total receiving antenna gain to be minimum 32 dBi, therefore beam must be quite narrow i.e. about 2°.

■ High Speed Tracking

Some mobile moves very fast like airplanes and trains they changes their direction rapidly. So mobile receiver must track the satellite very quickly, not only for beam steering but also for vehicles vibration compensation.

■ Height Limitation

The entire receiving configuration should be compact, light weight and not too high so that they can be easily accommodated on mobile. Generally the antenna unit on the mobile should be less than a foot high.

■ Countermeasure for Satellite Signal Interruption

In monopulse mobile tracking system, tracking is performed until signal from the satellite is being received. Therefore tracking becomes impossible when signal is interrupted by surrounding obstacle.

Hence to cope with satellite signal blocking several required techniques are there :

- Space diversity
- Holding the antenna direction by gyroscope or a magnetic sensor
- High speed mechanical searching for satellite signal for recapture.

Summary

1. The consideration of antenna transmission is inevitable in GMSC systems because their propagation characteristics are very much affected by different and changable local environments.
2. In 1895, the "Russian Professor of Physics Propov" designed first world's radio receiver with an antenna
3. First shipboard antenna were all made of haphazard lengths of wire string as high as possible.
4. In 1970's the idea came to virtually replace HF radio in Navy with the new flatsatcom military mobile system.
5. The first real global MSC system was the Marisat system which employed SES and L-band antenna systems
6. The general classification of MSA is based on service operator/provided.
7. Fundamentals classification of MSA according to gain values and characteristics.
8. The vehicle antenna has low gain, it has no need to perform tracking but capacity of communications link is limited.
9. The Imarsat-A SES is the inheritor of the first generation of Marisat. It is similar to SES.
10. In 1982 analog standard was operated. It uses inmarsat standard-A transceiving antenna system known as **Above Deck Equipment (ADU).**
11. Using second generation of Inmarsat satellites, at the end of 1993, Inmarsat-B digital standard started to be in service on ships.
12. MSA are transceiving antennas except for the Inmarsat-EGC receiver, which can also use a receiving antenna.
13. The Cospao-Sarsat system developed shipborne EPIRB with small built in VHF and UHF antennas.
14. Vehicleborne MSA were developed for installation on road and rail vehicles such as buses, cars, trucks, trailers and trains.
15. Like maritime application, the cospas-sarsat system has developed personal or vehicleborne PLB handheld satellite because with small built in VHF and UHF band transmitting antennas.
16. The new GEO and non GEO MSC systems developed small satellite antennas for GMPSC terminals.
17. The other types of MSA are receiving broadcast radio and TV MSA, receiving GNSS MSA, transceiving broadband MSA and antenna for military satellite communication and navigating equipment.
18. The omnidirectional antenna has low gain, generally from 0 to 4 dB L-band and they do not require the capability of satellite tracking.
19. The QHA low gain model consists of four identical helixes would equally spaced on a cylindrical surface.
20. The helical antenna gain depends on the number of turns (N) and typical gain and half power beam width are about 8 dBi and 50°.
21. Widely used dipole antenna has half wavelength $\left(\frac{\lambda}{2}\right)$.

22. A half wavelength dipole is a linear antenna whose current amplitude varies one half of a sine wave with a maximum at the center.
23. To optimize radiation pattern, a set of dipole antennas is bent toward the ground.
24. The CDDA serves as a transceiver antenna for L-band Inmarsat-C SES and VES applications mounted inside a radome.
25. A microstrip patch (disc) antenna has a very low profile and has mechanical strength.
26. To produce a circular polarized wave, a patch antenna is excited at two points orthogonal to each other and feel with signal in equal amplitude and a relative phase of 0° and 90°.
27. The resonant frequency excited by basic mode is given by :

$$f = \frac{\frac{184C}{2\pi a}}{\sqrt{\varepsilon_r}}$$

28. The medium gain directional MSA gives solution with a typical gain between 12 to 15 dBi.
29. The aperture reflector antennas are good solutions with medium gain characteristics used in MSC.
30. The SBF plane reflector antenna was developed by H.W Ehrenspeck in 1960s.
31. An SBF antenna is very attractive for gains in the order of 8 to 15 dBi peak RHCP.
32. The electrical characteristics of a simple SBF antenna can be improved by charging its main reflector from a flat disk to a conical or step plate and by adding a second small reflector.
33. Antenna stabilization is obtained by two-axis stabilized method and satellite pointing is carried out by a tracking program.
34. Wire antennas are monosyllabic construction or combination of elements.
35. An axial mode helical antenna has good circular polarization characteristics over wide frequency range.
36. A conical helix is important for L-band in the order of 100° and circular polarization without hybrid gain of 4 to 7 dBi.
37. The improved dipoles are known as V and U-type dipoles.
38. The crossed dipole can also produce circular polarization without using any external circuits like hybrid component.
39. Antenna resonance is obtained when its length is somewhat shorter than a free space half wavelength.
40. Crossed slot antennas are useful for L-band aeronautical satellite communications on high speed aircraft.
41. The slot antenna is circularly polarized and is complementary with the corresponding dipole antenna.
42. Conical spiral antenna is suitable for L-band LMSC and GPS applications while bifilar version is also used in KU-band satellite communications.
43. Various types of antenna can be arranged as array in space to make a directional pattern or one with desired radiation pattern.

44. The elements of MAA are arranged in 3 × 3 rows square array to obtain similar radiation pattern in different cut planes.

45. XSA antenna application is suitable for the MES Inmarsat-M maritime and land including Inmarsat-Aero solutions.

46. This helix integrated antenna is more suitable for shipborne inmarsat M standards.

47. Four element SBF array antenna is more suitable for maritime applications as a shipborne antenna.

48. Directional antennas for LMSC provide voice and HSD links for long have tracks and for private cars also.

49. In space, radiation pattern can be scanned by controlling the phase of the exciting current in each element of the array.

50. The key technology in AMSC is a directional medium gain antenna.

51. The microstrip phased array antenna is mounted on the top of the fuselage and has two planes with 16 circular patch elements.

52. High gain directional apart use antennas are very powerful transmission reflector and panels.

54. The first generating of parabolic dish antenna uses a reflector in diameter of max 1.2 m but new models reduces the size approximately 0.7 to 0.8 m.

55. Mostly parabolic reflector is used for high directivity of radio signals travelling in straight line same as light rays.

56. The antenna unit processor controls all ADE functions which include satellite tracking and electronic control.

57. The WQHA is highly resonant antenna and it consists of four helices placed at 90° to each other.

58. For resonant RHA (RQHA) it is assume that the QHA consists of four helical and four radial parts.

59. Satellite broadcasting at 12 GHz band becomes very popular and day by day the need for TV reception by relatively large mobile like ships, trucks, buses, trains and aircraft is rapidly growing.

60. In USA currently many services are performed by : Direct TV, USSB, Dish network and Alphastar Broadcast emissions.

Review Questions

1. What do you understand by GMSC ? Explain antenna configuration used for GMSC.
2. Discuss the evolution of antenna systems for mobile radio communications.
3. How many types of Mobile Satellite Antennas (MSA) are there ? Explain each of them briefly.
4. Briefly explain the important general requirements for mobile antenna solutions used in GMSC systems.
5. List the general relation ships of antenna systems which are required to describe the antenna functions.
6. Define the followings :

   (*a*) Radiation pattern

(*b*) Beamwidth
(*c*) Side lobes
(*d*) Antenna directivity.

7. What do you understand by polarization and axial ratio ? Write down the expression to calculate polarization of circular polarized radiated waves.

8. Classify mobile satellite antenna on the basis of transmission direction.

9. What do you understand by Quad Helix Antenna (QHA) ? With the help of diagram discuss its various components.

10. Draw and discuss the Crossed Drooping Dipole Antenna (CDDA) and Microstrip Patch Antenna (MPA).

11. What is medium gain directional antenna ? List the major characteristics of SBF, modified SBF and improved SBF.

12. Draw and discuss the various types of wire antennas.

13. Discuss the applications of crossed slot antenna, conical spiral antenna and planar spiral antenna.

14. Briefly explain the construction of array antennas. Explain its types and applications also.

15. What is wise quadrifilar helix antenna ? How a QHA is accommodated on hand held terminals ? Briefly explain.

16. Discuss the major electrical, mechanical and structural requirements of mobile antennas for MSB.

❑❑❑

APPENDICES

APPENDIX–I	:	What is Inside a Satellite
APPENDIX–II	:	Satellite Dish
APPENDIX–III	:	Comparison of Satellite Communication w.r.t. Fiber Optic Communication
APPENDIX–IV	:	Globalstars
APPENDIX–V	:	Ariane
APPENDIX–VI	:	Error Detection on Satellite Links
APPENDIX–VII	:	Multiple Access
APPENDIX–VIII	:	Parabolic Antenna

Appendix–I

WHAT IS INSIDE A SATELLITE ?

Satellites have a great deal of equipment packed inside them. A satellite has seven subsystems, and each one has its own work to do.

1. The **propulsion subsystem** includes the electric or chemical motor that brings the spacecraft to its permanent position, as well as small thrusters (motors) that help keep the satellite in its assigned place in orbit. Satellites drift out of position because of solar wind or gravitational or magnetic forces. When that happens, the thrusters are fired to move the satellite back into the right position in its orbit.

2. The **power subsystem** generates electricity from the solar panels on the outside of the spacecraft. The solar panels also store electricity in storage batteries, which provide power when the sun isn't shining on the panels. The power is used to operate the communications subsystem. A Boeing 702 generates enough power at the end of its service life to operate two hundred 75-watt light bulbs.

3. The **communications subsystem** handles all the transmit and receive functions. It receives signals from the earth, amplifies or strengthens them, and transmits (sends) them to another satellite or to a ground station.

4. The **structures subsystem** distributes the stresses of launch and acts as a strong, stable framework for attaching the other parts of the satellite.

5. The **thermal control subsystem** keeps the active parts of the satellite cool enough to work properly. It does this by directing the heat that is generated by satellite operations out into space, where it won't interfere with the satellite.

6. The **attitude control subsystem** maintains the communications "footprint" in the correct location. Satellites can't be allowed to jiggle or wander, because if a satellite is not exactly where it belongs, pointed at exactly the right place on the earth, the television program or the telephone call it transmits to you will be interrupted. When the satellite gets out of position, the attitude control system tells the propulsion system to fire a thruster that will move the satellite back where it belongs.

7. Operators at the ground station need to be able to transmit commands to the satellite and to monitor its health. The **telemetry and command subsystem** provides a way for people at the ground stations to communicate with the satellite.

Appendix–II

SATELLITE DISH

Fig. A.2.1. Special dish for up to 16 satellite positions (Ku-band)

Fig. A.2.2. Satellite dish antenna for C-Band

Fig. A.2.3. Satellite Dishes installed on an apartment complex

A **satellite dish** is a type of parabolic antenna designed with the specific purpose of transmitting signals to and/or receiving from satellites. A satellite dish is a particular type of microwave antenna. Satellite dishes come in varying sizes and designs, and are most commonly used to receive satellite television. Many of the offset type of satellite dishes are sections of a larger parabolic dish.

The parabolic shape of a dish reflects the signal to the dish's focal point. Mounted on brackets at the dish's focal point is a device called a feedhorn. This feedhorn is essentially the front-end of a waveguide that gathers the signals at or near the focal point and 'conducts' them to a low-noise block downconverter or LNB. The LNB converts the signals from electromagnetic or radio waves to electrical signals and shifts the signals from the downlinked C-band and/or Ku-band to the L-band range. Direct broadcast satellite dishes use an LNBF, which integrates the feedhorn with the LNB. (A new form of omnidirectional satellite antenna, which does not use a directed parabolic dish and can be used on a mobile platform such as a vehicle was announced by the University of Waterloo.

Modern dishes intended for home television use are generally 43 cm (18 in) to 80 cm (31 in) in diameter, and are fixed in one position, for Ku-band reception from one orbital position. Prior to the existence of Direct broadcast satellite services, home users would generally have a motorised C-band satellite dish of up to 3 metres in diameter for reception of channels from different satellites. Overly small dishes can still cause problems, however, including rain fade and interference from adjacent satellites.

Motorised satellite dishes are still popular with enthusiasts. There are three competing standards, which often are all supported by a set-top box, DiSEqC, USALS, and 36v positioners. They can only supply one receiver. A common misconception is that the LNBF (low-noise block/feedhorn), the device at the front of the dish, receives the signal directly from the atmosphere. See, for instance, this BBC News 24 countdown that shows a "red data stream" being received by the LNBF directly instead of being beamed to the dish, which because of its parabolic shape will collect the signal into a smaller area and deliver it to the LNBF.

In Europe the frequencies used by DBS services are 10.7 - 12.75 GHz on two polarisations H and V. This range is divided into a 'low band' with 10.7 - 11.7 GHz, and a 'high band' with 11.7 - 12.75 GHz. This results in four frequency bands, each with a bandwidth of about 1 GHz. In the LNB they become down converted to 950 - 2150 MHz, which is the frequency range allocated for the satellite service on the coaxial cable between LNBF and receiver. Lower frequencies are allocated to Cable and Terrestrial TV, FM radio, etc. Only one of these frequency bands fits on the coaxial cable, so each of these bands needs a separate cable from the LNBF to a switching matrix or the receiver needs to select one of the 4 bands at a time.

In a single receiver residential installation there is a single cable from receiver to LNB and the receiver uses different power supply voltages (14/18V) and pilot tones (22 kHz) to instruct the LNB to select one of the 4 bands. In larger installations each band is given its own cable, so there are 4 cables from the LNB to a switching matrix, which allows the connection of multiple receivers in a star topology using the same signalling method as in a single receiver installation.

The theoretical gain of a dish increases as the frequency increases. The actual gain depends on many factors including surface finish, accuracy of shape, feedhorn matching. With lower frequencies, C-Band for example, dish designers have a wider choice of materials. The large size of dish required for lower frequencies led to the dishes being constructed from metal mesh on a metal framework. At higher frequencies, mesh type designs are rarer though some designs have used a solid dish with perforations.

The dish is a reflector antenna and almost anything that reflects radio frequencies can be used as a reflector antenna. This has led to dustbin lids, woks and other items being used as "dishes". Coupled with low noise LNBs and the higher transmission power of DTH satellites, it is easier to get a usable signal on some of these "dishes".

Types

- Individual dishes serving one dwelling: Direct To Home (DTH).
- Collective dishes, shared by several dwellings: Satellite Master Antenna Television (SMATV) or Communal Antenna Broadcast Distribution (CABD).

Appendix–III

COMPARISON OF SATELLITE COMMUNICATION W.R.T. FIBER OPTIC COMMUNICATION

Fiber optic communication and the satellite communication are being developed continuously for the last two decades.

S. No.	Characteristics	Satellite	Fiber optic system
1.	Multipoint capability	Point to multipoint communication is easily implemented.	Basically a point to point communication.
2.	Security	Signals are encoded for security	It is difficult to tap without detection.
3.	Flexibility	Easy to reconfigure	Difficult to reconfigure to meet changing demand.
4.	Link durability	Storms can disable individual antennas but leave network intact	Storms can knock down overhead lines.
5.	Bandwidth	Most transponders have bandwidths of 36.54 or 72 MHz	The bandwidth of fiber is 1 terahertz and it is limited by electronics at terminals.
6.	Immunity to interference	Transmission subject to interference from various sources including microwaves	Immune to EM interference.
7.	Connectivity to customer size	Antennas are installed on customer premises with 14/12 GHz band and local loops are not required.	Local loops are required.

Appendix–IV

GLOBALSTAR

Globalstar is a low Earth orbit (LEO) satellite constellation for telephone and low-speed data communications, similar to (and competing with) the Iridium satellite system.

System Architecture

Globalstar differs from Iridium in several important ways :

- Globalstar satellites are simple "bent pipe" repeaters; there is no inter-satellite linking. A network of ground gateway stations provides connectivity from the 40 satellites to the public switched telephone network; users are assigned telephone numbers on the North American Numbering Plan in North America or the appropriate telephone numbering plan for the country that the overseas gateway is located in, except for Brazil, where the official globalstar country code (+8818) is used. Because there is no inter-satellite linking, a satellite must have a gateway station in view to provide service to any users it may see. Because there are no gateway stations to cover certain remote areas (such as oceans far from land), no service can be provided in those areas even though the satellites fly over them.
- Globalstar orbits have an inclination of 52 degrees, compared to the near-polar 86.4-degree orbits used by Iridium. Globalstar also does not cover polar areas due to the lower orbital inclination.
- The Globalstar system uses the Qualcomm CDMA air interface; however, the Ericsson and Telit phones accept standard GSM SIM cards, while the Qualcomm GSP-1600/1700 phones do not have a simcard interface, but use CDMA/IS-41 based authentication. Therefore the Globalstar gateways need to support both the CDMA/IS-41 and the GSM standards, which, however not all gateways do. This results in the lack of coverage for GSM-authentication-standard based phones in the Eastern Asian and Caribbean region.

While for Iridium only Telstra Australia offers GSM-roaming, most Globalstar providers have roaming agreements with local cellular operators, enabling the use of a cellular SIM card with a Globalstar handset and vice versa.

Despite (or because of) the limitations above, Globalstar's operational costs were significantly cheaper and can support somewhat higher data rates than Iridium. However, such cost differences later proved meaningless when both companies shed their multi-billion dollar debts through bankruptcy.

System Deployment

Like Iridium, Globalstar received its U.S. spectrum allocation from the FCC in January 1995, and continued to negotiate with various other sovereign nations for rights to use the same radio frequencies in their countries.

The first satellites were launched in February 1998, but system deployment was delayed through a series of embarrassing and costly launch failures, notably the September 1998 loss of 12 satellites in a launch by the Russian Space Agency. In February 2000, it launched the last of 52 satellites: 48 satellites and four spares (reduced from the original plan of eight spares).

The first call on the Globalstar system was placed on November 1, 1998, from Irwin Jacobs (chairman of Qualcomm) in San Diego to Bernard Schwartz (CEO and chairman of Loral Space and Communications) in New York.

In October 1999, the system began "friendly user" trials with 44 of 48 planned satellites. In December 1999, the system began limited commercial service (200 users) with the full 48 satellites (no spares in orbit). In February 2000, it began full commercial service with its 48 satellites and 4 spares in North America, Europe and Brazil. Initial prices were $1.79/minute vs. $9/minute for Iridium.

In 2005, some of the satellites began to reach the limit of their operational lifetime of 7.5 years. In December of 2005, Globalstar began to move some of its satellites into a graveyard orbit above LEO.

Recent Problems

According to documents filed with the SEC on January 30, 2007, Globalstar's previously identified problems with its S-band amplifiers used on its satellites for two-way communications are occurring at a rate faster than previously believed, possibly eventually leading to total loss of service in 2008.

In the filing, Globalstar made the following statements:

"Based on data recently collected from satellite operations, the Company has concluded that the degradation of the amplifiers is now occurring at a rate that is faster than previously experienced and faster than the Company had previously anticipated."

"Based on its most recent analysis, the Company now believes that, if the degradation of the S-band antenna amplifiers continues at the current rate or further accelerates, and if the Company is unsuccessful in developing additional technical solutions, the quality of two-way communications services will decline, and by some time in 2008 substantially all of the Company's currently in-orbit satellites will cease to be able to support two-way communications services."

Industry analysts speculate the problem is caused by radiation exposure the satellites receive when they pass through the South Atlantic Anomaly in their 800 km altitude orbits.

"The S-band antenna amplifier degradation does not affect adversely the Company's one-way "Simplex" data transmission services, which utilize only the L-band uplink from a subscriber's "Simplex" terminal to the satellites. "

"The Company is working on plans, including new products and services and pricing programs, and exploring the feasibility of accelerating procurement and launch of its second-generation satellite constellation, to attempt to reduce the effects of this problem upon its customers and operations. The Company will be able to forecast the duration of service coverage at any particular location in its service area and intends to make this information available without charge to its service providers, including its wholly owned operating subsidiaries, so that they may work with their subscribers to reduce the impact of the degradation in service quality in their respective service areas. The Company is also reviewing its business plan in light of these developments. "

"The Company's liquidity remains strong. At December 31, 2006, in addition to its credit agreement, the Company had unrestricted cash on hand and undrawn amounts under the Thermo Funding Company irrevocable standby stock purchase agreement of approximately $195 million."

Globalstar's Second-Generation of Satellite Constellation

In December 2006, Globalstar announced that Alcatel Alenia Space has been awarded a • 661 million contract for the Second- Generation 48 satellite constellation.

"The satellites are being designed with a life expectancy of 15 years, which is significantly longer than the design life of Globalstar's first generation constellation."

In addition, Globalstar announced on April, 03, 2007 that it has signed a EUR 9 million (approximately US$12 million) agreement with Alcatel Alenia Space to upgrade the Globalstar satellite constellation, including necessary hardware and software upgrades to Globalstar's satellite control network facilities.

"This Second-Generation of satellites are expected to provide service to Globalstar customers with satellite voice and data services until at least 2025."

Appendix–V

ARIANE

Ariane 5 is a European expendable launch system designed to deliver payloads into geostationary transfer orbit or low Earth orbit.

It is manufactured under the authority of the European Space Agency (ESA) with EADS Astrium Space Transportation as prime contractor, leading a consortium of sub-contractors. The rocket is operated and marketed by Ariane space as part of the *Ariane* programme. EADS Astrium Space Transportation builds the rockets in Europe and Arianespace launches them from the Guiana Space Centre.

It succeeded Ariane 4, but does not derive from it directly. Its development took 10 years and cost •7 billion. The ESA originally designed Ariane 5 to launch the manned mini shuttle Hermes, and thus intended it to be "human rated" from the beginning. After the ESA cancelled Hermes, the rocket became a purely robotic launcher.

Two satellites can be mounted using a Sylda carrier. Three main satellites are possible depending on size. Up to eight secondary payloads, usually small experiment packages or minisatellites, can be carried with an ASAP (Ariane Structure for Auxiliary Payloads) platform.

Components

Fig. A.5.1. ***Vulcain*** **engine.**

Ariane 5's cryogenic H158 main stage is called the EPC "Étage Principal Cryotechnique" (Cryotechnic Main Stage). It consists of a large tank 30.5 metres high with two compartments, one for 130 tonnes of liquid oxygen and one for 25 tonnes of liquid hydrogen, and a Vulcain engine at the base with thrust of 115 tonnes-force (1.13 meganewtons). This part of the first stage weighs about 15 tonnes when empty.

Attached to the sides are two solid propellant boosters, P238 (P241 for Ariane 5 ECA), each weighing about 277 tonnes full. Each delivers a thrust of about 630 tonnes-force (6.2 MN). These boosters can be recovered with parachutes, like the Space Shuttle Solid Rocket Boosters. They may have been retrieved for examination on early missions, but are not reused.

The second stage is on top of the main stage and below the payload. The Ariane 5G used the EPS (Étage à Propergols Stockables/Storable Propellant Stage), which is fueled by monomethylhydrazine (MMH) and nitrogen tetroxide, where as the Ariane 5 ECA uses the ESC (Étage Supérieur Cryotechnique/Cryogenic Upper Stage), which is fueled by liquid hydrogen and liquid oxygen. The payload and all upper stages are covered at launch by the fairing, which splits off once sufficient altitude has been reached.

Variants

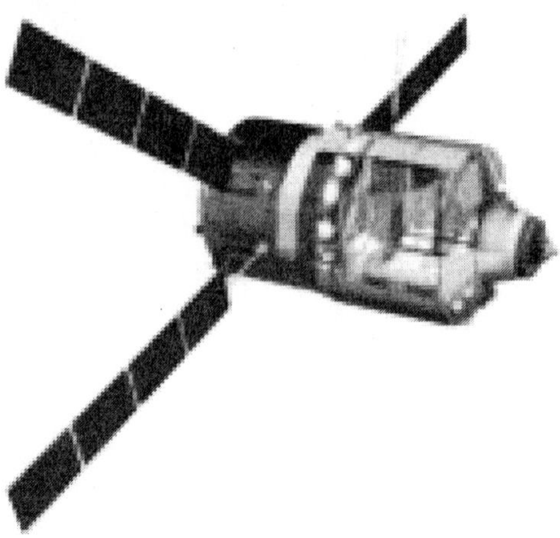

Fig. A.5.2. The ATV will be delivered into orbit by an Ariane 5 rocket

The original version is dubbed Ariane **5G** (Generic) with a launch mass of 737 tonnes. Its payload capability to Geostationary transfer orbit (GTO) was initially specified as 5970 kg, but was increased after the qualification flights to 6200 kg.

The Ariane **5G+** had an improved second stage, with a GTO capacity of 6950 kg for a single payload. It flew three times in 2004.

It was replaced in 2005 by the Ariane **5GS**, with the same solid boosters as the Ariane 5 ECA and a modified first Stage with a Vulcain 1B engine. It can carry a single payload of 6100 kg to GTO.

The Ariane 5 **ECA** has a GTO launch capacity of 10,000 kg for dual payloads or 10,500 kg for a single payload. This variant uses a new Vulcain 2 first-stage engine, and an ESC-A second stage, powered by an HM-7B engine, weighing 2100 kg and carrying 14,000 kg of cryogenic propellant. The second stage was previously used as the third stage of Ariane 4; in ECA use, the tanks are modified to shorten stage length. The revised Vulcain has a longer, more efficient nozzle with more efficient flow cycle and denser propellant ratio. The new ratio demanded length modifications to the first-stage tanks. Also, the solid booster casings have been lightened with new welds, and packed with more propellant. The ESC-A cryogenic second stage does not improve the performance to Low Earth orbit compared to Ariane 5G, and for this reason the Ariane 5 ECA will not be used to launch the Automated Transfer Vehicle (ATV).

The Ariane 5 **ES-ATV** will be used to launch the Automated Transfer Vehicle. It includes all the performance improvements of Ariane 5 **ECA** on EPC and EAP stages while the second stage is the EPS used on Ariane **5GS** variants.

Comparable rockets: Delta IV - Atlas V - Chang Zheng 5 - Angara - Proton - Falcon 9 - H-IIB

Future Developments

Ariane 5 **ES ATV** has been designed for launching the Automated Transfer Vehicle. It consists of the improved Vulcain 2-powered first stage and the upper stage from the Ariane 5G, and can put up to 21,000 kg in LEO.

Appendix–VI

ERROR DETECTION ON SATELLITE LINKS

Error detection is a user-defined service which allows the user to send and receive data with a greatly reduced probability of error, and a very high probability that uncorrected errors are identified and located within a block of data, so that the existence of an error is known even if the exact bit or word in error cannot be determined. The penalty for the user is a reduced transmission rate. The transmission rate is reduced only when errors occur.

Error Detection and Correction Methods

When a succession of pictures is transmitted, for example, weather maps, the previous value of a pixel can be used when a new value is in error. When an error detection code or CRC is used and the error must be corrected, a retransmission of the block of data containing the error must be made so that the correct data are acquired at the receiving terminal. If an error is detected in the block, a *not acknowledge* (NAK) signal is sent back to the transmit end of the link, which triggers a retransmission of the erroneous block of data. This is called an *automatic repeat request* (ARQ) system.

In terrestrial data communication systems, it is a common practice for the receiving terminal to send an *acknowledge* (ACK) signal to the transmitting end whenever it receives an error-free block of data. A receiving terminal receives data using a terrestrial protocol and generates the acknowledgements, then transmits data over the satellite link using a different protocol. This is called *spoofing.*

There are **three techniques for retransmission requests**, depending on the type of link used.

They are:

Simplex Link

Here, the ACK or NAK signal must travel on the same path as the data, so the transmitter must stop after each block and wait for the receiver to send back an NAK or ACK before it retransmits the last data block or sends the next one. With a one-way delay of 240 ms on a GEO satellite link, the data rate of this system is low. It is suitable for links in which data is generated slowly.

Stop-and-Wait System

Here the transmitting end sends a block of data and waits for the acknowledgement to arrive on the return channel. The delay is the same as in the simplex case , but implementation is simpler. Fig. a shows an example of a stop and wait sequence. In a continuous transmission system using the go-back-N technique, data are sent in blocks continuously and held in the buffer at the receive end of the link.

Each block is checked for errors as it arrives, and the appropriate ACK or NAK is sent back to the transmitting end, with the block number appended. When a NAK is received, the transmit end goes back to block N and retransmits all subsequent blocks as in Fig. A.VI.1.

This requires the transmitter on a satellite link to hold at least 480ms of data, to allow time for the data to reach the receive end and be checked for errors and for the acknowledgement to be sent to the transmit end.

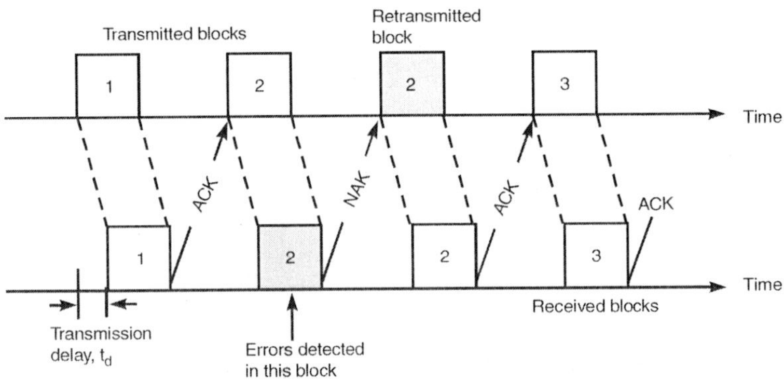

Fig. A.VI.1. Schematic of a Stop and Wait system

Selective repeat system

Here, sufficient buffering is provided at both ends of the link. Hence only the corrupted data is to be retransmitted. In fig b block 2 corrupted, and blocks 4,5,6,7 are transmitted before the NAK message is received. At this point, we could transmit block 3 only if blocks 4,5,6,7 are stored at the receive end. On receiving a correct version of block 3, the receive buffer substitutes it for the corrupted version and releases the data for retransmission.

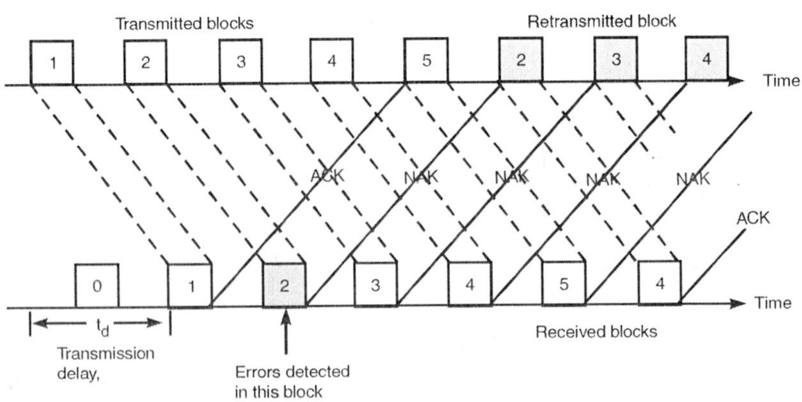

Fig. A.VI.2.

Appendix–VII

MULTIPLE ACCESS

Satellites are built with the intention that many users will share the bandwidth allocated to the satellite, allowing many separate communication links to be established through the satellite's transponders. The signals that earth stations transmit to a satellite may differ widely in their character- voice, video, data, facsimile-but they can be sent through the same satellite using multiplexing and multiple access techniques.

MULTIPLEXING

It is the process of combining a number of signals into a single signal, so that it can be processed by a single amplifier or transmitter over a single radio channel. Multiplexing can be done at baseband or at radio frequency. The corresponding technique that recovers the individual signals is called demultiplexing. Multiplexing is a key feature of all commercial long distance communication systems, and is part of the multiple access capability of all satellite communication systems.Mutiplexing is a transmission feature.

MULTIPLE ACCESS

Multiple access is the process by which a large number of earth stations interconnect their links through a satellite. It is the ability of the satellite to carry many signals at the same time. Multiple access allows the communication capacity of the satellite to be shared among a large number of earth stations, and to accommodate the different mixes of communication traffic that are transmitted by the earth stations. Multiple access schemes are employed in communication satellites with the use of many transponders. The individual transponders may carry one signal, a single analog television program, or hundreds of signals, as with mobile satellite telephone systems.

Difference between Multiplexing and Multiple access:

Multiplexing applies to signals that are generated at one location whereas; multiple access refers to signals from a number of different geographical locations.

MULTIPLE ACCESS TECHNIQUES

FDMA (Frequency Division Multiple Access)

Each radio station transmits at a unique allocated frequency.

TDMA (Time Division Multiple Access)

Each radio station is allocated a unique time slot at the satellite so that the signals pass through the transponder sequentially.

CDMA (Code Division Multiple Access)

Earth stations use spread spectrum transmissions with orthogonal codes to share a transponder without interference.

Appendix–VIII

PARABOLIC ANTENNA

The **parabolic antenna** is a high-gain reflector antenna used for radio, television and data communications, and also for radio location (RADAR), on the UHF and SHF parts of the electromagnetic spectrum. The relatively short wavelength of electromagnetic (radio) energy at these frequencies allows reasonably sized reflectors to exhibit the very desirable highly directional response for both receiving and transmitting.

Fig. AAVIII.1.1. Parabolic antennas at the Very Large Array Radio Telescope (image courtesy of NRAO/AUI).

A typical parabolic antenna consists of a parabolic reflector illuminated by a small feed antenna.

The reflector is a metallic surface formed into a paraboloid and usually truncated in a circular rim that forms the diameter of the antenna. This paraboloid possesses a distinct focal point by virtue of having the reflective property of parabolas, in that a point light source at this focus produces a parallel light beam aligned with the axis of revolution.

The feed antenna is placed at the reflector focus. This antenna is typically a low-gain type such as a half-wave dipole or a small waveguide horn. In more complex designs, such as the Cassegrain antenna, a sub-reflector is used to direct the energy into the parabolic reflector from a feed antenna located away from the primary focal point. The feed antenna is connected to the associated radio-frequency (RF) transmitting or receiving equipment by means of a coaxial cable transmission line or hollow waveguide.

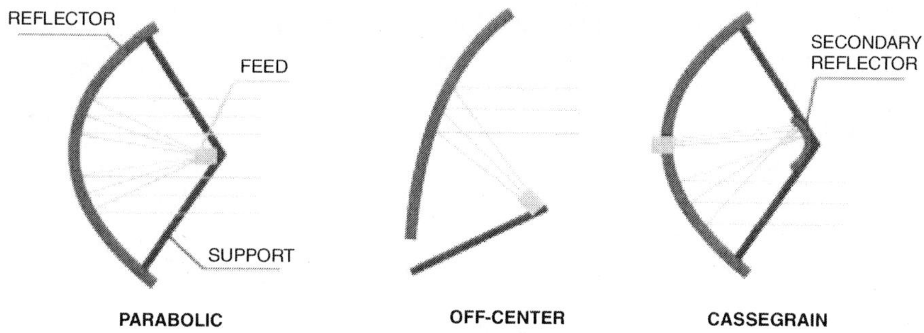

Fig. AAVIII.1.2. Main Types of Parabolic Antennas.

Considering the parabolic antenna as a circular aperture gives the following approximation for the maximum gain:

$$G \approx \left(\frac{\pi^2 D^2}{\lambda^2}\right)$$

or

$$G \approx \frac{9.87 D^2}{\lambda^2}$$

where :

G is power gain over isotropic

D is reflector diameter in same units as wavelength

λ is wavelength

Practical considerations of antenna effective area and sidelobe suppression reduce the actual gain obtained to between 35 and 55 percent of this theoretical value. For theoretical considerations of mutual interference (at frequencies between 2 and 30 GHz - typically in the Fixed Satellite Service) where specific antenna performance has not been defined, a *reference antenna* based on Recommendation ITU-R S.465 is used to calculate the interference, which will include the likely sidelobes for off-axis effects.

Applying the formula to just one of the 25-meter-diameter VLA antennas for a wavelength of 21 cm (1.42 GHz, a common radio astronomy frequency) yields an approximate maximum gain of 140,000 times or about 50 dBi (decibels above the isotropic level).

With the advent of TVRO and DBS satellite television, the parabolic antennas became an ubiquitous feature of urban, suburban, and even rural, landscapes. Extensive terrestrial microwave links, such as those between cellphone base stations, and wireless WAN/LAN applications have also proliferated this antenna type. Earlier applications included ground-based and airborne radar and radio astronomy. The largest "dish" antenna in the world is the radio telescope at Arecibo, PR, but, for beam-steering reasons, it is actually a spherical, rather than parabolic, reflector.

Structure

The reflector dish can be solid, mesh or wire in construction and it can be either fully circular or somewhat rectangular depending on the radiation pattern of the feeding element. Solid antennas have more ideal characteristics but are troublesome because of weight and high wind load. Mesh and wire types are easier to construct and have nearly ideal characteristics if the holes or gaps are kept under 1/10 of the wavelength.

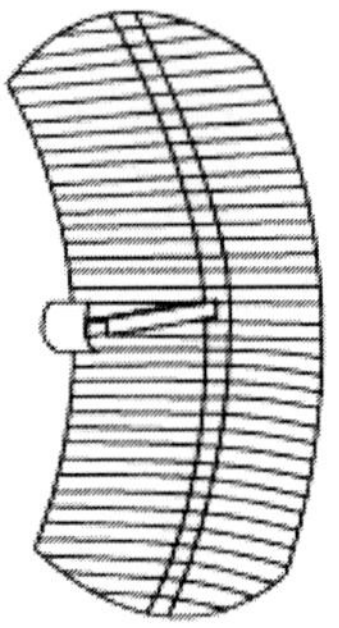

Fig. AVIII.1.3.

Wire-type parabolic antenna (Wi-Fi / WLAN antenna at 2,4 Ghz). Oriented to provide horizontal polarization: the reflector wires and the feed element are both horizontal. This antenna has a greater extent in the vertical plane and hence, a narrower beamwidth in that plane. The feed element has a wider beam in the vertical direction than the horizontal and hence matches the reflector by illuminating it fully.

More exotic types include the off-set parabolic antenna, Gregorian and Cassegrain types. In the off-set, the feed element is still located at the focal point, which because of the angles utilized, is usually located below the reflector so that the feed element and support do not interfere with the the main beam. This also allows for easier maintenance of the feed, but is usually only found in smaller antennas.

The Gregorian and Cassegrain types, sometimes generically referred to as "dual optics" antennas, utilize a secondary reflector, or "sub-reflector", allowing for better control over the collimated beam as well as allowing the antenna feed system to be more compact. These antennas are usually much larger where prime focus and off-set construction are not as practical. The feed element is usually located in a "feed horn" which protrudes out from the main reflector. This setup is used when the feed element is bulky or heavy such as when it contains a pre-amplifier or even the actual receiver or transmitter. Parabolic antenna theory closely follows optics theory. So a Gregorian antenna can be identified by the fact that it uses a concave sub-reflector, while a Cassegrain antenna uses a convex sub-reflector.

Feeding Parabolic Antennas

The actual 'antenna' in a parabolic antenna, that is, the device that interfaces the transmission line or waveguide containing the radio-frequency energy to free space, is the feed element. The reflector surface is entirely passive. This feed element should usually be at the center of the reflector at the focal point of that dish. The focal point is the point where all reflected waves will be concentrated, this point is calculate with the following equation.

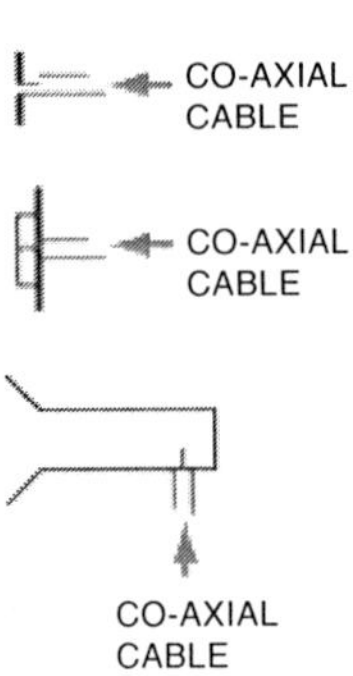

Fig. AVIII.1.4.

$$F = \frac{D^2}{16d}$$

where :

F is the Focal Point of the reflector

D is reflector diameter in same units as wavelength

d is depth of the reflector

The radiation from the feed element induces a current flow in the conductive reflector surface which, in turn, re-radiates in the desired direction, perpendicular to the directrix plane of the paraboloid. The feed element can be any one of a multitude of antenna types. Whichever type is used, it must exhibit a directivity that efficiently illuminates the reflector and must have the correct polarization for the application — the polarization of the feed determining the polarization of the entire antenna system. The simplest feed is a half-wave dipole which is commonly used at lower frequencies, sometimes in conjunction with a closely coupled parasitic reflector or "splash plate". At higher frequencies a horn-type becomes more feasible and efficient. To adapt the horn to a coaxial antenna cable, a length of waveguide is used to effect the transition.

❑❑❑

SHORT ANSWER QUESTIONS

1. What are the different applications of satellite systems ?

Ans.
- Largest International System(Intel sat)
- Domestic sat system in united states(Dom sat)
- U.S National Oceanography Atmospheric Administration (NOAA)

2. Mention the different services of satellite systems.

Ans.
- Fixed satellite services
- Broadcasting satellite services
- Mobile satellite services
- Navigational satellite services
- Meteorological satellite services.

3. Define Polar-orbiting Satellites.

Ans. Polar orbiting Satellites orbit the earth in such a way as to cover the north and south polar regions.

4. State Kepler's first law.

Ans. It states that the path followed by the satellite around the primary will be an ellipse. An ellipse has two focal points F_1 and F_2.The center of mass of the two body system, termed the barycenter is always centered on one of the foci.

$$E = V\frac{\sqrt{a^2 - b^2}}{a}$$

5. State Kepler's second law.

Ans. It states that for equal time intervals, the satellite will sweep out equal areas in its orbital plane, focused at the barycenter.

6. State Kepler's third law.

Ans. It states that the square of the periodic time of orbit is perpendicular to the cube of the mean distance between the two bodies.

$$a^3 = \frac{\mu}{n^2}$$

7. Define apogee and perigee.

Ans.
- The point farthest from the earth is known as apogee.
- The point closest from the earth is known as perigee.

8. What is line of apsides ?

Ans. The line joining the perigee and apogee through the center of the earth.

9. Define ascending and descending node.

Ans. The point where the orbit crosses the equatorial plane going from South North.

The point where the orbit crosses the equatorial plane going from South to North.

10. Define inclination.

Ans. The angle between the orbital plane and the earth's equatorial plane.It is measured at the ascending node from the equator to the orbit going from east to north.

11. Define mean anomaly and true anomaly.

Ans. **Mean anomaly:** It gives an average value of the angular position of the satellite with reference to the perigee.

True anomaly : It is the angle from perigee to the satellite position, measured at the earth's center.

12. Mention the apogee and perigee height.

Ans. $R_a = a(1 + e)$

$R_p = a(1 + e)$

$H_a = r_a - r_p$

$H_p = r_p - R_p$

13. Define Universal time.

Ans. It is the time used for all civil time keeping purposes and it is the time reference which is broadcast by the national bureau of standards as a standard for setting clocks.

UT day = l/24(hours + minutes/60 + seconds/3600)

UT° = 360 × UT day.

14. Mention the Julian dates.

JD = JD010 + day number + UT day

15. What is sideral time ?

Ans. Sideral time is time measured relative to the fixed stars. It will be seen that one complete rotation sideral time relative to the sun .This is because the earth moves in its orbit around the sun.

16. Define Sideral day.

Ans. It is defined as one complete rotation of the earth relative to the fixed stars.It is measured as $23^h56^m04^s$ mean solar time.

17. Defme Greenwich hour angle(GHA).

Ans. The angular distance from the I axis to the Greenwich merdian is measured directly as Greenwich sideral time ,also known as the Greenwich hour angle.

17. Write the formula for GST.

Ans. GST = 99.6910 + 36000.7689Xt + 0.0004Xt + 2 + UT deg

19. Define geocentric latitude.

Ans. As a point of interest for zero height, the angle

$$\Psi e\ (h = 0) = (1 - Ee^2)\tan \lambda E$$

20. What are the difference between the geodetic and geocentric latitudes ?

Ans. The latitudes reaches a maximum at a geocentric latitude of 45 deg, when the geodetic latitude is 45.192 deg.

21. What is meant by azimuth angle ?

Ans. It is defined as the angle produced by intersection of local horizontal plane and the plane passing through the earth station, the satellite and center of earth.

22. What are the conditions of location of the earth station and subsatellite point of the Azimuth angle ?

Ans. For ES in northern hemisphere:

A = 180deg – A(ES west of satellite)

A = 180deg + A′(ES east of satellite)

For ES in northern hemisphere:

A = A (ES west of satellite)

A = 360 – A′ (ES east of satellite)

23. Write short notes on altitude control system.

Ans. It is the system that achieves and maintains the required altitudes.The main functions of altitude control system include maintaining accurate satellite position throughout the life span of the system.

24. What is an polar antenna?

Ans. A single actuator is used which moves the antenna in a circular arc ie known as polar mount antenna.

25. What is declination?

Ans. The angle of tilt is often referred to as the declination which must not be confused with the magnetic declination used in correcting compass readings.

26. Define the terms in Eclipse.

Ans. During euinox periods.the earth the sun and the satellite are in alignment with the result that earth's shadow eclipses tha satellite and the sunlight fails to reach the satellite solar cells.The eclipse effect is noticeable for periods of about four weeks and the maximum daily eclipse duration is about 1.20 hours.

27. What is meant by payload?

Ans. The payload refers to the equipment used to provide the service for which the satellite has been launched.

28. What is meant by transponder ?

Ans. In a communication satellite,the equipment which provides the connecting link between the satellite's transmit and receive antennas is referred to as the transponder.

29. Write short notes on station keeping.

Ans. It is the process of maintenance of satellite's altitude against different factors that can cause drift with time. Satellites need to have their orbits adjusted from time to time because the satellite initially placed in the correct orbit, natural forces induce a progressive drift.

30. What is mean by Pitch angle ?

Ans. Movement of a spacecraft about an axis which is perpendicular to its longitudinal axis.It is the deg of elevation or depression.

31. What is an propellant ?

Ans. A solid or liquid substance burnt in a rocket for the purpose of producing thurst.

32. What is an Yaw?

Ans. Yaw is the rotation of a vechicle about its vertical axis.

33. What is an Zero 'g'?

Ans. Zero 'g' is a state when the gravitational attraction is opposed by equal and opposite inertial forces and the body experiences no mechanical stress.

34. Describe the spin stabilized satellites.

Ans. In a, spin stabilized satellites,the body of the satellite spins at about 30 to 100 rpm about the axis perpendicular to the orbital plane. The satellites are normally dual spin satellites with a spinning section and a despun section on which antennas are mounted. These are kept stationary w.r.to earth by counter rotating the despunsection.

35. What is meant by frequency reuse?

Ans. The carrier with opposite senses of polarization may overlap in frequency this technique is known as frequency reuse.

36. What is meant by spot beam antenna?

Ans. A beam generated by a communication satellite antenna of sufficient size that the angular spread of sufficient size that the angular spread of the energy in the beam is very small with the result that a region that is only a few hundred km in diameter is illuminated on earth.

37. What is an TWTA?

Ans. The TWTAS are widely used in transponder to provide the final output power required to the transtube and its power supplies.

38. What is meant by Intermodulation distortion?

Ans. The AM/PM conversion is then a complicated function of carrier amplitudes,but in addition,the nonlinear transfer characteristic introduces a more serious form of distortion known as intermodulation distortion.

39. Define input backoff.

Ans. In order to reduce the intermodulation distortion,the operating point of the TWT must be shifted closer to the linear portion of the curve, the reduction in input power being referred to as i/p backoff.

40. Define diplexer.

Ans. The transmit and receives signals are separated in a device known as diplexer

41. What is an OMT?

Ans. The polarization seperation takes place in a device known as an orthocoupler or orthogonal mode transducer.

42. What is an polarization interleaving?

Ans. Overlap occurs between channels, but these are alternating polarized left hand circular and righthand circular to reduce interference to acceptable levels.This is referred to as polarization interleaving.

43. What is an SCPC?

Ans. In a thin route circuit, a transponder channel (36 MHz) may be occupied by a no. of single carriers, each associated with its own voice circuit.

44. Defme S/N ratio.

Ans. The S/N introduced in the preceding section is used to refer to the ratio of signal power to noise power at the receiver output. This ratio is sometimes referred to as the post detector.

45. What is noise weighting?

Ans. Improve the post detection signal to noise ratio is referred to as noise weighting.

46. What is an EIRP?

Ans. It is a measure of radiated or transmitted power of an antenna.It can be completed from the antenna gain and the power fed to the antenna input.

47. Write the equations of losses for clear sky conditions.

Ans. Losses = (FSL) + (RFL) + (AML) + (AA) + (PL)

48. What is an noise power spectral density ?

Ans. Noise power per unit BW is termed the NPS density.

$$NO = PN/BN=KTN \text{ joules}$$

49. What is an Intermodulation noise?

Ans. Intermodulation distortion in high power amplifier can result in signal products which appear as noise and in fact is referred to as Intermodulation noise.

50. What are the types of antenna losses?

Ans.
- Sky noise
- Antenna losses

51. What is an antenna losses?

Ans. It is add to noise received as radiation and the total antenna noise temperature is in the sum of the equivalent noise temperature of all these sources.

52. Define sky noise.

Ans. It is a term used to describe the microwave radiation which is present throughtout universe and which appears to originate from matter in any form, at finite temperature.

53. Define noise factor.

Ans. An alternative way of representing amplifier noise is by means of its noisefactor. In defining the NF of an amplifier, denoted by to usually taken as 290k

$N_{0,out}$ = FGKTO

54. What is an Apsorptive n/w?

Ans. It is one which contains resistive elements.These introduce losses by absorbing energy from the signal and converting it to heat. Resistive attenuators, transmission lines and waveguides are all examples of absorptive networks.

55. Write the equation of system noise factor.

Ans. TS = Tant+ Tel + (L – l)TO/Gl+L(F – l)TO/Gl

56. Define saturation flux density.

Ans. The flux density required at the receiving antenna to produce saturation of TWTA is termed the saturation flux density.

57. A satellite downlink at 12 GHz operates with a transmit power of 6 W and an antenna gain of 48.2 dB. Calculate the EIRP in dBw.

EIRP = $10 \log_6 + 48.2 = 56$ dBw.

58. Calculate the gain of a 3 m parabolidal antenna operating at a frequency of 12 GHz. Assume an aperture efficiency of 0.5.

Ans. G = $10 \log 78168 = 48.9$ dB

59. The range between a ground station and a satellite is 42000 km. Calculate the free space loss a frequency of 6 GHz.

Ans. (FSL) = $32.4 + 20\log_4 2000 + 20 \log 6000 = 200.4$ dB.

60. An antenna has a noise temperature of 35 K and its matched into a receiver which has a noise temp of 100 K. Calculate the noise power density and the noise power for a BW of 36 MHz.

Ans. $NO = (35 + 100) \times 1.38 \times 1\,Q^{-23} = 1.86 \times 10^{-21} J$

$PN = 1.86 \times 10^{-21} J \times 36 \times 10^6 = 0.067$ PW

61. What is a single mode of operation?

Ans. A transponder channel aboard a satellite may be fully loaded by a single transmission from an earthstation. This is referred to as a single acess mode of operation.

62. What are the methods of multiple acess techniques?

Ans. FDMA (Frequency division multiple techniques) TDMA (Time division multiple techniques)

63. What is in CDMA7 and its types?

Ans. In this method each signal is associated with a particular code that is used to spread the signal in frequency and or time.

- Spread spectrum multiple acess
- Pulse address multiple acess

64. What is a thin route service?

Ans. SCPC systems are widely used on lightly loaded routes, this type of service being referred to as a thin route service.

65. What is an important feature of Intelsat SCPC system?

Ans. The system is that each channel is voice activated. This means that on a two way telephone conversation only one carrier is operative at any one time.

66. What is an TDMA? What are the Advantage?

Ans. Only one carrier uses the transponder at anyone time,and therefore intermodulation products, which results from the nonlinear amplification of multiple carriers are absent.

Merits : The transponder traveling wave tube can be operated at maximum power o/p or saturation.

67. What is preamble ?

Ans. Certain time slots at the beginning of each burst are used to carry timing and synchronizing information.These time slots collectively are referred to as preamble.

68. Define guard time.

Ans. It is necessary to between bursts to prevent the bursts from overlapping. The guard time will vary from burst to burst depending on the accuracy with which the various bursts can be positioned within each frame.

69. What is meant by decoding quenching?

Ans. In certain phase detection systems the phase detector must be allowed time to recover from one burst before the next burst is received by it.This is know as decoding quenching.

70. What is meant by direct closed loop feedback?

Ans. The timing positions are reckoned from the last bit of the unique word in the preamble .The loop method is also known as direct closed loop feedback.

71. What is meant by feedback closed loop control?

Ans. The synchronization information is transmitted back to an earth station from a distant,that is termed feedback closed loop control.

72. Define frame efficiency.

Ans. It is a measure of the fraction of frame time used for the transmission of traffic. Frame efficiency is defined as

ηf = traffic bits/total bits.

ηf = 1 – overhead bits/total bits.

73. What is meant by telephone load activity factor?

Ans. The fraction of time a transmission channel is active is known as the telephone load activity factor.

74. What is meant by digital speech interpolation ?

Ans. The point is that for a significant fraction of the time the channel is available for other transmissions, and advantages is taken of this in a form of demand assignment known as digital speech interpolation.

75. What are the types of digital speech interpolation?

Ans.
- Digital time assignment speech interpolation
- Speech predictive encoded communications

76. What is meant by freeze out?

Ans. It has assumed that a free satellite channel will be found for any incoming speech spurt, but of course there is a finite probability that all channels will be occupied and the speech spurt lost. Losing a speech spurt in this manner is referred to as freeze out.

77. What is DSI?

Ans. The DSI gain is the ratio of the number of terrestrialchannels to number of satellite channels.It depends on the number of satellite channels provided as well the design objectives stated above.

78. What are the advantage of SPEC method over DSI method?

Ans. The SPEC method over DSI method is that freezout does not occur during overload conditions.

79. What is ratio of bit rate IF bandwidth?

Ans. R_b/BiF = m/l+p 'm' is the roll of factor m = 1 forBPSK M = 2 for QPSK.

80. What are the demerits of conventional approach method?

Ans.
- Excessive size and weight
- Power consumption.

81. Define space division multiplexing.

Ans. The satellites in Geostationary orbit can be achieved through the use of antenna spot beams. The use of spot beam is also known as space division multiplexing.

82. Define satellite switched TDMA?

Ans. Space division multiplexing can be realized by switching the antenna interconnections in synchronism with the TDMA frame rate,this being know as satellite switched TDMA.

83. What is SS/TDMA?

Ans. A modern pattern is a repetitive sequence of satellite switch modes, also referred to as SS/TDMA.

84. What is processing gain?

Ans. The jamming or interference signal energy is redused by a factor known as the processing gain.

85. What are the applications of Radarsat?

Ans.
- Shipping and fisheries
- Ocean feature mapping
- Oil pollution monitoring
- Iceberg detection
- Crop monitoring

86. What is ECEF?

Ans. The geocentric equatorial coordinate system is used with the GPS system, where it is called the earth centered, earth fixed coordinate system.

87. What is dilution of precision?

Ans. Position calculations involve range differences, and where the ranges are nearly equal, any error is greatly magnified in the difference. This effect, brought about as known as a result of the satellite geometry is known as dilution of precision.

88. What is PDOP?

Ans. With the GPS system dilution of position is taken into account through a factor known as the position dilution of precision.

89. What is burst code word ?

Ans. It is a binary word, a copy of which is stored at each earth station.

90. Define SIC.

Ans. This identifies the Transmitting station.

91. What is an start of receiving frame?

Ans. At any given traffic station, detection of the unique word in the reference burst signals the start of receiving frame.

92. What is meant by burst position acquisition and burst position synchronization?

Ans. A station just entering, or reentering after a long delay to acquire its correct slot position.

93. What is an single access?

Ans. A transponder channel aboard a satellite may be fully loaded by a single transmission from earth station.

94. What is an multiple access technique?

Ans. A transponder to be loaded by a number of carriers. These may originate from a number of earth station may transmit one or more of the carriers. This mode of operation known as multiple access technique.

95. What are the two types of multiple access technique?

Ans.
- FDMA
- TDMA

96. What is meant by frequency reuse?

Ans. The satellite as a whole to be accessed by earth stations widely separated geographically but transmitting on the same frequency i.e., known as frequency reuse.

97. What is meant by space division multiple access?

Ans. The satellite as a whole to be accessed by earth stations widely separated geographically but transmitting on the same frequency i.e., known as frequency reuse. This method of access known as space division multiple access.

98. Write the equations of C/N ratio.

Ans. C/NO = (EIRP) + (G/T) – LOSSES – (K) dBHz.

99. What is an error detecting code?

Ans. A code which allows for the detection of errors is termed an error detecting code.

❑❑❑

GLOSSARY

A

AC : Assignment channel. A channel carrying assignment information in the Intelsat TDMA system.

Access Line : A local loop.

Access : The general term for the ability of a telecommunications user to make use of a network.

ACK : Acknowledge signal. This is sent when data are received correctly in an ARQ system.

Active Sensor : A measuring instrument in the earth exploration satellite service or in the space research service.

Adaptive Differential : A variant of PCM which encodes only the difference between Pulse Code. The value of the current sample and the previous one modulation (ADPCM).

Addressability : The ability of a network, especially a satellite or cable system, to individually address and thus control (usually to enable decryption) user's receivers.

ADM : Adaptive delta modulation. Delta modulation that varies its step size in response to the behavior of the input signal.

Advanced Data Communication Control Procedure (ADCCP) : A bit-oriented datalink protocol similar to HDLC.

Aeronautical Earth Station : An earth station in the fixed satellite service, or, in some cases, in the aeronautical mobile satellite service, located at a specified point on land to provide a feeder link to a satellite.

Aeronautical Mobile Satellite Service : A mobile satellite service in which mobile earth stations are located on board aircraft. Emergency position indicating radio beacon stations may also participate in this service.

Aeronautical Radio Incorporated (ARINC) : A not-for-profit corporation owned by the airline, intended to coordinate their use of telecommunications used in the airline industry.

Aeronautical Radio Navigation Satellite Service : A radio navigation satellite service in which earth stations are located on board aircraft.

Aircraft Earth Station : A mobile earth station in the aeronautical mobile satellite service located on board an aircraft.

Allocation (of a Frequency Channel) : Entry in the Table of Frequency Allocations of a given frequency band for the purpose of its use by one or more terrestrial or space radio communications services under specified conditions. This term also is applied to the frequency band concerned.

Allotment (of a Radio Frequency Channel) : Entry of a designated frequency channel in a plan, adopted by a competent conference, for use by one or more administrations for a terrestrial or space radio communications service in one or more countries or geographical areas and under specified conditions.

ALOHA : A random access system.

Alternate Routing : A communication path used if the primary path is not available.

Amplifier : An electronic device that increases the power of a receiver signal.

Amplitude Modulation (AM) : The technique of varying the strength of a carrier wave in proportion to the strength of a signal.

Analog : A signal that can take on any of a continuous range of values between some minimum and some maximum. Compare digital.

Analog Channel : A channel on which the carrier of the information can take on any of a range of values.

Analog Data : Information conveyed by continuously varying some property.

Analog Transmission : Transmission of a carrier wave which varies continuously and can take on any of a range of value of one or more of the properties of the wave.

Antenna : A metal structure or wire that picks up or transmits electromagnetic energy through space.

Antenna gain : An increase in flux density in a given direction produced by an antenna relative to flux density produced by an isotropic antenna transmitting the same total power.

Aperture antenna : A microwave antenna using a horn or a feed and reflector.

Aperture efficiency : The ratio of effective aperture of an antenna to its physical area, typically in range 50 to 75 percent. A lossless, uniformly illuminated aperture has an aperture efficiency of 100 percent.

Aperture : The area across which an antenna radiates or receives energy.

Apogee : The point in an orbit of greatest distance" from the earth.

APT : Automatic picture transmission. A format for sending weather data from a satellite.

ARM : Apogee kick motor. A rocket engine that boosts a satellite from the Space Shuttle orbit into a transfer orbit.

ASCII code : A standardized binary code to represent alphabetic letters, numbers, and sym-bols. ASCII stands for American Society for Computer Information Interchange.

Assigned Frequency Band : The frequency band within which the emission of a station is authorized; the width of the band equals the necessary bandwidth plus twice the absolute value of the frequency tolerance. Where space stations are concerned, the assigned frequency band includes twice the maximum Doppler shift that may occur in relation to any point of the earth's surface.

Assignment (of a Radio Frequency or Radio Frequency Channel) : Authorization given by an administration for a radio station to use a radio frequency channel under specified conditions.

Asynchronous : A transmission technique in which each group of bits to be sent is preceded by a "start" character(s) and followed by "stop" characters.

AT&T : American Telephone and Telegraph Company.

Attenuation : Time-varying fading.

Attitude and : A system that keeps a satellite in the correct orbit and

Audio Frequency (AF) : The frequencies of sound waves that can be perceived by the human ear, roughly 30 to 20,000 Hz. Also sometimes called voice frequencies (VF), but actually broader bandwidth.

Audio subcarrier : A carrier that is modulated by the audio component of a television signal.

Authorized Power : The maximum signal strength allowed for a particular class of licensed transmitting station.

Automatic Request For Reptition (ARQ) : An error correction system in which the receiver asks the sender to repeat information received incorrectly.

Autotrack : Automatic tracking of satellite motion by earth station antenna.

Azimuth : The horizontal angle measured east from north to the line from an observer to a satellite.

B

B Channel : A 64 kbps channel, part of the ISDN system, usable for voice or data.

Backoff : The process of reducing the input power level of a traveling wave tube to obtain more linear operation.

Bandwidth : with-out distortion.

Base Station : A stationary station in the land mobile service which can communicate with other mobile stations.

Baseband Signalling : Transmission of information at its original range of frequencies.

Baseband : The frequency band that a signal occupies when initially generated. BCH codes : Bose-Chaudhuri-Hocquenghem codes, named after discoverers. These are linear block codes.

Baud : A measure of the signaling speed of a circuit, the baud rate is the number of discrete signal conditions per second.

BEL : A unit of strength of a signal, equal to 10 decibels (q.v.)

BER : Bit error rate. The fraction of a sequence of message bits that are in error. Binary cyclic codes : Block codes with cyclic properties. These can be generated with shift registers. Block code : A binary data transmission format in which message bits and parity check bits are formed into blocks, each with a predetermined number of bits. Blockage : The loss of energy in a reflector antenna caused by the presence of obstacles in the aperture (feed, feed supports, etc.).

Binary Coded Decimal : An encoding scheme in which each decimal digit is encoded as a 4-bit binary number.

Binary Synchronous : A character oriented synchronous datalink protocol that allows frames of data of arbitrary length, ASCII or EBCDIC is used for encoding the characters.

Bit Rate : The rate at which bits are transmitted, expressed as bits per second.

Bit Stuffing : A technique of interesting a 0-bit in a transmitted datastream used in bit-oriented protocols to ensure that six is never appear consecutively.

Bit : Binary digit, the smallest piece of information possible, usually expressed as either a "1" of a "0."

Block Up Converter (BUC) : Translate L-band input frequency to either C-, X- or Ku-band.

Boresight : The center of an antenna beam, usually the direction of maximum gain. Boltzmann's constant : The constant used to obtain noise power from noise temperature. Value is 1.38×10^{-23} J/K, or –228.6 dBW/KHz.

BPSK : Binary (or bipolar) phase shift keying. A digital modulation technique in which the carrier phase takes on one of two possible values.

Broadband : Referring to a bandwidth greater than the baseband bandwidth, or greater than a voice frequency bandwidth.

Broadcast Satellite Service (BSS) : The satellite service designed to bring primarily video entertainment from satellites directly to consumers via high power satellites and small user antennas. More commonly referred to as DBS, Direct Broadcast Service.

Broadcasting Service : A radio communications service in which the transmissions are intended for direct reception by the general public. This service may include sound and television transmission.

Broadcasting Station : A station in the broadcasting service.

Burst errors : Errors that occur in adjacent bits.

Byte : Eight bits.

C

C band : Frequency range 4 GHz to 8 GHz.

Carrier : A continuous (usually high frequency) electromagnetic wave which can be modulated by a signal to carry information.

Carrier System : A technique of frequency multiplexing in which individual signals are carried on separate carrier waves of different frequencies.

Carrier-to-noise Ratio (C/N) : The ratio of the strength of the carrier signal to that of the noise in a channel, usually expressed in decibels.

Carson's rule : The theoretical bandwidth needed to transmit an FM signal

Cassegrain antenna : A two-reflector antenna with a configuration originally used by William Cassegrain for an optical telescope.

Cassegrain Feed : An optical system used in telescopes and satellite antennas in which the feed is placed behind the dish, and the radio waves are brought to a focus there after being reflected first from the main dish and then from a subreflector placed near the dish's prime focus.

CATV : Originally, this stood for community antenna television, but now it is used as a general abbreviation for cable television.

C-Band : Frequencies in the 4 to 6 GHZ range used both by terrestrial microwave links and for satellite links.

CCIR : International Radio Consultative Committee.

CCITT : International Telegraph and Telephone Consultative Committee.

CDMA : Code division multiple access. A multiple access scheme in which stations use spread-spectrum modulation and orthogonal codes to avoid interfering with one another.

Cellular Mobile Radio : A land mobile telephone service provided by a grid of lowpower computer connected and controlled cells, rather than by one highpower station. The user may travel from ceil to cell, and the system maintains track and switches the links accordingly.

Channel Group : A block of 12 telephones channels, transmitted together.

Channel : An electrical path between two devices. Also called link, channel, path, line.

Character : A of the letters, numbers, and other symbols used as part of a written message.

Circuit Switching : A network technique in which a physical connection is made between the sender and receiver for the duration of the transmission.

Circular Polarization : A mode of transmission in which signals travel in a rotating corkscrew-like pattern as they are downlinked to earth. By utilizing both right-hand and left-hand circular polarization, the satellite's transmission capacity is doubled.

Circumscribed circle : A circle that is tangential to an orbit and that encloses the orbit.

Class Of Emmision : The set of characteristics of an emission, designated by standard symbols, e.g., type of modulation of the main carrier, modulating signal, type of information to be transmitted, and also if appropriate, any additional signal characteristics.

Clear sky : Conditions under which atmosphere does not cause excess attenuation or de-polarization of RF signals.

Clip : The time delay between the onset of sound and the completion of a channel assign-ment and connection in a demand-assignment system.

Clustuer Controller : A device used to control a number of terminals.

Code vector : A codeword.

Codec : Coder/Decoder, a device which converts an analog signal into and/or from a digital signal.

Codeword : A combination of message bits and parity check bits into a word of fixed length.

Coma : Distortion of plane wave in antenna aperture caused by movement of the feed away from the focus in the focal plane.

Collocation : The placing of several satellites near each other in orbit. Co-location allows a single fixed receiving antenna to receive signals from all the satellites without moving from one satellite to another (tracking).

Communications Satellite : An artificial satellite, usually placed in geostationary orbit, used to relay radio transmission.

Community Reception (in the Broadcast Satellite Service) : The reception of emissions from a space station in the braodcasting satellite service equipment, which in some cases may be complex and have antennas larger than those used for individual reception. Intended for use by a group of the general public at one location or through a distribution system covering a limited area.

Companding : The process by which the dynamic range of speech is compressed before transmission and expanded after detection. The name is a contraction of compression and expanding.

COMSAT : The Communications Satellite Corporation, the U.S. signatory to the Intelsat and Inmarsat organizations.

Conical scan : Autotrack technique that derives angular error of a satellite in two planes simultaneously by rotation.

Convolutional code : A code in which there is not a one-to-one correspondence between data bits and coded bits.

Corrugated horn : A microwave antenna with high efficiency. It is used as a feed for reflector antennas because of its symmetrical beam shape and low cross-polarization.

Costal Earth Station (CES) : One of the land-based nodes in the maritime satellite services which is connected to the public telephone network.

CR/BTR : Carrier recovery/bit timing recovery.

Crosstalk : Transfer of signal between two separate channels or circuits.

CSC : A common signaling channel in the SPADE system.

CSSB : Companded single sideband. An analog FDM technique using companding and single sideband amplitude modulation.

D

Data Communication : The transfer of digital data by electronic or electrical means.

Data Set : A device that provides the modem and control functions enabling data communications between terminals over some communications facilities.

Data Terminal Equipment (DTE) **:** A device capable of transmitting digital data over a communications circuit.

Date Circuit : Termination Equipment. Device that is the interface between a data transmission facility and data terminal equipment.

dBK : Noise temperature unirexpressed in decibels greater than 1 K.

dBK^{-1} : Units of G/T ratio, decibels per Kelvin.

dBm : Decibels relative to one milliwatt.

dBp : Power in dB above 1 picowatt. It may be weighted or unweighted, depending on context.

DBS : Direct broadcast satellite.

dBW : Decibels relative to one watt.

Decoder : A device that reconstructs an encrypted signal so that it can be clearly received.

Deemphasis : The removal of some high-frequency noise from a demodulated FM signal. See *preemphasis*.

Delay Distortion : Distortion of a transmitted signal due to different frequencies traveling through a transmission path at different speeds.

Delay Equalizer : A device to eliminate the effects of delay distortion in a transmission path.

Delta modulation : A process for digitally encoding and transmitting an analog waveform that avoids the use of analog-to-digital and digital-to-analog converters.

Demand Assignment Multiple Access (DAMA) : Any system whereby many users may access a network on demand, rather than be permanently connected.

Demodulation : The process of extracting the original signal from a modulated carrier.

Demultiplexer : A device that recovers individual signals that had been multiplexed for transmission.

Despun antenna : An antenna, mounted on a satellite with a spinning body, which is rotated in the opposite direction to the body rotation so that the antenna beam points in a fixed direction.

Differential modulation : Digital modulation in which the change in the carrier phase or amplitude from one symbol to the next is determined by the modulating signal. This is in contrast to direct modulation, in which the state of the carrier phase (or amplitude) is determined by the modulating signal.

Digital Speech Interpolation : A statistical multiplexing system which time-assigns channels to telephone calls only when a user is actually speaking.

Direct Broadcasting Satellites (DBS) : Satellite operating in the 12GHz band range and beyond that provide direct-to-home transmissions.

Direct modulation : A digital modulation scheme in which the phase (or amplitude) of the carrier is determined by the modulating signal. This is in contrast to differential modulation, where the change in carrier amplitude (or phase) is determined by the modulating signal.

DM : Delta modulation.

DOMSAT : Domestic communications satellite.

Doppler shift : The change in radio frequency that results from motion of the transmit or receiver.

Downconverter : That portion of a satellite television receiver that translates signals in a certain frequency or block of frequencies to a lower portion of the electromagnetic spectrum, e.g., from microwave to UHF of VHP.

Downlink : The signal sent from a communications satellite to the Earth station.

DSI : Digital speech interpolation. The digital implementation of TASI, a technique for reassigning channels during speech pauses.

Duplex link : A link capable of simultaneous two-way transmission of data.

E

Earth Exploration Satellite Service : A radio communications service between an earth station and one or more space stations. The service may include links between space stations in which information relating to the characteristics of the earth and its natural phenomena is obtained from active or passive sensors on earth satellites. Similar information is collected from airborne or earth-based platforms and such information may be distributed to earth stations within the systems concerned. Platform interrogation may be included and the service may also include feeder links necessary for its operation.

Earth Station : The earth-based link in a communications satellites transmission system.

Eccentricity : A measure of the ellipticity of an orbit. If the eccentricity is zero, the orbit is circular.

Echo Canceller : A device which eliminates the echo in a channel.

Echo Check : An error-checking technique in which data sent to a receiver is sent back to the sender for verification.

Echo Suppressor : A device which reduces the signal coming back to a transmitter to reduce echo.

Echo : The unwanted refection of a signal to its originator.

Eclipse : A satellite in the shadow of the earth.

Effectvie Monopoleradiated Power (EMRP) (in a Given Direction) : The product of the lower supplied to the antenna and its gain relative to a short vertical antenna in a given direction.

EIRP : Effective isotropically radiated power. This is equal to antenna gain multiplied by transmitted power.

Elevation : The angle measured up from local horizontal to the line from an observer to a satellite.

ELV : Expendable launch vehicle.

Encryption : Systematic modification of a signal to prevent unauthorized use.

Energy dispersal : The process of modulating a lightly loaded link to maintain a low power spectral density and avoid interference with terrestrial links.

Equalization : The technique of compensating for difference in attenuation of a signal at different frequencies.

Equalizer : A filter with an amplitude or phase characteristic designed to correct for amplitude or phase distortion in another part of a communication channel.

Equivalent Satellite Link Noise Temperature : The noise station temperature of the receiving antenna of the earth station corresponding to the radio frequency noise power that produces the total observed noise at the ouput of the satellite link excluding noise due to interference coming from satellite links using other satellites and form terrestrial systems.

Erlang B model : The preferred equation for calculating the number of channels required to carry a given amount of traffic.

Erlang : The unit of traffic flow in a communication system.

Error correction : The correction of bits in a digital data stream that have been corrupted during transmission.

Error detection : The detection of bits in a digital data stream that have been corrupted during transmission.

Error syndrome : A word that indicates whether a codeword is correct or in error.

Error-detecting Code : A technique by which a message is sent along with additional information derived from the message by which means the receiver can determine if any errors occurred during transmission.

ESA : European Space Agency.

Even (odd) parity : Addition of one or more bits to a data signal to permit error detection or correction. In even parity, the binary sum of the bits in a character or block is even. In odd parity, the binary sum is odd.

F

Far field : A region sufficiently distant from a transmitting antenna that the observed wave is locally plane.

FAW : Frame alignment word. A pattern of bits that identifies the start of a frame.

FCC : Federal Communications Commission. The government agency in the United States that regulates radio communications and allocates frequencies.

FDM : Frequency division multiplexing. A technique whereby several signals from the same earth station share a transponder by using different frequencies.

FDMA : Frequency division multiple access. A technique whereby signals from several earth stations share a satellite or transponder by using different frequencies.

Feeder Link : A radio link from an earth station at a specified fixed point to a space station, or vice versa, that conveys information for a space radio communications service other than for the fixed satellite service.

FH : Frequency hopping. A spread-spectrum technique in which the transmitter frequency is "hopped" over a wide bandwidth.

Fiber Optics : The technology in which a modulated beam of light carries information through a thin glass or plastic fiber.

Flux density : Power per unit area when an electromagnetic wave is incident on a surface.

FM improvement : Increase in (S/N) at the output of an FM demodulator relative to the (C/N) at its input.

Forward Error Correction : A technique for transmitting with an item of information enough additional information to enable the receiver to reconstruct the original information if it is received with errors.

FPD : Focal plane distribution. Spatial variation of energy in the focal plane of an antenna.

Frame (TDM) : A portion of a digital transmission that contains one word from each channel and sufficient synchronization information to identify the start of the frame.

Frame (TDMA) : A portion of a digital transmission that contains one burst from each earth station plus synchronization and housekeeping information.

Frequency coordination : A process by which interference is avoided when planning an earth station installation.

Frequency Derived Channel : A channel within a frequency divided multiplexed group of channels.

Frequency Modulation (FM) : A technique whereby a carrier wave is made to carry information by changing its frequency in portion to variations in strength of a lower frequency signal.

Frequency reuse : Transmitting to or receiving from a satellite two independent data channels at the same radio frequency. The channels are separated by the spacecraft antenna using directional beams in spatial frequency reuse and by orthogonal polarizations in polarization frequency reuse.

Frequency Shift Keying (FSK) : A FM technique in which different states of the original signal are transmitted as different frequencies of the carrier wave.

Front-fed antenna : An antenna with a single reflector and a radiating feed at the focus of the reflector.

FSS : Fixed satellite service. An FCC term describing satellite communications with fixed earth stations.

G

G/T : Antenna gain to noise temperature ratio. Used to characterize earth stations.

GaAsFET : Gallium Arsenide Field Effect Transistor.

Gain of an Antenna : The ratio, usually expressed in decibels, of the power required at the input of a loss-free reference antenna to the power supplied to the input of the given direction, the same field strength or the same power fluxdensity at the same distance. When not specified otherwise, the gain refers to the direction of maximum radiation. The gain may be considered for a specified poloarization.

GCE : Ground control equipment (at earth station).

Geostationary arc : A circle of radius 42,242 km lying in the plane of the equator and having its center at the center of the earth.

Geostationary orbit : A circular orbit of radius 42,242 km lying in the plane of the equator. To an observer on the ground, a satellite in geostationary orbit remains at the same fixed position in the sky.

Geosynchronous orbit : A circular orbit of radius 42,242 km that is not in the earth's equa-torial plane. A satellite in geosynchronous orbit has an orbital period equal to the earth's rotational period, but its inclination with respect to the equatorial plane makes its position with respect to an observer on the ground change with time.

Global beam : A satellite antenna beam covering the whole of earth, as seen from the satellite.

Grating lobe : An unwanted sidelobe caused by two radiating or receiving antennas or elements that are separated by more than one half wavelength.

Gregorian antenna : A two-reflector antenna with a configuration originally used by James Gregory for an optical telescope.

Guard time : An empty time interval used to separate TDMA bursts.

Guardband : An empty frequency band used to separate FDM signals.

H

Hamming code : A simple block code that is useful for correcting or detecting a small number of errors per block.

Housekeeping : This term refers to systems used on a satellite to keep it operating, but that do not form part of the payload.

HPA : High-power amplifier.

I

I channel : The bit stream that modulates the in-phase carrier of a QPSK system.

IF : Intermediate frequency.

Illumination efficiency : A measure of the efficiency with which an aperture collects energy from an incident plane wave. Uniform illumination of the aperture gives 100 percent efficiency when the antenna transmits or receives.

Inclination of an Orbit (of an Earth Satellite) : The angle determined by the plane containing the orbit and the plane of the Earth's equator.

INMARSAT : The international Maritime Satellite Organization, a body which international maritime satellite telephone traffic via satellites.

Integrated Receiver Decoder (IRD) : A satellite receiver which contains a decoding circuit for scrambled signals.

INTELSAT : The International Telecommunications Satellite Organization, a body which international telecommunications traffic, especially from the public telephone networks, via satellites placed over regions.

Intermodulation noise : Noise generated by the interaction of signals in a nonlinear device (usually a TWT in the context of this text).

Inter-satellite Service : A radio communications service that provides links between artificial earth satellites.

ISI : Intersymbol interference. This is interference between different symbols in a serial digital transmission.

Isotropic source (antenna) : An antenna that radiates equal power in all directions.

J

Julian date : The number assigned toll day in a standard astronomical dating system.

Jumbogroup Multiplex : A frequency division mulitplexing of three jumbo groups, totaling only, 10,800 telephone channels.

Jumbogroup : A frequency division multiplexing of six mcfsTergroups, totaling of 3,600 telephone channels.

K

Ka- BAND : Frequencies in roughly the 18 GHz to 31 GHz range.

Kelvin : Unit of noise temperature.

Key : The information needed to successfully decode an encrypted data sequence.

Ku-band : Frequency range of 12 GHz to 18 GHz

L

Land Mobile Satellite Service : A mobile satellite service in which mobile earth stations are located on land.

L-band : Frequency range of 1 GHz to 2 GHz.

LDM : Linear delta modulation. Delta modulation that uses a fixed step size.

LHCP : Left-hand circular polarization.

Link power budget : Method for calculation of received power in a satellite link.

LNA : Low noise amplifier.

Loading factor : A factor that is multiplied by the rms test-tone deviation to yield the rms deviation of an FDM/FM multiplexed telephone signal.

Loading : A measure of the traffic carried by a multiplex telephone link.

Look angles : The coordinates to which an earth station antenna must be pointed to communicate with a satellite.

Luminance : The amount of white light at a point in a television picture.

M

Margin : The amount (usually in decibels) by which a received signal exceeds a predeter-mined lower limit.

Maritime Satellite Service (MSS) : The satellite telecommunications service designed for communications between ships on the high seas and coastal Earth Stations which are connected to the switched telephone network.

Mastergroup Multiplex : A frequency division multiplexing of 3 mastergroups, totaling 1800 telephone channels.

Mastergroup : A frequency division multiplexing of 10 supergroups, totaling 600 telephone channels.

Meteorological Satellite Service : An earth exploration satellite service for meteorological purpose.

Microterminal : A very small Earth station, often called a VSAT.

Microwave : The portion of the electromagnetic spectrum above 1 GHz in frequency.

Moblie Earth Service : A radio communications services between mobile earth stations and one or more space stations, or between mobile earth station by means of one or more space stations. This service may also include feeder links necessary for its operation.

Moblie Earth Station : An earth station in the mobile satellite service intended to be used while in motion or during halts at unspecified points.

Modem : A modulator/demodulator.

Modulation : A process of having a signal change the characteristic(s) of another (usually higher frequency) wave.

Momentum wheel : A heavy wheel, mounted within a satellite, which rotates at high speed to provide angular momentum in one direction. It is used in the attitude control system.

Monopulse : An autotrack technique that derives angular error of a satellite in two planes simultaneously.

Most Economical Routing Scheme : A system in which the network attempts to send a message first via the least expensive and so on until a connection is made.

MTBF : Mean time before failure.

Multiple Access : Any technique which allows many users to access a given transmission facility.

Multiple access : The sharing of a transponder or satellite by signals from several earth stations.

Multiplexer : A device that combines signals for transmission.

Multiplexing : The sharing of a transponder by several signals from the same earth station.

Multipoint Distribution Service : A one day microwave radio service from a central hub to several receiving stations.

Multi-satellite Links : A radio link between a transmitting earth station and a receiving earth station through two or more satellites, without any intermediate earth station. A multisatellite link comprises one uplink, one or more satellite-to-satellite links and one downlink.

N

(*n*, *k*) code : An arrangement of binary data such that in a block of *n* bits, *k* bits are the message and $(n - k)$ are parity check bits.

NAK : Not acknowledge signal. It is sent when data are received incorrectly in an ARQ system.

NASA : The National Aeronautics and Space Administration.

NOAA : National Oceanographic and Atmospheric Agency. U.S. Government body that operates weather satellites.

Noise figure : A measure of the noise power generated by a device or system.

Noise power budget : A method for calculation of total noise power in a receiving system.

Noise temperature : A measure of noise power that is independent of measurement bandwidth.

Noise : Any unwanted contribution to a signal. May be natural or interference from other signals.

North-South, East-West, stationkeeping maneuver : Movement of the satellite perpendicular to (N-S) and along (E-W) the geostationary orbit to correct orbital errors.

NPR : Noise power ratio. A measure of intermodulation noise in a multiplexed telephone channel.

NRZ : Non-return-to-zero. A baseband digital transmission scheme in which the logical t and 0 signals correspond to equal-amplitude positive and negative voltages.

NTSC : The 525-line/60-Hz television transmission standard used in North America and Japan.

Nyquist filter : A filter that produces zero 1SI waveforms when driven by an impulse.

O

Occasional-use Transponder : A satellite transponder that invented on a periodic, typically hourly, basis.

Offset reflector : A reflector 'that is not a rotationally symmetric conic section.

OMT : Orthogonal mode transducer. This separates polarizations in an antenna feed.

OQPSK : Offset QPSK modulation.

Orbit Control : pointing in the correct direction. Made up of gas jets

Orbit : The path, relative to a specified frame of reference, described by the center of mass of a satellite or other object in space subjected primarily to natural forces, mainly the force of gravity.

Orbital elements : A set of six constants that are sufficient to specify an orbit.

Osculating orbit : The orbit that a spacecraft would follow if all perturbing forces were removed.

OTS : Orbital Test Satellite, built by the European Space Agency.

Outage : Loss of communication.

Overdeviation : Intentionally modulating an FM transmitter so that the Carson's rule band-width of the uplink signal will be significantly greater than the transponder (and pos-sibly the IF receiver) bandwidth.

P

P channel : The bit stream that modulates the in-phase carrier of a QPSK system.

Packet Assembly And Disassembly (PAD) : The device or process of splitting a data message into packets for transmission over packet network.

Packet : A unit of data to be transmitted over a packet network.

PAL : The European 625-line/50-Hz television transmission system.

PAM : Payload assist module. A rocket engine used in Space Shuttle launches to move a satellite from low orbit to geostationary orbit.

Parabolic torus reflector : A reflector that is circular in one plane and parabolic in the orthogonal plane. Beam can be scanned by feed movement in the plane of circular curvature.

Parallel Transmission : A system in which bits making up a character are sent simultaneously along several channels.

Paramp : Parametric amplifier.

Parity check matrix : A method for calculating the error syndrome in a block codeword.

Parity : An error detection technique in which an extra bit is added to the bits of an encoded character so that the character sent over a transmission line contains either an even number (even parity) or an odd number (odd parity) of 1's.

Passive Sensor : A measuring instrument in the earth exploration satellite service or in the space research service by means of which information is obtained by reception of radio waves of natural origin.

Path loss : The apparent loss of power caused by spherical expansion of a transmitted wave. Equal to $(1/4pR)^2$.

PCM : Pulse code modulation. Any modulation scheme that transmits digitally encoded quantized samples of an incoming signal.

Peak factor : A factor that is multiplied by the rms deviation to yield the true deviation of an FDM/FM multiplexed telephone signal.

Perigee : The point in the orbit where a satellite is closest to the earth.

Period (of a Satellite) : The time elapsing between two consecutive passages of a satellite through a characteristics point on its orbit.

Personal Earth Station (PES) : M/Acom term for a microterminal.

Phase Modulation : The technique of changing the phase of a carrier wave in response to changes in state of a signal.

PN sequence : A sequence of apparently random ones and zeros that actually repeats after a finite time. The spectrum of a PN sequence is similar to that of white noise.

PN : Pseudonoise. This is also the abbreviation for a spread-spectrum scheme in which a transmitted signal is spread through multiplying it by a PN sequence.

Polarizer : A device to generate wanted polarization in an antenna feed.

Preamble : The part of a TDMA burst that contains synchronization and other house-keeping information.

Preemphasis : A process for increasing the signal-to-noise ratio of a demodulated FM signal by filtering out some of the high-frequency noise. The signal is distorted before modulation by a preemphasis filter, and the distortion is removed along with some of the noise by a deemphasis filter.

Principal planes : The planes in an antenna that align with reference axes such as the direc-tion of linear polarization and its orthogonal plane.

Propagation Delay : The time it takes for a signal to go from a sender to a receiver.

PSK : Phase shift keying. A modulation technique in which the phase of the RF carrier is varied.

Psophometric weighting : A correction factor used in signal-to-noise ratio calculation that accounts for the nonuniform response of the human ear to white noise.

Pulse Code Modulation (PCM) : A modulation scheme whereby the strenght of the modulating signal is expressed by a digital code.

Pulse Modulation : Any modulation scheme in which the modulating signal is carried by a series of pulses. Includes pulse amplitude, pulse duration (pulse width), pulse position, and pulse code techniques.

Pulse stuffing : Insertion of dummy words into a multiplexed digital transmission. It is done to allow an input channel with a low transmission rate to catch up with the rest of the system,

pWp : Picowatts psophometrically weighted. Power in picowatts corrected for the nonuni-form response of the human ear to white noise for use in signal-to-noise ratio calcu-lation.

Q

Q channel : The bit stream that modulates the quadrature channel of a QPSK system.

QPSK : Quadrature phase shift keying. A digital modulation technique in which the carrier phase takes on one of four possible values.

Quantization : The process of resolving analog samples of a signal into one of a finite number of possible values.

R

Radio Frequency Interference (RFI) : Interference with a receiver by a signal inadvertently generated at one near the wavelength that the receiver is attempting to receive. The Federal Communications Commission establishes RFI standards for electronic devices (such as video games) to minimize this interface.

Radio Naviagation Satellite Service : A radiodetermination satellite service used for the purpose of radio navigation. This service may also include feeder links necessary for its operation.

Radio Spectrum : The portion of the electromagnetic spectrum within which frequencies can be generated and detected by electronic currents. Signals on such frequencies can be propagated through space.

Radiodetermination : The determination of the position, velocity and/or other cnaracteristtcs of an object, or the obtaining of information relating to these parameters, by means of propagation properties of radio waves.

Radiodetermination Satellite Service (RDSS) : A satellite telecommunications service intended to provide navigation and message service to mobile users on land.

RARC : Regional Administrative Radio Conference. Sets frequency allocations for a region of the world.

RCA : Radio Corporation of America.

Realtime : A term used to describe any system which operates such that input, processing, and output take place over *a* short period of time and without any long delays or storage of input or of intermediate or final results.

Redundancy : The provision of spare components in a satellite or earth station to ensure continuous operation when one component fails.

Redundant bits : Extra bits inserted in a digital data stream to enable error detection or correction to be carried out.

Reference burst : The burst that marks the start of a TDMA frame.

Reflecting Satellite : A satellite intended to reflect radio communications signals.

Reliability : Measure of the likelihood of a component or system continuing to operate correctly after a given time.

Repeater : Any device which receives signal and repeats it to parts of a network farther down the line, usually amplified or restored to proper shape and intensity. On a communications satellite, a transponder.

RF : Radio frequency.

RHCP : Right-hand circular polarization.

rms multicarrier deviation : The rms deviation of an FDM/FM telephone signal.

Roll, Pitch, Yaw : Rotational motion about three Cartesian axes centered in the spacecraft.

S

S band : Frequency range of 2 GHz to 4 GHz.

SAM : Simple attenuation model.

Satellite Link : A radio link between a transmitting earth station and a receiving earth station through one satellite. A satellite link consists of one uplink and one downlink.

Satellite Master Antenna Television (SMATV) : A cable television system, sometimes serving a large area, but more commonly referring to service to an apartment building, housing complex, etc.

Satellite Network : A satellite system or part of a satellite system, consisting of only one satellite and the cooperating earth stations.

Satellite System : A space system that consists or more than one artificial earth satellites.

SBS : Satellite Business Systems. A U.S. company that operates satellite networks, primarily digital.

SC : Satellite channel. The bits representing 16 terrestrial channels (TC) in the Intelsat TDMA system.

Scalar feed : Feed for front-fed reflector antennas with high efficiency and symmetry.

Scintillation : Rapid fluctuations in signal strength.

SCPC : Single channel per carrier. An FDM technique in which each signal is modulated onto its own carrier for transmission.

SER : Symbol error rate. The fraction of a sequence of message symbols that are in error.

Serial : A transmission technique in which the bits or a character are sent one after another.

SES : Ship earth station in the Inmarsat system.

Shannon capacity : The maximum possible data rate of a communication channel for a given signal-to-noise ratio, with no bit errors.

Shaped reflector : Antenna reflector with modified profile for improved efficiency.

Ship Earth Station : A mobile earth station in the maritime mobile satellite service located on board a ship.

SIC : Station identification code.

Sideband : The band of frequencies either side of a carrier frequency produced by the process of modulating it by other frequencies.

Signal-to-nosie Ratio (SNR) : The ratio of the strength of the signal to the strength of noise in a channel, usually expressed in decibels.

Simplex link : A link that is capable of only one-way transmission of data.

Site-diversity : A scheme in which two or more redundant earth stations located a few kilometers apart communicate with the same satellite.

Sky noise : Electromagnetic radiation from galactic sources and thermal agitation of atmospheric gases and particles.

Slot : A grouping of bits from one channel (in TDM) or from one earth station (in TDMA).

SOF : Start of frame in a TDMA system.

Solar sails : Long flat panels covered with solar cells, projecting from a satellite. The panels are rotated to face the sun and generate the spacecraft's electrical power from incident sunlight.

SORF : Start of receive frame. The reference time for an earth station's TDMA downlink.

SOTF : Start of transmit frame. The reference time for an earth station's TDMA uplink.

Space Operation Service : A radio communications service in which spacecraft, in particular space telemetry and space telecommand. These functions will normally be provided within the service in which the space station is operating.

Space qualification : Special test procedures to ensure high reliability of spacecraft parts.

Space Research Service : A radio communications service in which spacecraft or other objects in space are used for scientific or technological research purposes.

Space Station : A station located on a object which is beyond, the major portion of the earth's atmosphere.

Space System : Any group of cooperating earth stations and/or space stations employing space radio communications for specific purposes.

Space Telemetry : They use telemetry for the transmission from a space station of results of measurements made in a spacecraft, including those relating to the functioning of the spacecraft.

Space Tracking : Determination of the orbit, velocity or instantaneous position of an object in space by means of radiodetermination, excluding primary radar, for the purpose of following the movement of the object.

Spacecraft : Any vehicle in orbit. All communication satellites are spacecraft.

SPADE : Single channel per carrier PCM multiple-access demand assignment equipment. An Intelsat demand-access system.

Sparklies : White flashes or dots appearing on a TV screen in FM TV systems. They are caused by noise impulses from the FM demodulator.

Specialized Common Carrier (SCC) : Carrier that provides point-to-point common carrier services other than telephone services.

Spectrum : The entire range of frequencies of waves, such sound waves or electromagnetic waves, or the graphical depiction of such the strength of the waves as a function of frequency.

Spherical reflector : A reflector having a spherical surface, which can employ a movable feed to scan the beam.

Spillover : Energy lost in an antenna that does not contribute to the illumination efficiency.

Spinner : A satellite with a body that rotates to provide gyroscopic stabilization of attitude.

Spot beam : Very narrow beam covering a small part of the earth's surface.

Spread Spectrum Multiple Access : Also called CDMA, Code Division Multiple Access, a digital technique which allows multiple users access to a transmission path by assigning each a unique code which is combined with the data to produce the transmitted signal.

Spread Spectrum : A modulation technique in which the signal is spread over a wider than the minimum bandwidth in order to provide better signal-to-noise ratio and/or communications security.

SQPSK : Staggered QPSK modulation.

SS : Spread-spectrum. A modulation technique in which a signal's energy is spread over a bandwidth much wider than that required for transmission.

SSPA : Solid-state power amplifier.

SS-TDMA : Satellite switched TDMA. A TDMA system in which the satellite can interconnect different uplink and downlink spot beams.

Standard Frequency and Time Signal Satellite Service : A radio communications service using space stations on earth satellites for the same purposes as those of the standard frequency and times signal service. This service may also include feeder links necessary for its operation.

Standard Frequency and Time Signal Service : A radio communications service for the scientific, technical and other purposes, providing the transmission of specified frequencies, time signal or both stated high precision, intended for general reception.

Statistical Multiplexing : A technique of time-division multiplexing that allows the combining of many lowspeed signals onto fewer highspeed channels in order to use them more efficiently. Examples are TASI and DSI.

Stop Bit : A bit sent in asynchronous transmission indicating the end of the character.

Store And Forward : The technique of temporarily stopping the flow of a message from the sender to receiver some time later,

STS : Space Transportation System. The Space Shuttle.

Subdish : Secondary reflector in a Cassegrain or Gregorian antenna.

Subsatellite point : The place where a line drawn from the center of the earth to a satellite passes through the earth's surface.

Superframe : The number of TDM or TDMA frames required to transmit all housekeeping and synchronization information once.

Supergroup : A frequency division multiplexing of five 12 channel groups, totaling 60 telephone channels.

Surface errors : Departure of reflector surface from ideal curvature due to manufacturing tolerance, alignment errors, etc.

Symbol : One of the unique states of the carrier in digital modulation. A symbol may carry one or more bits.

Synchronous Transmission : System for data transmission in which all stations operate at the same frequency determined by a master clock.

System (AOCS) : and/or momentum wheels.

System Network Architecture (SNA) : A network architecture from IBM comprising protocols and standards intended for data communications.

System noise temperature : The total apparent noise temperature at a reference point in a receiving system.

T

T1 : A carrier system in which a 24 digitized telephone channels (12 2-way calls) are combined and transmitted at 1.544 Mbps.

Taper : Distribution of field strength across an aperture or reflector.

TASI : Time-assigned speech interpolation. A technique for reassigning channels during speech pauses.

TC : Terrestrial channel. The bits containing 16 samples of one voice channel carried in the Intelsat TDMA system.

TDM : Time division multiplexing. A technique whereby several signals from the same earth station share a transponder by using it at different times.

TDMA : Time division multiple access. A technique whereby signals from several earth stations share a satellite or transponder by using it at different times.

Telemetry, Tracking, and Command (TT&C) : A system that provides two-way communication between an earth station and the satellite to monitor spacecraft systems and to send instructions for changes.

Test tone : A 1-mW (0-dBm) 1-kHz signal used to adjust telephone equipment.

Three-axis stabilization (body stabilization) : A technique to maintain the body of a satellite in the same orientation relative to earth at ail times. The body of the satellite does not rotate.

Threshold : A level at which the operation of a system changes suddenly.

Thruster : A small rocket used to provide fine control of the velocity or attitude of a spacecraft.

Tl carrier : A 24-channel TDM system developed by the Bell System for telephony.

Transceiver : A device that can both transmit and receive.

Transfer orbit : An intermediate orbit used in the process of launching a geostationary satellite.

Translator : A device which converts information expressed in one system to the proper expression in another system.

Transparent transponder : A transponder that does not modify the signal, other than by amplification and frequency conversion.

Transponder hopping : One TDMA earth station working several transponders.

Transponder : Basically a receiver followed by a transmitter. A transponder receives an uplink signal at one frequency, amplifies it, and retransmits it on another frequency.

Trellis diagram : A technique for locating errors in convolutional code codewords.

TRT : The timing reference transponder for an Intelsat TDMA network.

TVRO : Television receive-only terminal.

TW'T : Traveling wave tube.

TWTA : Traveling wave tube amplifier.

U

Universal time (UT) : Standard time for space operations and scientific observations.

Unprotected-use Transponder : A transponder whose protection is not offered by the common carrier. Should the transponder fail, the programmer is not guaranteed replacement or service restoration.

Uplink : The communications channel from an earth station to a satellite.

UW : Unique word. A sequence of bits used in TDMA to mark each frame or superframe.

V

Very Small Aperture Terminal (VSAT) : A satellite dish between 1.2 an 1.8 meters in diameter, typically located at a retail or other remote locations as an endpoint of a VAST network.

Viterbi codes : Convolution codes named for A. J. Viterbi.

Voice Frequency (VF) : The range of frequencies carrier in the public telephone network without special line conditioning, from about 300 Hz. Sometimes called Audio Frequencies (AF), but actually smaller bandwidth.

W

WARC : World Administrative Radio Conference. This sets frequency allocations for the whole world.

Windowing : A process in which a TDMA earth station searches for the unique word only during some time window.

X

X band : Frequency range of 8 GHz to 12 GHz.

XPD : Cross-polarization discrimination. A measure of the conversion of power from one wave polarization to the orthogonal polarization.

XPI : Cross-polarization isolation. A measure of the coupling between nominally orthogonally polarized channels.

Z

Z-axis intercept : Intersection of the satellite Z axis and Earth's surface. Defines the antenna pointing direction.

Zero Bit Insertion : A technique used in bit-oriented protocols to ensure than six 1's never appear consecutively in the data stream.

Zone beam : Satellite antenna beam covering a part of the earth's surface.

❑❑❑

MODEL QUESTION PAPERS

MODEL QUESTION PAPER—1

SATELLITE COMMUNICATION

Time : 3 Hours *Total Marks : 100*

1. Attempt any **four** parts of the following : **(5 × 4 = 20)**

(*a*) What are the elements of a satellite communication ? Explain each of them with a suitable block diagram.

(*b*) What is the difference between geosynchronous and geostationary orbit ? Discuss the advantages and disadvantages of geostationary orbit.

(*c*) Explain, what is meant by apogee and perigee heights. The COSMOS 1675 satellite has apogee height of 39,342 Km and a perigee height of 613 Km. Determine the semimajor axis and ecentricity. Assume a mean earth's radius is 6371 Km.

(*d*) Explain how a satellite is placed into geostationary orbit from earth ?

(*e*) An earth station is located at 30° W longitude and 60° latitude. Determine the earth station azimuth and elevation angle with respect to geostationary satellite located at 50° W longitude. If the orbital radius and earth radius are 42 & 64 Km and 6378 Km respectively.

(*f*) Derive the equation of a satellite orbit.

2. Attempt any **four** parts of the following : **(5 × 4 = 20)**

(*a*) Explain the working of a transponder with the help of a block diagram. Why does the conversion in a transponder use RF to RF conversion instead RF to IF to RF ?

(*b*) Draw the blockdiagram of a satellite showing various subsystems and explain the function of each block in brief.

(*c*) What is the attitude and orbit control subsystems ? Explain as to how these perform their functions .

(*d*) Explain how intermodulation noise originates in a satellite link ? Describe how it may be reduced ?

(*e*) What is the system noise temperature ? Derive the expression for equivalent noise temperature.

(*f*) The following parameters apply to a satellite downlink :

Saturation EIRP = 22.5 dB W
Free space Loss = 195 dB
Other losses and margins = 1.5 dB
Earth station (G/T) = 37.5 dB/K

Calculate the C/N at earth station. Assuming an output backoff of 6 dB is applied. What is the new value of (C/N) ?

3. Attempt any **two** parts of the following : **(10 × 4 = 20)**

(*a*) What is meant by error rate performance of a system ? How do you calculate the error rate performance for MSK , BPSK, QPSK, orthogonal FSK and DPSK system.

(*b*) Briefly discuss the FM pre-emphasis and de-emphasis circuit.

(*c*) Explain the DA-TDMA burst structure. In what way it is different from a simple TDMA burst structure.

(*d*) Briefly describe how demand assignment may be implemented in a TDMA network. What are the advantages of TDMA over FDMA in this respect.

4. Attempt any **two** parts of the following : **(10 × 2 = 20)**

(*a*) Compare linear block codes, cyclic codes and convoluation codes by giving their advantages and disadvantages.

(*b*) What is channel capacity ? Derive an expression of channel capacity in terms of bandwidth.

(*c*) The generator polynomial of a (7, 14) cyclic code is $G(P) = P^3 + P + 1$. Obtain all the code vector for the code in nonsystematic and systematic form.

(*d*) Discuss the troposheric multipath and scientillation effects.

5. Attempt any **two** parts of the following : **(10 × 2 = 20)**

(*a*) What do you understand by site diversity ? Derive an expression for diversity gain.

(*b*) Draw the network structure of mobile satellite communication (MSAT) communication. How many types of communications are possible through this network.

(*c*) List the various factors to use KU band for direct broadcast satellite service.

(*d*) Discuss the various application areas of VSAT and data broadcast satellite.

MODEL QUESTION PAPER—2

Time : 3 Hours *Total Marks : 100*

1. Attempt any **four** parts of the following : **(5 × 4 = 20)**

(*a*) Briefly explain the satellite frequency spectrum why is uplink frequency kept different from downlink frequency ?

(*b*) What do you understand by look angle ? Explain with reference to a geostationary satellite.

(*c*) Explain, how a satellite is placed into geostaionary orbit from earth ?

(*d*) What are the elements of satellite communication system ? Explain each with a suitable block diagram.

(*e*) How can one determine the earth coverage and slant range for geostationary satellite ? What are the maximum values of two parameters ?

(*f*) The semimajor axis and the semiminor axis of an elliptical satellite orbit are 20,000 Km and 1,600 Km respectively. Determine the apogee and perigee distances.

2. Attempt any **four** parts of the following : **(5 × 4 = 20)**

(*a*) Describe the telemetry, tracking and command subsystem with reference to an earth station. Do they are a part of space segment or ground segment ?

(*b*) Derive general link equation. Findout experission for C/N and G/T ratio. Explain the importance of the satellite link design.

(*c*) Discuss the antenna requirements for large and small earth station.

(*d*) What is redundancy ? Why is it necessary for the satellite subsystem particularly in repeaters ?

(*e*) What are the various interferences that may affect the satellite link performance ? Explain briefly.

(*f*) Two amplifiers are connected in cascades having a gain of 20 dB each. If the noise temperature is 200 K, determine the overall gain and effective noise temperature refered to input.

3. Attempt any **two** parts of the following : **(10 × 2 = 20)**

(*a*) Define TDMA, FDMA and CDMA. Discuss their relative merits and demerits.

(*b*) What do you understand by process gain and Jam margin ? Explain the importance of these parameters in spread spectrum communication system.

(*c*) What are the various methods of digital modulation techniques and which is mostly used in digital satellite communication ?

(*d*) What is time division multiplexing ? Now does it differ from frequency division multiplexing.

4. Attempt any **two** parts of the following : **(10 × 2 = 20)**

(*a*) What do you understand by cyclic codes ? List their advantages and disadvantages also.

(*b*) Briefly discuss the various propagation effects that are not associated with hyrometers.

(*c*) What is attenuation ? Derive an expression of attenuation probability for SAM model.

(*d*) A (7, 4) linear block code whose generator matrix is given as

$$a = \begin{bmatrix} 1 & 0 & 0 & 0 & : & 1 & 0 & 1 \\ 0 & 1 & 0 & 0 & : & 1 & 1 & 1 \\ 0 & 0 & 1 & 0 & : & 1 & 1 & 0 \\ 0 & 0 & 0 & 1 & : & 0 & 1 & 1 \end{bmatrix}$$

Determine,

(*i*) All codevectors of this code.

(*ii*) Parity check matrix of this code.

5. Attempt any **two** parts of the following : **(10 × 2 = 20)**

(*a*) Discuss the various applications of satellite and explain each of them in briefly.

(*b*) Explain the important features of RADARSAT.

(*c*) Draw the block diagram of outdoor unit of a DBS home receiver ? Explain the function of each block briefly.

(*d*) State and explain the various segments of a global positioning satellite system.

❑❑❑

PREVIOUS YEARS EXAMINATION PAPERS

B.TECH

(SEM. VII) EXAMINATION, 2007-08

SATELLITE COMMUNICATION

Time : 3 Hours *Total Marks : 100*

Note : *Attempt **all** questions. All questions carry **equal** marks.*

1. Attempt any **four** parts of the following : **(5 × 4 = 20)**

(*a*) In what way satellite has an edge over communication compared to other methods ? What are the future trends in satellite communications.

(*b*) Explain the following with respect to satellites :

(*i*) Ascending and Descending modes.

(*ii*) Satellite axis.

(*iii*) Look, azimuths, elevation angles.

(*iv*) Inclination perigee, apogee.

(*c*) A satellite is orbiting in a geosynchronous orbit of radius 41500 km. Find the velocity and time of orbit. What will be the change in velocity if the radius reduces to 36000 km ? Assume g_0 = 398600.5 km^3 s^2.

What are the torques that affect the position of a geostationary satellite ?

(*d*) Explain the effect of a solar eclipse on the performance of a geostationary satellite. In what way it is related to fixing the parking place of a satellite.

A low orbititing satellite has an 8-hour progade orbit. How long during each orbit will an earth station be able to communicate with it above an elevation angle of 15% ?

(*e*) Explain how attitude and orbit control is achieved from an earth station. A satellite receives sun rays at 7°6' and the duration of the eclipse is 56 min. Calculate the radius of orbit and height of satellite.

(*f*) Explain the master control facility including the satellite control earth station, satellite control center and the UPS.

2. Attempt any **four** parts of the following : **(5 × 4 = 20)**

(*a*) Explain why some satellites employ cylindrical solar arrays whereas other employ solar cell arrays for the production of primary power. State the typical power output to be expected from each type. Why is it necessary to carry batteries in addition to solar cell arrays ?

(*b*) (*i*) Briefly explain the various methods of satellite stabilisation.

(*ii*) What is a despun antenna ? How is despping achieved ? Explain.

(*c*) In a satellite receiving system the equivalent noise picked up by antenna is 18°K. The antenna is connected to the received through a cable having a loss factor of 0.15 dB/m and of 15 m length. The physical temperature for the design is to be considered as 290 K. The stages of the receiver is shown in fig. below. Calculate the overall noise temperature and G/T ratio if the received power is –96 dB.

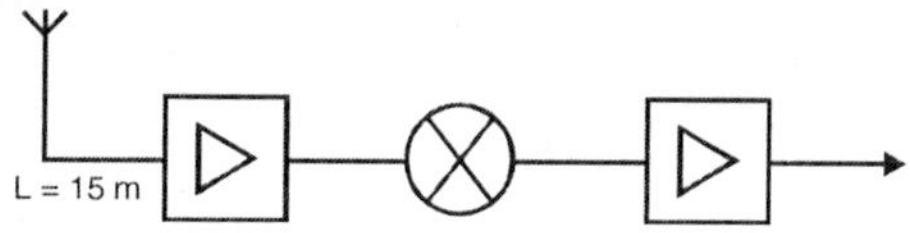

T_0 = 290 K, GLNA = 15 dB, G_m = –dBH, G_{if} = 40 dB, T_a = 29 K, T_m = 300 K, T_{if} = 800 K.

(*d*) Explain the following briefly :

(*i*) TT&C subsystem. (*ii*) Transponder.

(*e*) The EIRP of a 240 W transponder is 57 dBW. Calculate the approximate gain of the antenna. If the transponder is switched to 120 W, calculate the new [EIRP], assuming that the same antenna is used.

(*f*) Show that C/N of a satellite receiver system is directly proportional to G/T ratio.

3. Attempt any **two** parts of the following : **(10 × 2 = 20)**

(*a*) Distinguish between multiplexing and multiple access. Compare FDMA, TDMA, CDMA and DAMA.

(*b*) Describe the concept of threshold in an FM demodulator and how this is useful in satellite communication system. An FM system has a receiver threshold of 15 dB. How much received carrier power and RF bandwidth is needed to transmit a 4 kHz baseband signal with a demodulated SNR of 40 dB ? Take N_0 = 10 to the power of –10 W/Hz.

(*c*) In a digtital transmission E_b/N_0 = 11 dB for a polar NRZ transmission over BPSK. The system uses 8 bits per level. Calculate the S/N ratio in dBS. With a block diagram, explain a typical PCM/TDM system.

4. Attempt any **two** parts of the following : **(10 × 2 = 20)**

(*a*) Explain the difference between error detection and correction. Why would FEC normally be used in satellite circuits ? A transmission takes place where the average probability of error is 10^6. Given that a message containing 10^8 bits is transmitted, what is the average number of bit errors to be expected ?

(*b*) A convolutional code is described by g_1 = [1 0 0], g_2 = [1 0 1], g_3 = [1 1 1].

(*i*) Draw the encoder corresponding to this code.

(*ii*) Draw the state transition diagram and the trellis diagram for this code.

(*iii*) Find the transfer function and the free distance of this code.

(*c*) Explain the various propagation effects and their impact on satellite-earth links. Also write note on depolarisation effects and cloud fire effects on these links.

5. Attempt any **two** parts of the following : **(10 × 2 = 20)**

(*a*) Write short notes on :

(*i*) VSAT (*ii*) Non-geostationary satellites

(*iii*) LEO satellites for internet transmission.

(*b*) Explain different types of satellite based mobile systems. What is the role of IMMARSAT ?

(*c*) Explain how global positioning is achieved. With a neat diagram, explain a GPS receiver.

B.TECH
(SEM. VII) EXAMINATION, 2008-09
SATELLITE COMMUNICATION

Time : 3 Hours *Total Marks : 100*

Note : (1) *Attempt **all** questions.*

(2) ***All** questions carry **equal** marks.*

1. Attempt any **four** parts of the following : **(5 × 4 = 20)**

(*a*) What are the elements of satellite communication ? Explain each of them with a suitable block diagram.

(*b*) What are the factors that affect the uplink design and downlink design in geostationary satellite communication ? Discuss in detail.

(*c*) A satellite at a distance of 50,000 km from a point on the earth's surface radiates a power of 2.5 W from an antenna with a gain of 15 dB in the direction of observer. Find the flux density at the receiving point, and the power received by an antenna with an effective area of 11 m^2. If the operating frequency of satellite is 12.5 GHz and the receiving antenna has a gain of 50 dB, find the received power.

(*d*) Explain different types of antenna used in satellite system.

(*e*) Explain the parameters that control the design of earth station. What is the optimum G/T ratio for a standard earth station ?

2. Attempt any **two** parts of the following : **(10 × 2 = 20)**

(*a*) Draw the block diagram of tracking, telemetry command and monitoring system. Explain the function of each block.

(*b*) Describe the KU-band uplink and Downlink design.

(*c*) What are the advantages of a communication system, that uses digital signal transmission ? Explain as to how the digital satellite communication has reduced the size of the earth station.

3. Attempt any **two** parts of the following : **(10 × 2 = 20)**

(*a*) What is the difference between multiplexing and multiple access techniques ? What is TDMA super frame ? Explain its structure.

(*b*) What is the function of Demand Assignment control in DAMA system ? Explain various such control systems.

(*c*) What is meant by TDMA frame acquisition and frame synchronization ? What is frame delay ? How does it helps in carrying out TDMA frame acquisition techniques.

4. Attempt any **two** parts of the following : **(10 × 2 = 20)**

(*a*) Why convolutional codes provide better error correction performance ? Explain it.

(*b*) Describe the Rain and ice effects on propagation.

(*c*) What is the perceived polarization lilt angle at an earth station located at 52°N, 1°E for vertical polarized signals transmitted from a GEO satellite located at 60°E ?

5. Write short notes on any **four :** **(5 × 4 = 20)**

(*a*) VSAT (*b*) GPS (*c*) Satellite Television and Radio

(*d*) CDMA (*e*) FM Video transmission.

B. TECH

(SEM. VII) ODD SEMESTER THEORY EXAMINATION 2010-11

SATELLITE COMMUNICATION

Time : 3 Hours *Total Marks : 100*

Note : Attempt **all** the questions.

1. Attempt any **four** parts : **(5 × 4 = 20)**

   (*a*) Explain different types of antenna used in Satellite Communication.

   (*b*) What is optimum G/T ratio for a standard earth station ?

   (*c*) What are the factors that affect the uplink and downlink design ?

   (*d*) Write short note on :

   (*i*) Look, azimuths, elevation angle (*ii*) Satellite axis.

   (*e*) The semimajor and the semiminor axis of an elliptical satellite orbit are 20,000 km and 1600 km respectively. Determine the apogee and perigee distance.

   (*f*) Explain geostationary satellite.

2. Attempt any four parts : **(5 × 4 = 20)**

   (*a*) Derive general link equation. Find out expression for *C/N* and *G/T* ratio.

   (*b*) Discuss the antenna requirements for large and small earth station.

   (*c*) What are the various interference that may affect the satellite link performance ? Explain.

   (*d*) Two amplifiers are connected in cascades having a gain of 20 dB each. If the noise temperature is 200 K, determine the overall gain.

   (*e*) The EIRP of a 240 W transponder is 57 dBW. Calculate the approximate gain of the antenna if the transponder is switched to 120W, calculate the new [EIRP], assuming that the same antenna is used.

   (*f*) Explain TT and C subsystem briefly.

3. Attempt any **two** parts : **(2 × 10 = 20)**

   (*a*) In a digital transmission E_b/N_b = 11dB for a polar NRZ transmission over BPSK. The system uses 8 bits per level. Calculate the S/N ratio in dBS with a block diagram, explain a typical PCM/TDM system.

   (*b*) What is function of Demand Assignment control in DAMA system ? Explain various such control systems.

   (*c*) What is the difference between multiplexing and multiple access techniques ? What is TDMA super frame ? Explain its structure.

4. Attempt any **two** parts : **(2 × 10 = 20)**

   (*a*) Describe the Rain and ice effects on propagation.

   (*b*) The generator matrix for a (6, 3) block code is shown below. Obtain all code of words of their code :

$$G = \begin{bmatrix} 1 & 0 & 0 & : & 0 & 1 & 1 \\ 0 & 1 & 0 & : & 1 & 0 & 1 \\ 0 & 0 & 1 & : & 1 & 1 & 0 \end{bmatrix}$$

   (*c*) The parity check matrix of a (7, 4) linear block code is expressed as :

$$H = \begin{bmatrix} 1 & 1 & 1 & 0 & : & 1 & 0 & 0 \\ 1 & 1 & 0 & 1 & : & 0 & 1 & 0 \\ 1 & 0 & 1 & 1 & : & 0 & 0 & 1 \end{bmatrix}$$

   Obtain the generator matrix (*G*) and list of all code vectors.

5. Attempt any **two** parts : **(2 × 10 = 20)**

   (*a*) Explain DBS home receiver with block diagram.

   (*b*) State and explain the various segments of GPS system.

   (*c*) Write short note on any two :

   (*i*) VSAT

   (*ii*) LEO satellites for internet transmission

   (*iii*) Non Geostationary satellites.

B. TECH

(SEM. VII) ODD SEMESTER THEORY EXAMINATION 2011-12

SATELLITE COMMUNICATION

Time : 3 Hours *Total Marks : 100*

Note : (*i*) Attempt **all** the questions.

(*ii*) All questions carry equal marks.

(Given Kepler's constant = 3.9861×10^5 km^3/s^2, Boltzmann's constant, k = 1.38×10^{-23} J/K)

1. Attempt any **four** parts of the following : **(5 × 4 = 20)**

(*a*) What is meant by look angles ? Explain them with reference to a geostationary satellite and the earth station.

(*b*) Satellite-1 in an elliptical orbit has the orbit semi major axis equal to 18000 km and satellite-2 in an elliptical orbit has semi-major axis equal to 24000 km. Determine the relationship between their orbital periods.

(*c*) With the aid of a neat sketch, explain what is meant by each of the angles : inclination; argument of perigee; right ascension of the ascending node.

(*d*) Explain briefly what is meant by sun transit outage ?

(*e*) Explain as to how a satellite is placed into geostationary orbit from the earth ?

(*f*) What is meant by station keeping of satellite ? Explain its significance and also the methods to achieve it.

2. Attempt any **four** parts of the following : **(5 × 4 = 20)**

(*a*) With the help of a simple block diagram, explain the function of telemetry, tracking and command (TTC) sub-system.

(*b*) Give the name of space craft antennas and explain any two in detail with diagrams.

(*c*) A satellite carrying a 11.7 GHz continuous wave (CW) beacon transmitter is located in geosynchronous orbit 38000 km from an earth station. The beacon's output power is 200 mW, and it feeds an antenna with an 18.9 dB gain towards the earth station. The earth station receiving antenna aperture efficiency of 50 percent.

(A) Calculate the satellite EIRP in W, dBW and dBm,

(B) Calculate the path-loss in dB.

(C) Calculate the receiving antenna gain in dB.

(*d*) Derive the G/T ratio of earth station. What is the standard value for good quality of earth station ?

(*e*) How is the uplink design different than the downlink design ? In what conditions a complete satellite link become down link limited ?

(*f*) A 6/4 GHz band communication satellite transponder requires a saturation flux density of –67.5 dBW/m^2, when an input back-off of 11 dB is employed. The spacecraft has a G/T of –11/6 dB/K. If the downlink saturated EIRP is 26.6 dBW, estimate the back-off required if all overall (C/N0) of 92.6 dBHz has to be maintained. The earth station G/T is 40.7 dB/K and total losses may be assumed to be 196.7 dB.

3. Attempt any **two** parts of the following : **(10 × 2 = 20)**

(*a*) What are the factors that affect the uplink design and the downlink design in geostationary satellite communication ? Discuss in detail.

(*b*) Write brief notes on the advantages and disadvantages of using satellite in LEO, MEO and GEO for mobile satellite communications.

(*c*) With the help of suitable diagram explain the satellite communication system architecture using VSAT. Also write the applications of VSAT.

4. Attempt any **two** parts of the following : **(10 × 2 = 20)**

(*a*) Explain the position location principle of GPS. Also draw the block diagram of a GPS receiver and explain its working.

(*b*) With the help of suitable diagram describe the working of home receiver units in direct broadcast satellite television.

(*c*) Write short notes on GPS navigation message and GPS timing accuracy.

5. Attempt any **four** parts of the following : **(5 × 4 = 20)**

(*a*) Describe the main features of the Iridium system, and comment briefly on how this differs from the Orbcomm system.

(*b*) Describe briefly about the types of low-gain directional mobile satellite antennas. Also name their applications.

(*c*) Briefly describe the types of antennas used for hand held terminals.

(*d*) Describe the features of phased-array antenna.

(*e*) Write short note on antenna systems for mobile satellite broadcasting.

(*f*) Write a short note on vehicle earth station antennas.

B. TECH

(SEM. VII) ODD SEMESTER THEORY EXAMINATION 2012-13

SATELLITE COMMUNICATION

Time : 3 Hours *Total Marks : 100*

Note : (*i*) Attempt **all** the questions.

(*ii*) All questions carry equal marks.

1. Attempt any **four** parts of the following : **(5 × 4 = 20)**
   (*a*) Write short note on history of development of satellite launch vehicles.
   (*b*) Satellite-1 in an elliptical orbit has the orbit semi major axis equal to 18000 km and satellite-2 has semi major axis equal to 24000 km. Determine the relationship between their orbital periods.
   (*c*) Explain briefly what is meant by sun transit outage.
   (*d*) What is meant by look angles ? Explain them with reference to a geostationary satellite and the earth station.
   (*e*) The semi major and semi minor axis of an elliptical satellite orbit are 20000 km and 1600 km respectively. Determine the apogee and perigee distance.
   (*f*) Explain geostationary orbit and geostationary satellite in short.
2. Attempt any **four** parts of the following : **(5 × 4 = 20)**
   (*a*) Derive general link equation. Find out expression for C/N and G/T ratio.
   (*b*) What are the various interferences that may affect the satellite link performance ? Explain in short.
   (*c*) The EIRP of a 240 W transponder is 57 dBW. Calculate the approximate gain of the antenna if the transponder is switched to 120 W. Also calculate the new EIRP assuming that the same antenna is used.
   (*d*) How is the uplink design different from the downlink design ?
   (*e*) Three amplifiers are connected in cascade having a gain of 20 dB each. If the noise temperature is 100 K, determine the overall gain.
   (*f*) With the help of a suitable block diagram explain the function of telemetry, tracking and command (TTC) sub-system.
3. Attempt any **two** parts of the following : **(10 × 2 = 20)**
   (*a*) Describe in detail the rain, cloud and ice effect on propagation in satellite communication.
   (*b*) With the help of suitable diagram explain the satellite communication system architecture using VSAT. Also write the applications of VSAT.
   (*c*) A satellite carrying a 11.7 GHz continuous wave (CW) beacon transmitter is located in a geostationary orbit 38000 km from an earth station. The beacon's output power is 200 MW and it feeds an antenna with a 18.9 dB gain towards the earth station. The earth station receiving antenna aperture efficiency is 50 percent. Calculate the path loss in dB and receiving antenna gain in dB.
4. Attempt any **two** parts of the following : **(10 × 2 = 20)**
   (*a*) With the help of suitable diagram describe the working of home receiver units in DBS television.
   (*b*) Explain the position location principle of GPS. Also draw the block diagram of a GPS receiver and explain its working.
   (*c*) Write short note on satellite signal acquisition.
5. Attempt any **two** parts of the following : **(5 × 4 = 20)**
   (*a*) Write short note on antenna systems for mobile satellite broadcasting.
   (*b*) Describe in brief different types of antennas used in satellite communication.
   (*c*) Write short note on Wire Quadrifilar Helix Antenna (WQHA) for hand held terminals.

B. TECH

(SEM. VII) ODD SEMESTER THEORY EXAMINATION 2013-14

SATELLITE COMMUNICATION

Time : 3 Hours *Total Marks : 100*

Note : (*i*) Attempt **all** questions.

(*ii*) All questions carry equal marks.

(Given Kepler's constant = 3.986×10^5 km^3/s^2, Earth's radius = 6360 km, Boltzmann's constant, k = 1.38×10^{-23} J/K)

1. Attempt any **four** parts of the following : **(5 × 4 = 20)**

(*a*) Explain Kepler's laws of planetary rotation. How are these applied to the case of geostationary satellite ?

(*b*) A satellite moving in a highly eccentric Molniya orbit having the farthest and the closest points as 35000 km and 500 km respectively from the surface of the earth. Determine the orbital time period and the velocity at the apogee and perigee points.

(*c*) List the orbital elements of a satellite and briefly explain them.

(*d*) What is the mechanism of launching a satellite ? Briefly explain each step of the launch sequence.

(*e*) An earth station is located at 30° West longitude and 60° North latitude. Determine the earth station azimuth and elevation angles with respect to geostationary satellite located at 50° West longitudes. (Assume orbital radius = 42164 km and earth's radius 6360 km)

(*f*) Explain what is meant by the geostationary orbit. How do the geostationary orbit and a geosynchronous orbit differ ?

2. Attempt any **four** parts of the following : **(5 × 4 = 20)**

(*a*) A satellite TV signal occupies the full 36 MHz transponder bandwidth and is desired to provide a carrier-to-noise (C/N) ratio of 22 dB at the earth station. If a downlink frequency of 4 Ghz is employed and other losses amount to 3.4 dB, what must be the G/T of the earth station, if EIRP is 37 dBW ? The path length may be assumed to be 4000 km.

(*b*) With the help of a simple block diagram, explain the function/working of a satellite transponder.

(*c*) Explain how intermodulation noise originates in a satellite link, and describe how it may be reduced. In a satellite link, and describe how it may be reduced. In a satellite circuit the carrier-to-noise ratios are uplink-25 dB; intermodulation 13 dB. Calculate the overall carrier-to-noise ratio.

(*d*) Describe the east-west and north-south station-keeping manoeuvres required in satellit estation keeping. What are the angular tolerances in station keeping that must be achieved ?

(*e*) The following parameters apply to a satellite downlink: saturation [EIRP] 22.5 dBW, free-space loss 195 dB, other losses are 1.5 dB, earth station G/T is 37.5 dB/K. Calculate the carrier-to-noise density (C/N_0) at the earth station. Assuming an output back-off of 6 dB is applied, what is the new value of (C/N_0) ratio ?

(*f*) Explain what is meant by satellite attitude and briefly describe two forms of attitude control.

3. Attempt any **two** parts of the following : **(10 × 2 = 20)**

(*a*) Describe about different types of atmospheric losses in satellite link.

(*b*) Describe the operation of a typical VSAT system. State briefly where VSAT systems have widest application.

(*c*) (*i*) What is Ice Effect ? Explain how depolarization is caused by rain.

(*ii*) Write short note on Non-geostationary satellite system.

4. Attempt any **two** parts of the following : **(10 × 2 = 20)**

(*a*) Explain the working of a Global Positioning System (GPS) receiver. Explain why a minimum of four satellites must be visible at an earth location utilizing the GPS system for position determination. What does the term dilution refer to ?

(*b*) Explain the working of Direct Broadcast Satellite (DBS) television network.

(*c*) (*i*) Briefly describe the video compression process used in MPEG-2 used in direct broadcast satellite television.

(*ii*) Write short notes on the GPS codes and Signal Levels.

5. Attempt any **two** parts of the following : **(5 × 4 = 20)**

(*a*) With the help of suitable diagram, explain the working of Iridium system.

(*b*) (*i*) Describe briefly about the types of medium-gain directional mobile satellite antennas. Also name their applications.

(*ii*) Describe the features of Wire Quadrifilar Helix Antenna. Why this antenna is used for handheld terminals ?

(*c*) Briefly describe the types of antennas used for mobile satellite broadcasting. What are their technical requirements ?

❑❑❑

Index

INDEX

A

B

C

D

❑❑❑

Win Prizes

ATTENTION READERS

Win Prizes

We request for your frank assessment regarding various aspects of the book as given below :

SATELLITE COMMUNICATION

By : SAPNA KATIYAR **(UPTU)**

Please Fill up the Sheet in Neat and Clean Writing

(i) What topics of your syllabus, important for your examination point of view, are not covered in this book ?

....

....

....

....

(ii) What are the chapters wherein the treatment of the subject matter is not systematic or organised or updated ?

....

....

....

....

(iii) Have you come across misprints/mistakes/factual inaccuracies in the book ? Please specify the chapters and the page numbers.

....

....

....

....

(iv) What other books on the same subject have you found/heard better than the present book ? Please specify in terms of price and quality.

....

....

....

....

(v) Further suggestions and comments for the improvement of this book.

....

....

....

....

(PS : If need be please attach a separate sheet to write your views)

OTHER DETAILS

(i) Who recommended you this book ? (Please tick (✓) in the box near the option relevant to you.

☐ Teacher ☐ Friends ☐ Bookseller

....

....

....

....

(ii) Name of the recommending teacher, his designation and address:

....

....

....

....

(iii) Name and address of the bookseller you purchased the book from :

....

....

....

....

(iv) Name and address of your institution (Please mention the University or Board as the case may be)

....

....

....

....

(v) Your name and complete address :

....

....

....

....

(vi) Course you are enrolled in :

....

....

....

....

The best assessment will be awarded each month : The award will be in the form of our publications as decided by the Editorial Board.

Please mail the filled up coupon at your earliest to :

S.K. KATARIA & SONS®

4885/109, Prakash Mahal,
Dr. Subhash Bhargav Lane,
Opposite Delhi Medical Association,
Daryaganj, New Delhi-110002
Phone: +91-11-43551243
Mobile: +91-9871775858
Email: katariabook@gmail.com